Computational Intelligence in Sustainable Reliability Engineering

Scrivener Publishing
100 Cummings Center, Suite 541J
Beverly, MA 01915-6106

Sustainable Computing and Optimization

Series Editor: Prasenjit Chatterjee, Morteza Yazdani and Dilbagh Panchal

Scope: The objective of "Sustainable Computing and Optimization" series is to bring together the global research scholars, experts, and scientists in the research areas of sustainable computing and optimization from all over the world to share their knowledge and experiences on current research achievements in these fields. The series aims to provide a golden opportunity for global research community to share their novel research results, findings, and innovations to a wide range of readers, present globally. Data is everywhere and continuing to grow massively, which has created a huge demand for qualified experts who can uncover valuable insights from data. The series will promote sustainable computing and optimization methodologies in order to solve real life problems mainly from engineering and management systems domains. The series will mainly focus on the real life problems, which can suitably be handled through these paradigms.

Publishers at Scrivener
Martin Scrivener (martin@scrivenerpublishing.com)
Phillip Carmical (pcarmical@scrivenerpublishing.com)

Computational Intelligence in Sustainable Reliability Engineering

Edited by

S. C. Malik

Department of Statistics, M.D. University, Rohtak India

Deepak Sinwar

Department of Computer and Communication Engineering, Manipal University, Jaipur, India

Ashish Kumar

Department of Mathematics and Statistics, Manipal University, Jaipur, India

S. R. Gadde

Department of Statistics, The University of Dodoma, Tanzania

Prasenjit Chatterjee

Department of Mechanical Engineering, MCKV Institute of Engineering, West Bengal, India

and

Bui Thanh Hung

Faculty of Information Technology, Artificial Intelligence Laboratory, Ton Duc Thang University, Ho Chi Minh City, Vietnam

Scrivener Publishing

WILEY

This edition first published 2023 by John Wiley & Sons, Inc., 111 River Street, Hoboken, NJ 07030, USA and Scrivener Publishing LLC, 100 Cummings Center, Suite 541J, Beverly, MA 01915, USA

For more information about Scrivener publications please visit www.scrivenerpublishing.com.

Wiley Global Headquarters
111 River Street, Hoboken, NJ 07030, USA

For details of our global editorial offices, customer services, and more information about Wiley products visit us at www.wiley.com.

Library of Congress Cataloging-in-Publication Data

ISBN 978-1-119-86501-8

Cover image: Pixabay.Com
Cover design by Russell Richardson

Set in size of 11pt and Minion Pro by Manila Typesetting Company, Makati, Philippines

Printed in the USA

The editors would like to dedicate this book to their parents, family members, friends and readers.

Contents

Preface

With the design of every new product, the world is witnessing the continuous development brought on by cross-disciplinary technologies. Instead of taking raw materials and sending them through a real manufacturing process that repeatedly combats tolerances, errors, and energy consumption to arrive at the final product, the assembly details can be directly input into the computation model in order to obtain the material characteristics as output to reduce effort and process costs. To ensure maximum reliability of product development, it is desired that the manufacturing process be driven by optimization. However, even though optimization has previously been applied for various fields, over the past two decades, computational optimization has become very popular for industrial optimizations. Computational intelligence-based optimization is one of several computational techniques that help achieve sustainability in product design and development phases. Among computational intelligence-based techniques, metaheuristic optimization is found to be specifically suitable for industrial optimizations. There are mainly two types of metaheuristic approaches; single-solution based and population based. As per the applications in the field of industrial optimization, this book mainly focuses on population-based (swarm intelligence) metaheuristic approaches.

Swarm intelligence is an important sub-area of optimization that helps develop sustainable materials at nano-, micro-, meso- and macro-levels by identifying the optimum values for different parameters. With the exponential rise in demand for sustainable materials for various purposes, optimization has played an important role over the last few years. Not only is materials data available for researchers and scientists, but sufficient processing resources are also available, which need to be optimized through AI techniques.

Traditional techniques employed by researchers are often cumbersome, expensive and lack sustainability. Hence, there is always a need for having recourse to time-efficient, fail-safe, cheaper intelligent technologies to address problems and ensure long-term sustainability. Since the existing literature available in this respect is nonexistent, this book is proposed to serve as a treatise and knowledge base for the community to inspire them to adapt environment-friendly and sustainable solutions for the future.

This book focuses on developing advanced computational intelligence algorithms for the analysis of data involved in reliability engineering, material design, and manufacturing with the objective of ensuring sustainability. It reveals applications of different models of evolutionary algorithms in the field of optimization with the objective of solving problems to help the manufacturing industries. Some special features of this book include a comprehensive guide for utilizing computational models for reliability engineering, state-of-the-art swarm intelligence methods for solving manufacturing processes and developing sustainable materials, high-quality and innovative research contributions, and a guide for applying computational optimization to reliability and maintainability theory. A chapter-wise summary of the information presented herein follows.

Chapter 1 presents a stochastic model for reliability indices of a computer system with priority and server failure. The model is analyzed by using the semi-Markov process and regenerative point technique. The reliability indices, such as mean time to system failure (MTCSF), availability, busy period of the server due to hardware repair and software upgradation, expected number of treatments given to the server, expected number of hardware repair, and software upgradation, are obtained for arbitrary values of the parameters. The profit analysis of the system model has also been carried out to discern the usefulness of the system under different parametric situations.

Chapter 2 presents a study that optimizes the availability of a turbine unit (TU) of a steam turbine power plant (STPP) using mathematical modeling and a genetic algorithm. The mathematical model is developed using the Markovian birth-death process (MBDP) and Chapman-Kolmogorov differential equations derived for the proposed model. The analytical solution of the mathematical model is derived for a particular case by considering exponential distribution for random variables associated with failure and repair rates. By using a nature-inspired algorithm (NIA), namely a genetic algorithm (GA), an effort is made to attain the global solution of the TU.

Chapter 3 covers the development of the Laplacian artificial bee colony (LABC) algorithm for effective harmonic estimator design. For designing the estimator, a hybrid approach based on least square error minimization with the help of a new version of the artificial bee colony algorithm is proposed. The proposed version employs a Laplacian factor-based update equation in the scout bee phase. For proving the modification meaningful, first the proposed algorithm is tested on several standard benchmark problems, and then it is applied to the estimator design problem. Results reported in on both parts indicate that the proposed modification is meaningful and the performance of the LABC algorithm is comparable with that of many other state-of-the-art algorithms.

Chapter 4 discusses the applications of the cuckoo search algorithm in reliability optimization, which is a novel nature-inspired algorithm that is used to solve complex optimization problems. The algorithm depends on the brood-parasitic strategy of cuckoo species. The usage of Lévy flights is used to produce new candidate resolutions. It can improve the relationship between exploration and exploitation towards the potential of searching. It can also be used in solving engineering problems such as embedded systems, distribution of networks, and scheduling problems. In this chapter, a study of the reliability of the software at static and runtime is performed and the results are also discussed.

Chapter 5 carries out a performance evaluation of the series-parallel computer system with a Gumbel-Hougaard copula family. To analyze the reliability of the system, the partial differential equations are derived from the system's schematic diagram in which reliability measures of system strength, such as reliability, availability, mean time to failure (MTTF), and cost function, are computed. The MTTF of devices, such as workstation, hub, and router, obeys exponential distribution whereas the corresponding repair time follows two different distributions, namely general and copula distribution. The findings of the study are depicted with the help of suitable diagrams and tabular representations.

Chapter 6 covers the applications of artificial intelligence (AI) in sustainable energy development and utilization. To combat the energy and environmental crises, clean and renewable fuels like biofuels are popular as petrodiesel replacement fuels. Biofuels can be obtained from different feedstocks and are successfully tested in diesel engines. However, several parameters influence the output results during their production and engine testing. The accurate prediction of end results is considered challenging with the traditional techniques. Therefore, AI techniques have emerged as being the most successful in solving nonlinear problems and achieving a high success rate in prediction. In this chapter, different AI techniques that have been successfully used in finding a feasible solution for complex problems in biodiesel production and engine testing are discussed in detail.

Chapter 7 introduces a new joint reliability achievement worth (JRAW), joint reliability reduction worth (JRRW), and joint reliability Fussell-Vesely (JRFV) measure for three multistate components of a multistate system. This is a new approach to detect the joint effect of a group of components in improving system reliability. The differencing technique is used in the proposed measures. A steady-state performance level distribution restricted to the component's states is used to evaluate the proposed measures. The universal generating function (UGF) technique is applied for the evaluation of proposed joint importance measures with suitable examples.

Chapter 8 presents some inferences about inverse Rayleigh distribution based on joint progressive Type-II censoring. The maximum likelihood estimation and the corresponding asymptotic confidence interval estimation are used as the classical estimation methods. The Bayes estimates are calculated under the squared error loss function (SELF) using Tierney-Kadane's approximation and Metropolis-Hastings algorithm, along with the construction of Bayes estimates highest posterior density credible intervals. A Markov chain Monte Carlo simulation study is carried out to compare different estimation methods and a real-life problem is discussed for illustrative purposes.

Chapter 9 deals with component reliability estimation through competing risk analysis of fuzzy lifetime data. In many cases, the lifetimes of systems are not precisely observed, or they are reported in "vague" terms. This imprecision or vagueness in data can be dealt with more accurately by incorporating fuzzy concepts. In this chapter, a competing risk analysis of lifetime data is performed by considering lifetimes as fuzzy numbers. Using different membership functions, the authors provide procedures for maximum likelihood and a Bayesian estimation of component reliability. They also evaluate bootstrap confidence intervals and the highest posterior density intervals. To observe the impact of various membership functions on the considered estimators, a comprehensive simulation study has been carried out. Finally, a real data set of small electric appliances has been analyzed.

Chapter 10 discusses the cost-benefit analysis of a redundant system with the provision of refreshment. Sometimes, due to some system-oriented snags and glitches, system performance may be hindered that can be overcome by repair. The goal of this chapter is to look at the survey of cost-benefit of a two-unit system with a single unit that can operate the system and another unit held as a spare in case of server failure, with refreshment provided to the server on demand.

Chapter 11 introduces a few novel inequalities of fuzzy measures and establishes the bounds in terms of triangular discrimination. Some new relations between new and existing fuzzy divergence measures are obtained with the help of the properties of a convex function and a new f-divergence measure. The utility of new fuzzy divergence measures in multi-criteria decision-making problems is also presented for better understanding.

Chapter 12 discusses the contribution of refreshment provided to the server during the job of repairing a cold standby system. The concept of probabilities of state transitions is presented followed by mean sojourn time and mean time to failure of the system. When calculating steady-state availability, a busy period of the server due to repair of the failed unit and a busy period of the server due to refreshment is computed followed by estimated visits made by the server. Novel conclusions are drawn based on considering particular cases and profit analysis.

Chapter 13 deals with stochastic modeling and availability optimization of a heat recovery steam generator using a genetic algorithm. The study presented in this chapter proposes a novel mathematical model for a heat recovery steam generator (HRSG) system to assess its availability. For this purpose, a state transition diagram is developed using the Markov birth-death process by considering all time-dependent failure and repair rates as exponentially distributed. The Chapman-Kolmogorov differential-difference equations are derived for the proposed model. The availability of the proposed model is optimized using a genetic algorithm to attain the global solution.

Chapter 14 investigates the reliability and maintainability of a piston manufacturing plant. For this analysis, data on time to repair and the number of failures was collected over two years. A descriptive analysis of the subsystems was performed along with trend and serial correlation testing. The best-fitted repair and failure time distributions among Weibull, normal exponential and lognormal distributions were investigated. The useful parameters corresponding to best-fitted distribution were estimated using U-statistics methodology and non-homogeneous Poisson process–power law process (NHPP-PLP). The reliability, availability, and hazard rates of the entire plant were calculated. The results were stored in numerical and graphical order concerning time to highlight the importance of the study. The model is useful not only in assessing the anticipated time for planning a maintenance schedule of a plant but also in terms of identifying the occurrence of failures in manufacturing plants.

Editors
Prof. S. C. Malik
Department of Statistics
M.D. University, Rohtak, India
Dr. Deepak Sinwar
Department of Computer and Communication Engineering
Manipal University Jaipur, India

Dr. Ashish Kumar
Department of Mathematics and Statistics
Manipal University Jaipur, India
Prof. S. R. Gadde
Department of Statistics
The University of Dodoma, Tanzania
Dr. Prasenjit Chatterjee
Department of Mechanical Engineering
MCKV Institute of Engineering, West Bengal, India
Dr. Bui Thanh Hung
Faculty of Information Technology, Artificial Intelligence Laboratory
Ton Duc Thang University, Ho Chi Minh City, Vietnam
December 2022

Acknowledgment

The editors wish to express their sincere thanks and appreciation to those who provided valuable support, constructive suggestions and assisted in editing this book.

This book would not have been possible without the valuable scholarly contributions of the authors.

The editors avow the endless support and motivation from their family members and friends.

Mere words cannot express the editors' deep gratitude to the entire Scrivener Publishing team, particularly Mr. Martin Scrivener for keeping faith and showing the right path to accomplish this very timely book.

Finally, the editors take this opportunity to thank all the readers and expect that this book will continue to inspire and guide them in high end researches.

The Editors

1

Reliability Indices of a Computer System with Priority and Server Failure

S.C. Malik*, R.K. Yadav and N. Nandal

Department of Statistics, M.D. University, Rohtak, India

Abstract

A stochastic model for a computer system has been described by considering theconcepts of failure of service facility, priority, and redundancy. The computer system comprises hardware and software components, which work together to perform the desired goals. The failure of the components is assumed as independent, which follow some probability distributions. There is a single server that has the responsibility to rectify the snags that occur during the operation of the components. The server is subjected to failure while performing jobs related to software upgradation but can resume the same after getting treatment. The server repairs the hardware components at failure while the software components are upgraded from time to time on a need basis. The preference to the software upgradation is given over the hardware repair. The computer system is considered as a single unit and an additional identical unit (redundant) is provided in order to meet out the emergency requirements. The distributions for repair time, treatment time and upgradation time are considered as negative exponential. The system model is analyzed by using semi-Markov process and regenerative point technique. The reliability indices, such as MTCSF, availability, busy period of the server due to hardware repair and software upgradation, expected number of treatments given to the server, expected number of hardware repair and software upgradation are obtained for arbitrary values of the parameters. The profit analysis of the system model has also been carried out to see the usefulness of the system under different parametric situations.

Keywords: Computer system, cold standby, failure of service facility, reliability indices, priority, software upgradation

**Corresponding author*: sc_malik@rediffmail.com

S. C. Malik, Deepak Sinwar, Ashish Kumar, S. R. Gadde, Prasenjit Chatterjee and Bui Thanh Hung (eds.) Computational Intelligence in Sustainable Reliability Engineering, (1–22) © 2023 Scrivener Publishing LLC

Mathematical Subject Classification: AMS: 60K10

1.1 Introduction

The entire responsibility for completing a task by a computer system lies on its constituents called h/w and s/w components. Therefore, the failure of any one component makes the computer system fail. These failures give rise to data loss, error in computer programs, natural disasters, economy loss, etc. Thus, it becomes very essential to maintain the reliability of the components of the computer system in order to provide effective services to the users. Friedman and Tran [4] gave the reliability techniques for combined hardware/software systems. The reliability modeling of hardware/software system has been discussed by Welke *et al.* [19]. Lai *et al.* [12] analyzed the availability of the distributed software/hardware system model.

The other way to intact the reliability of a computer system is to provide unit wise redundancy in it. The unit wise redundancy means to make a provision of another identical computer system which can be used as and when required at the failure of operating computer system. This provision of unit-wise redundancy may be made in cold standby or warm standby or in hot standby modes. The researches available in reliability theory indicate that the provision of cold standby redundancy in a system is much better than that of the others so far as reliability of the system is concerned. Sridharan & Mohanavadivu [18] and Meng *et al.* [15] analyzed the two unit cold standby systems. The stochastic analysis of cold standby system with two stage repair and waiting time has been carried out by El-Said and Sl-Sherbeny [3]. Bao and Cui [1] studied reliability for a cold standby Markov repairable system with neglected failures. Kumar *et al.* [5] checked the behavior of cold standby system with maximum repair time stochastically. Kumar and Goel [10] obtained profit of a two-unit cold standby system for general distribution. In the development of computer system model, the cold standby redundancy (unit wise/component wise) technique has also been used by the researchers, including Malik & Anand [13] and Munday & Malik [16].

Further, the idea of priority in repair discipline has been suggested by the researchers which can help in making the system more profitable to use and also to avoid unnecessary expenses on undesirable repair activities. In this direction, many authors, including Kumar & Malik [6], Kumar *et al.* [11], Kumar & Saini [8] and Kumar & Yadav [9], have worked and developed system models stochastically under different repair and maintenance strategies. But they failed to address the application of this idea in the stochastic modeling of computer systems.

The studies referred in this paper has no proof to authenticate the proper functioning of the service facility while conducting repair activities like s/w upgradation and h/w repair. In the computer system, upgradation of software means an improvement (newest version) in existing software, which is unable to meet out the assigned job. They considered the failure of the service facility negligible. But in realistic environment, the breakdown of the server is possible while repairing or upgrading the components of the operating systems. However, some authors including Bhardwaj & Singh [2], Nandal & Rathee [17] and Kumar & Saini [7] have tried to develop the system models by considering the failure of service facility. For the computer system, the service facility (server) can be the repairman or Internet or an individual (manually), which does h/w repair and s/w upgradation. The failure of service facility available for the repair of the computer system cannot be ignored as there may be the chances of its failure due to interruption in Internet connection or slow Internet speed or less knowledge of individual (manually) about that software. The failure of service facility has been considered by Yadav & Malik [20] and Malik *et al.* [14] while constructing stochastically the reliability models for a computer system.

The purpose of this chapter is to develop a stochastic model for a computer system by considering the ideas of failure of service facility, priority, and redundancy. The computer system comprises hardware and software components, which work together to perform the desired goals. The failure of the components is assumed as independent which follow some probability distributions. There is a single server who has the responsibility to rectify the snags, which occur during operation of the components. The server is subjected to failure while performing jobs related to software upgradation but can resume the same after getting treatment. The server repairs the hardware components at failure while the software components are upgraded from time to time whenever required. The preference to the software upgradation is given over the hardware repair. The computer system is considered as a single unit and redundancy in it with the identical unit is provided in order to meet out the emergency requirements. The distributions for repair time, treatment time and upgradation time are considered as negative exponential. The system model is analyzed by using semi-Markov process and regenerative point technique. The reliability indices, such as MTCSF, availability, busy period of the server due to hardware repair and software upgradation, expected number of treatments given to the server, expected number of hardware repair, and software upgradation, are obtained for arbitrary values of the parameters. The profit analysis of the system model has also been carried out to examine the usefulness of the system. The behavior of some reliability indices is presented graphically for arbitrary values of the parameters.

1.2 Some Fundamentals

Here, we shall describe in brief the following fundamentals.

1.2.1 Reliability

In broad term, reliability is considered as the probability of no failure of the system. If "T" is the lifetime of the system, then the system reliability is defined as:

$$R(t) = P_r[T > t] = \int_t^{\infty} f(u)du = 1 - F(t) = \bar{F}(\mathrm{t})$$

Where $f(t)$ is a probability density function of life time "T" and $F(t)$ is the cumulative density function of life time "T" or unreliability of the system, and $R(t)$ is the probability that the item does not fail in the time interval (0 , 1], and is still functioning at time "t."

1.2.2 Mean Time to System Failure (MTSF)

The expected time before the system completely fail is called mean time to system failure. Let $f(t)$ be the failure density function, then

$$MTSF = E(T) = \int_0^{\infty} tf(t)dt, \text{where, T is the time to failure}$$

$$= \lim_{t\to\infty} \int_0^t R(t)dt = \lim_{s\to 0} R^*(s), \text{where } R^*(s) \text{ is the Laplace transform of } R(t).$$

1.2.3 Steady State Availability

The probability that the system is operating successfully at time "t" is called availability of the system which is given by

$$\text{Availability } A(t) = \frac{\text{SystemUpTime}}{\text{SystemUpTime} + \text{SystemDownTime}}$$

The expected fraction of time that the system operates satisfactorily in the long run is known as steady state availability. Thus, steady state availability is

$$A(\infty) = \lim_{t \to \infty} A(t)$$

1.2.4 Redundancy

Redundancy is a common approach to improve the reliability and availability of a system. The provision of parallel paths (or alternative means) in a system for performing a given task such that all means must fail before causing the system failure, is called redundancy. It is mainly of two types: active redundancy and standby redundancy. The redundancy in which all spare units operate simultaneously is known as active redundancy, while the standby redundancy is that in which failed unit is replaced manually or automatically by its similar spare unit, and this process will continue until all the spare units (standby) have been exhausted.

For example, the system of power supply through electric transformer and generator is a case of standby system where a generator is kept as a spare (called redundant) and can be switched on as and when power supply through electric transformer is interrupted.

1.2.5 Semi-Markov Process

The semi-Markov process is a process in which transition from one state to another is governed by the transition probabilities of a Markov process but the time spent in each state before a transition occurs is a random variable depending upon the last transition made.

Mathematically, we assume that the process is time homogeneous, i.e., $Pr\{X_{n+1} = j, t_{n+1} - t_n < t | X_n = i\} = Q_{ij}(t)$, $i, j \in s$ is independent of n, then there exist limiting transition probabilities.

Here, $Q_{ij}(t)$ is the cdf of passage time from regenerative state S_i to a regenerative state S_j or to a failed state S_j without visiting any other regenerative state in (0, t]

$$p_{ij} = \lim Q_{ij}(t) = \Pr\{X_{n+1} = j | X_n = i\},$$

then $\{X_n$, n = 0, 1, 2, … … ..} constitute a Markov chain with state space E and transition probability matrix (t. p. m.)

$$P = [p_{ij}]$$

1.2.6 Regenerative Point Process

Regenerative stochastic process was defined by Smith (1955) and has been crucial in the analysis of complex system. In this, we take time points at which the system history prior to the time points is irrelevant to the system conditions. These points are called regenerative points. Let $X(t)$ be the state of the system of epoch. If t_1, t_2, … .. are the epochs at which the process probabilistically restarts, then these epochs are called regenerative epochs and the process {X(t), t = t_1, t_2, … … …} is called **regenerative process.**

1.3 Notations and Abbreviations

MTCSF	Mean Time to Computer System Failure
SMP	semi-Markov process
RPT	regenerative point technique
MST	mean Sojourn time
O/Cs	the unit is operative/in cold standby
a/b	probability of hardware/software failure
$x_1/x_2/\mu$	hardware/software/server failure rates
HFUr/HFWr	the failed hardware is under/waiting for repair
HFUR/HFWR	the failed hardware is continuously under/waiting for repair from prior state
SFUg/SFWg	the failed software is under/waiting for upgradation
SFUG/SFWG	the failed software is continuously under/waiting for up-gradation from prior state
SUt	the failed server (service facility) is under treatment
SUT	the failed server (service facility) is continuously under treatment from prior state
h(t)/H(t)	pdf/cdf of hardware repair time

$u(t)/U(t)$	pdf/cdf of software repair time
$s(t)/S(t)$	pdf/cdf of server treatment time
$m(t)/M(t)$	pdf/cdf of hardware preventive maintenance time
$q_{ij}(t)/Q_{ij}(t)$	pdf/cdf of first passage time
m_{ij}	contribution to MST(μ_i) in state S_i when system transits directly to state S_j
$M_i(t)$	probability that the system up initially in regenerative state S_i is up at time t without visiting any other regenerative state
$W_i^H(t)$	probability that the server is busy in the state S_i due to hardware failure up to time "t" without making any transition to any other regenerative state or returning to the same state via one or more non-regenerative states
$W_i^S(t)$	probability that the server is busy in the state S_i due to software up-gradation up to time "t" without making any transition to any other regenerative state or returning to the same state via one or more non-regenerative states
Ⓢ/©	standard notation for Laplace-Stieltjes convolution/Laplace convolution
*/**	symbol for Laplace Transform (LT)/Laplace Stieltjes Transform (LST)
P	profit function of the system model
K	system revenue per unit up-time
C/E	busy period cost of the server per unit time due to hardware repair/software up-gradation
L/M	repair/upgradation cost per unit time due to failure of hardware/software
N	treatment cost of the server per unit time

1.4 Assumptions and State Descriptions

The present stochastic model is developed under the following assumptions:

- There are two similar units (computer systems) in which one unit (computer system) is working and the other unit (computer system) is in cold standby.
- There is a single server to handle the h/w repair and s/w upgradation.
- The server may fail while upgrading the s/w.
- After the repair of components and treatment of server, the system will work with full efficiency.
- The preference is given to upgradation of s/w over h/w repair.
- The failure rates of the components are considered as constant while repair rates of components and treatment rate of server follow arbitrary distributions.

The descriptions of the states are as follows:

S_0	Good state as one unit is in operation and another unit is in cold standby.
S_1 & S_2	Good state as one unit is in operation and another unit is failed

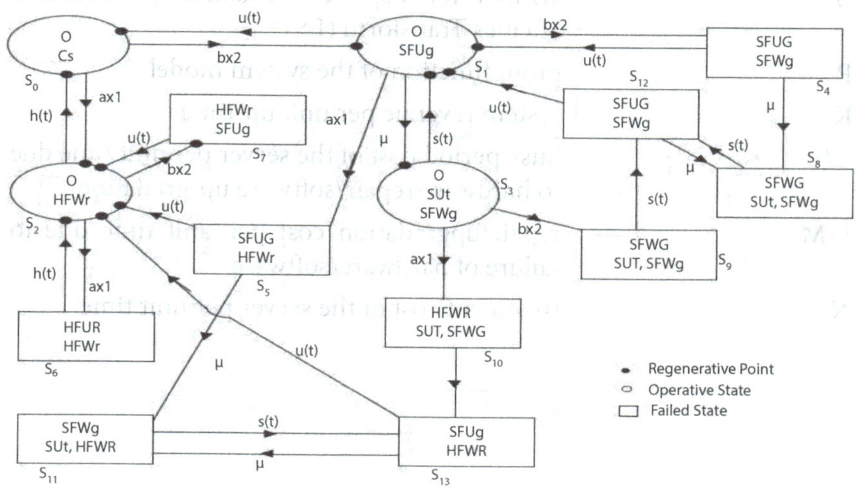

Figure 1.1 State transition diagram.

S_3	Good state as one unit is in operation, another unit and server is failed
$S_4 - S_7$, S_{12}& S_{13}	Failed state as both units are failed
S_8S_{11}	Failed state as both units and service facility are failed

The state transition diagram is shown in Figure 1.1.

1.5 Reliability Measures

1.5.1 Transition Probabilities

The differential transition probabilities for state S_0 are given by

$$dQ_{01}(t)=bx_2e^{-(ax_1+bx_2)t}dt, dQ_{02}(t)=ax_1e^{-(ax_1+bx_2)t}dt$$

Taking LST of above equations and using the following result

$$p_{ij}=\lim_{s\to 0}\varnothing_{ij}^{**}(s)=\varnothing_{ij}^{**}(0)=\int_0^{\infty}\mathrm{d}Q_{ij}(t)=\int_0^{\infty}q_{ij}(t)\mathrm{d}t, \text{we get}$$

$$p_{01}=\int_0^{\infty}bx_2e^{-(ax_1+bx_2)t}dt=\frac{bx_2}{ax_1+bx_2}, p_{02}=\int_0^{\infty}ax_1e^{-(ax_1+bx_2)t}dt=\frac{ax_1}{ax_1+bx_2}$$

Similarly, the remaining transition probabilities are calculated as:

$$p_{10}=u^*(ax_1+bx_2+\mu), p_{13}=\frac{\mu}{ax_1+bx_2+\mu}\{1-u^*(ax_1+bx_2+\mu)\},$$

$$p_{14}=\frac{bx_2}{ax_1+bx_2+\mu}\{1-u^*(ax_1+bx_2+\mu)\}, p_{15}=\frac{ax_1}{ax_1+bx_2+\mu}\{1-u^*(ax_1+bx_2+\mu)\},$$

$$p_{20}=h^*(ax_1+bx_2), p_{26}=\frac{ax_1}{ax_1+bx_2}\{1-h^*(ax_1+bx_2)\}, p_{27}=\frac{bx_2}{ax_1+bx_2}\{1-h^*(ax_1+bx_2)\}, p_{31}=s^*(ax_1+bx_2),$$

$$p_{39}=\frac{bx_2}{ax_1+bx_2}\{1-s^*(ax_1+bx_2)\}, p_{3,10}=\frac{ax}{ax_1+bx_2}\{1-s^*(ax_1+bx_2)\}, p_{41}=p_{52}=p_{12,1}=p_{13,7}=u^*(\mu),$$

$$p_{48}=p_{5,11}=p_{12,8}=p_{13,11}=1-u^*(\mu), p_{62}=h^*(0). p_{72}=u^*(0), p_{8,12}=p_{9,12}=p_{10,13}=p_{11,13}=s^*(0),$$

$$p_{11.4}=p_{14}p_{41}, p_{12.5}=p_{15}p_{52}, p_{22.6}=p_{26}, p_{31.9,12}=p_{39}p_{12,1}, p_{32.10,13}=p_{3,10}p_{13,2}, p_{11.4(8.12)n}=p_{14}p_{48},$$

$$p_{12.5(11,13)^n}=p_{15}p_{5,11}, p_{31.9,12(8,12)^n}=p_{39}p_{12,8}, p_{32.10,\ 3(11,13)^n}=p_{3,10}p_{13,11}$$

1.5.2 MST

The MST (μ_i) in state S_i are calculated by the following relations

$$m_{ij} = \left| -\frac{d}{ds} Q_{ij}^{**}(s) \right|_{s=0} = -Q_{ij}^{**\prime}(0) \text{ and } \mu_i = \sum_j m_{ij} \text{ where } Q_{ij}^{**}(s) = \int_0^\infty e^{-st}\, dQ_{ij}(t). \text{ Thus, we have}$$

$$\mu_0 = m_{01} + m_{02}, \mu_1 = m_{10} + m_{13} + m_{14} + m_{15}, \mu_2 = m_{20} + m_{26} + m_{27}, \mu_7 = m_{72}, \mu_3 = m_{31} + m_{39} + m_{3,10},$$

$$\mu_1' = m_{10} + m_{13} + m_{11.4} + m_{12.5} + m_{11.4(8,12)^n} + m_{12.5(11,13)^n}, \mu_2' = m_{20} + m_{22.6} + m_{27}, \mu_3' = m_{31} + m_{31.9,12} + m_{32.10,13}$$

$$+ m_{31.9,12(8,12)^n} + m_{32.10,13(11,13)^n}$$

1.5.3 Reliability and MTCSF

Let $\varnothing_i(t)$ be the c.d.f. of first passage time from regenerative state S_i to a failed state. Regarding the failed state as absorbing state, we have following recursive relations for $\varnothing_i(t)$:

$$\varnothing_i(t) = \sum_j Q_{ij}(t) \circledS \varnothing_j(t) + \sum_k Q_{ik}(t)$$

where S_j is an unfailed regenerative state to which the given regenerative state S_i can transit and S_k is a failed state to which the state S_i can transit directly. Thus, we have

$$\varnothing_0(t) = Q_{01}(t) \circledS \varnothing_1(t) + Q_{02}(t) \circledS \varnothing_2(t)$$
$$\varnothing_1(t) = Q_{10}(t) \circledS_0(t) + Q_{13}(t) \circledS \varnothing_3(t) + Q_{14}(t) + Q_{15}(t)$$
$$\varnothing_2(t) = Q_{20}(t) \circledS \varnothing_0(t) + Q_{26}(t) + Q_{27}(t)$$
$$\varnothing_3(t) = Q_{31}(t) \circledS \varnothing_1(t) + Q_{39}(t) + Q_{3,10}(t)$$

Taking LST of above expressions, we get

$$\varnothing_0^{**}(s) = Q_{01}^{**}(s)\varnothing_1^{**}(s) + Q_{02}^{**}(s)\varnothing_2^{**}(s)$$
$$\varnothing_1^{**}(s) = Q_{10}^{**}(s)\varnothing_0^{**}(s) + Q_{13}^{**}(s)\varnothing_3^{**}(s) + Q_{14}^{**}(s) + Q_{15}^{**}(s)$$
$$\varnothing_2^{**}(s) = Q_{20}^{**}(s)\varnothing_0^{**}(s) + Q_{26}^{**}(s) + Q_{27}^{**}(s)$$
$$\varnothing_4^{**}(s) = Q_{41}^{**}(s)\varnothing_1^{**}(s) + Q_{46}^{**}(s) + Q_{48}^{**}(s)$$

By the use of Cramer's Rule, we determine $\varnothing_0^{**}(s)$ as follows:

$$\varnothing_0^{**}(s)=\frac{\Delta_1}{\Delta}$$

$$\text{where } \Delta_1=\begin{vmatrix} 1 & -Q_{01}^{**}(s) & -Q_{02}^{**}(s) & 0 \\ -Q_{10}^{**}(s) & 1 & 0 & -Q_{13}^{**}(s) \\ -Q_{20}^{**}(s) & 0 & 1 & 0 \\ 0 & -Q_{31}^{**}(s) & 0 & 1 \end{vmatrix} \text{ and}$$

$$\Delta=\begin{vmatrix} 0 & -Q_{01}^{**}(s) & -Q_{02}^{**}(s) & 0 \\ Q_{14}^{**}(s)+Q_{15}^{**}(s) & 1 & 0 & -Q_{13}^{**}(s) \\ Q_{26}^{**}(s)+Q_{27}^{**}(s) & 0 & 1 & 0 \\ Q_{39}^{**}(s)+Q_{3,10}^{**}(s) & -Q_{31}^{**}(s) & 0 & 1 \end{vmatrix}$$

Now, we have $R^*(s)=\dfrac{1-\varnothing_0^{**}(s)}{s}$

The reliability is obtained by taking inverse LT of $R^*(s)$ as follows:

$$R(t) = L^{-1}\,[R^*(s)]$$

MTCSF is given by

$$MTCSF=\lim_{s\to 0}R^*(s)=R^*(0)=\frac{N_1}{D_1},\text{where } N_1=(1-p_{13}p_{31})(p_{02}\mu_2+\mu_0)$$

$$+p_{01}(p_{13}\mu_3+\mu_1)\text{ and } D_1=(1-p_{13}p_{31})(1-p_{02}p_{20})-p_{01}p_{10}$$

1.5.4 Availability

Let $A_i(t)$ be the probability that the system is in up-state at epoch "t" given that the computer system entered regenerative state S_i at $t = 0$. The recursive relations for $A_i(t)$ are given as

$$A_i(t)=M_i(t)+\sum_j q_{ij}^{(n)}(t)©A_j(t)$$

where S_j is any successive regenerative state to which the regenerative state S_i can transit through n transitions. Thus, we have

$$A_0(t)=M_0(t)+q_{01}(t)©A_1(t)+q_{02}(t)©A_2(t)$$
$$A_1(t)=M_1(t)+q_{10}(t)©A_0(t)+\left[q_{11.4}(t)+q_{11.4(8,12)^n}(t)\right]©A_1(t)$$
$$+\left[q_{12.5}(t)+q_{12.5(11,13)^n}(t)\right]©A_2(t)+q_{13}(t)©A_3(t)$$
$$A_2(t)=M_2(t)+q_{20}(t)©A_0(t)+q_{22.6}(t)©A_2(t)+q_{27}(t)©A_7(t)$$
$$A_3(t)=M_3(t)+\left[q_{31}(t)+q_{31.9,12}(t)+q_{31.9,12(8,12)^n}(t)\right]©A_1(t)$$
$$+\left[q_{32.10,13}(t)+q_{32.10,13(11,13)^n}(t)\right]©A_2(t)$$
$$A_7(t)=q_{72}(t)©A_2(t)$$

Where

$$M_0(t)=e^{-(ax_1+bx_2)t}, M_1(t)=e^{-(ax_1+bx_2+\mu)t}\overline{U}(t), M_2(t)=e^{-(ax_1+bx_2)t}\overline{H}(t), M_3(t)$$
$$=e^{-(ax_1+bx_2)t}\overline{S}(t)$$

We can obtain $A_0^*(s)$ (calculation is similar as given in Sec. 1.4.1), and thus steady state availability is given by

$$A_0(\infty)=\lim_{s\to 0} sA_0^*(s)=\frac{N_2}{D_2} \text{ where}$$
$$N_2=(p_{13}\mu_3+\mu_1)p_{20}p_{01}+\left(p_{20}\mu_0+\mu_2\right)\left(p_{10}+p_{15}+p_{13}p_{3,10}\right)-\mu_2 p_{01}p_{10}$$
$$D_2=\left(p_{13}\mu_3'+\mu_1'\right)p_{20}p_{01}+\left(p_{20}\mu_0+\mu_2'\right)\left(p_{10}+p_{15}+p_{13}p_{3,10}\right)-\mu_2 p_{01}$$
$$-\left(\mu_2'-\mu_7 p_{27}\right)p_{01}p_{10}$$

and $\mu_i=M_i^*(0), i=0,1,2,3$

1.5.5 Expected Number of Hardware Repairs

Let $R_i(t)$ be the expected number of the hardware repairs by the server in the interval (0,t] given that the computer system entered regenerative state S_i at $t = 0$. The expected number of the hardware repairs is given by

$$R_0(\infty) = \lim_{s \to 0} sR_0^{**}(s)$$

The recursive relations for $R_i(t)$ are given as:

$$R_i(t) = \sum_j Q_{ij}^{(n)}(t) ⓢ [\delta_j + R_j(t)]$$

where S_j is any regenerative state to which the regenerative state S_i can transit through n transitions and $\delta j = 1$, if S_j is the regenerative state where server does job afresh, otherwise $\delta j = 0$. Thus, we have

$$R_0(t) = Q_{01}(t) ⓢ R_1(t) + Q_{02}(t) ⓢ R_2(t)$$

$$R_1(t) = Q_{10}(t) ⓢ R_0(t) + \left[Q_{11.4}(t) + Q_{11.4(8,12)^n}(t)\right] ⓢ R_1(t)$$
$$+ \left[Q_{12.5}(t) + Q_{12.5(11,13)^n}(t)\right] ⓢ R_2(t) + Q_{13}(t) ⓢ R_{33}(t)$$

$$R_2(t) = Q_{20}(t) ⓢ [1 + R_1(t)] + [+Q_{21.7}(t) ⓢ [1 + R_1(t)] + Q_{27}(t) ⓢ R_7(t)$$

$$R_3(t) = \left[Q_{31}(t) + Q_{31.9,12}(t) + Q_{31.9,12(8,12)^n}(t)\right] ⓢ R_1(t)$$
$$+ \left[Q_{32.10,13}(t) + Q_{32.10,13(11,13)^n}(t)\right] ⓢ R_1(t)$$

$$R_7(t) = Q_{72}(t) ⓢ R_2(t)$$

We can obtain $R_0^{**}(s)$ (calculation is similar as given in Sec. 1.4.1), and thus the expected no. of the h/w repairs is given by

$$R_0(\infty) = \lim_{s \to 0} sR_0^{**}(s) = \frac{N_3}{D_2}$$

where $N_3 = (1 - 1p_{27})(p_{15} + p_{13}p_{3,10} + p_{02}p_{10})$ and

$$D_2 = (p_{13}\mu_3' + \mu_1')p_{20}p_{01} + (p_{20}\mu_0 + \mu_2')(p_{10} + p_{15} + p_{13}p_{3,10}) - \mu_2 p_{01}$$
$$- (\mu_2' - \mu_7 p_{27})p_{01}p_{10}$$

1.5.6 Expected Number of Software Upgradations

Let $U_i(t)$ be the expected number of the software upgradations by the server in the interval (0,t] given that the computer system entered regenerative state S_i at $t = 0$. The expected number of the software upgradations is given by

$$U_0(\infty)=\lim_{s\to 0} sU_0^{**}(s)$$

The recursive relations for $U_i\,(t)$ are given as:

$$U_i(t)=\sum_j Q_{ij}^{(n)}(t)\circledS[\delta_j+U_j(t)]$$

where S_j is any regenerative state to which the regenerative state S_i can transit through n transitions and $\delta j = 1$, if S_j is the regenerative state where server does job afresh, otherwise $\delta j = 0$. Thus, we have

$$\begin{aligned}
U_0(t)&=Q_{01}(t)\circledS U_1(t)+Q_{02}(t)\circledS U_2(t)\\
U_1(t)&=Q_{10}(t)\circledS\ \ 1+U_0(t)\ \ +\left[Q_{11.4}(t)\ \ +Q_{11.4(8,12)^n}(t)\ \ \right]\circledS[1+U_1(t)]\\
&\quad+\left[Q_{12.5}(t)+Q_{12.5(11,13)^n}(t)\right]\circledS[1+U_2(t)]+Q_{13}(t)\circledS U_3(t)\\
U_2(t)&=Q_{20}(t)\circledS U_0(t)+Q_{21.7}(t)\circledS U_1(t)+Q_{27}(t)\circledS U_7(t)\\
U_3(t)&=Q_{31}(t)\circledS U_1(t)+\left[Q_{31.9,12}(t)+Q_{31.9,12(8,12)^n}(t)\right]\circledS[1+U_1(t)]\\
&\quad+\left[Q_{32.10,13}(t)+Q_{32.10,13(11,13)^n}(t)\right]\circledS[1+U_1(t)]\\
U_7(t)&=Q_{72}(t)\circledS[1+U_2(t)]
\end{aligned}$$

We can obtain $U_0^{**}(s)$ (calculation is similar as given in Sec. 1.4.1), and thus the expected no. of the software upgradations is given by

$$U_0(\infty)=\lim_{s\to 0} sU_0^{**}(s)=\frac{N_4}{D_2}$$

where $N_4=(1-p_{13}p_{31})p_{01}p_{20}+p_{27}(p_{15}+p_{13}p_{3,10}+p_{02}p_{10})$ and

$$\begin{aligned}
D_2&=(p_{13}\mu_3'+\mu_1')p_{20}p_{01}+(p_{20}\mu_0+\mu_2')(p_{10}+p_{15}+p_{13}p_{3,10})-\mu_2 p_{01}\\
&\quad-(\mu_2'-\mu_7 p_{27})p_{01}p_{10}
\end{aligned}$$

1.5.7 Expected Number of Treatments Given to the Server

Let $T_i\,(t)$ be the expected number of the treatments given by the server in the interval (0,t] given that the computer system entered regenerative state S_i at $t = 0$. The expected number of the server treatments is given by

$$T_0(\infty) = \lim_{s \to 0} sT_0^{**}(s)$$

The recursive relations for $T_i(t)$ are given as:

$$T_i(t) = \sum_j Q_{ij}^{(n)}(t) \circledS [\delta_j + T_j(t)]$$

where j is any regenerative state to which the regenerative state i can transit through n transitions and δj = 1, if j is the regenerative state where server does job afresh, otherwise δj = 0. Thus, we have

$$T_0(t) = Q_{01}(t) \circledS T_1(t) + Q_{02}(t) \circledS T_2(t)$$

$$T_1(t) = Q_{10}(t) \circledS T_0(t) + Q_{11.4}(t) \circledS T_1(t) + Q_{11.4(8,12)^n}(t) \circledS [1 + T_1(t)]$$
$$+ Q_{12.5}(t) \circledS T_2(t) + Q_{12.5(11,13)^n}(t) \circledS [1 + T_2(t)] + Q_{13}(t) \circledS T_3(t)$$

$$T_2(t) = Q_{20}(t) \circledS T_0(t) + Q_{21.7}(t) \circledS T_1(t) + Q_{27}(t) \circledS T_7(t)$$

$$T_3(t) = [Q_{31}(t) + Q_{31.9,12}(t) + Q_{31.9,12(8,12)^n}(t)] \circledS [1 + T_1(t)]$$
$$+ [Q_{32.10,13}(t) + Q_{32.10,13(11,13)^n}(t)] \circledS [1 + T_1(t)]$$

$$T_7(t) = Q_{72}(t) \circledS T_2(t)$$

We can obtain $T_0^{**}(s)$ (calculation is similar as given in Sec.1.4.1), and thus the expected number of the treatments given to the server is given by

$$T_0(\infty) = \lim_{s \to 0} sT_0^{**}(s) = \frac{N_5}{D_2}$$

where $N_5 = [p_{48}(p_{14} + p_{15}) + p_{13}]\, p_{20} p_{01}$ and

$$D_2 = (p_{13}\mu_3' + \mu_1')\, p_{20} p_{01} + (p_{20}\mu_0 + \mu_2')(p_{10} + p_{15} + p_{13} p_{3,10})$$
$$- \mu_2 p_{01} - (\mu_2' - \mu_7 p_{27})\, p_{01} p_{10}$$

1.5.8 Busy Period of Server Due to H/w Repair

Let $B_i^H(t)$ be the probability that server is busy in repairing the unit due to hardware failure at epoch "t" given that the computer system entered state S_i at $t = 0$. The recursive relations for $B_i^H(t)$ are given as:

$$B_i^H(t) = W_i^H(t) + \sum_j q_{ij}^{(n)}(t) © B_j^H(t)$$

where S_j is any successive regenerative state to which the regenerative state S_i can transit through n transitions. Thus, we have

$$B_0^H(t) = q_{01}(t) © B_1^H(t) + q_{02}(t) © B_2^H(t)$$
$$B_1^H(t) = q_{10}(t) © B_0^H(t) + \left[q_{11.4}(t) + q_{11.4(8,12)^n}(t)\right] © B_1^H(t)$$
$$+\left[q_{12.5}(t) + q_{12.5(11,13)^n}(t)\right] © B_2^H(t) + q_{13}(t) © B_3^H(t)$$
$$B_2^H(t) = W_2^H(t) + q_{20}(t) © B_0^H(t) + q_{22.6}(t) © B_2^H(t) + +q_{27}(t) © B_7^H(t)$$
$$B_3^H(t) = \left[q_{31}(t) + q_{31.9,12}(t) + q_{31.9,12(8,12)^n}(t)\right] © B_1^H(t)$$
$$+\left[q_{32.10,13}(t) + q_{32.10,13(11,13)^n}(t)\right] © B_2^H(t)$$
$$B_7^H(t) = q_{72}(t) © B_2^H(t)$$

where $W_2^H(t) = \left[e^{-(ax_1+bx_2)t} + \left(ax_1 e^{-(ax_1+bx_2)t} © 1\right)\right]\bar{H}(t)$ and we can obtain $B_0^{H^*}(s)$ (calculation is similar as given in Sec. 1.4.1), and thus busy period of the server due to h/w repair is given by
where

$$B_0^H(\infty) = \lim_{s\to 0} sB_0^{H^*}(s) = \frac{N_6}{D_2},$$

where $N_6 = W_2^{H^*}(0)(p_{15} + p_{13}p_{3,10} + p_{02}p_{10})$ and

$$D_2 = (p_{13}\mu_3' + \mu_1')p_{20}p_{01} + (p_{20}\mu_0 + \mu_2')(p_{10} + p_{15} + p_{13}p_{3,10})$$
$$-\mu_2 p_{01} - (\mu_2' - \mu_7 p_{27})p_{01}p_{10}$$

1.5.9 Busy Period of Server Due to Software Upgradation

Let $B_i^S(t)$ be the probability that server is busy in repairing the unit due to software upgradation at epoch "t" given that the computer system entered state S_i at $t = 0$. The recursive relations for $B_i^S(t)$ are given as:

$$B_i^S(t) = W_i^S(t) + \sum_j q_{ij}^{(n)}(t) © B_j^S(t)$$

where S_j is any successive regenerative state to which the regenerative state S_i can transit through n transitions. Thus, we have

$$B_0^S(t) = q_{01}(t)©B_1^S(t) + q_{02}(t)©B_2^S(t)$$

$$B_1^S(t) = W_1^S(t) + q_{10}(t)©B_0^S(t) + \left[q_{11.4}(t) + q_{11.4(8,12)^n}(t)\right]©B_1^S(t)$$
$$+\left[q_{12.5}(t) + q_{12.5(11,13)^n}(t)\right]©B_2^S(t) + q_{13}(t)©B_3^S(t)$$

$$B_2^S(t) = q_{20}(t)©B_0^S(t) + q_{22.6}(t)©B_2^S(t) + +q_{27}(t)©B_7^S(t)$$

$$B_3^S(t) = \left[q_{31}(t) + q_{31.9,12}(t) + q_{31.9,12(8,12)^n}(t)\right]©B_1^S(t)$$
$$+\left[q_{32.10,13}(t) + q_{32.10,13(11,13)^n}(t)\right]©B_2^S(t)$$

$$B_7^S(t) = W_7^S(t) + q_{72}(t)©B_2^S(t)$$

where

$$W_1^S(t) = \Big[e^{-(ax_1+bx_2+\mu)t} + \left(ax_1e^{-(ax_1+bx_2+\mu)t}©1\right) + \left(ax_1e^{-(ax_1+bx_2+\mu)t}©\mu e^{-\mu t}©s(t)©1\right) +$$
$$\left(bx_2e^{-(ax_1+bx_2+\mu)t}©1\right) + \left(bx_2e^{-(ax_1+bx_2+\mu)t}©\mu e^{-\mu t}©s(t)©1\right)\Big]\bar{U}(t) \text{ and } W_7^S(t) = \bar{U}(t)$$

We can obtain $B_0^{S^*}(s)$ (calculation is similar as given in Sec. 1.4.1 and thus busy period of the server due to s/w upgradation is given by

$$B_0^S(\infty) = \lim_{s\to 0} sB_0^{S^*}(s) = \frac{N_7}{D_2}, \text{where}$$

$$N_7 = W_1^{S^*}(0)p_{20}p_{01} + W_7^{S^*}(0)p_{27}(p_{15} + p_{13}p_{3,10} + p_{12}p_{10}) \text{ and}$$

$$D_2 = (p_{13}\mu_3' + \mu_1')p_{20}p_{01} + (p_{20}\mu_0 + \mu_2')(p_{10} + p_{15} + p_{13}p_{3,10})$$
$$-\mu_2 p_{01} - (\mu_2' - \mu_7 p_{27})p_{01}p_{10}$$

1.6 Profit Analysis

The profit of the computer system model can be evaluated by following formula

$$P = KA_0(\infty) - LR_0(\infty) - MU_0(\infty) - NT_0(\infty) - CB_0^H(\infty) - EB_0^S(\infty)$$

1.7 Particular Case

Let us consider $h(t) = \alpha e^{-\alpha t}$, $u(t) = \beta e^{-\beta t}$ and $s(t) = \gamma e^{-\gamma t}$ then transition probabilities are given by

$$p_{01} = \frac{bx_2}{ax_1 + bx_2}, p_{02} = \frac{ax_1}{ax_1 + bx_2}, p_{10} = \frac{\beta}{ax_1 + bx_2 + \mu + \beta}, p_{13} = \frac{\mu}{ax_1 + bx_2 + \mu + \beta}, p_{15} = \frac{ax_1}{ax_1 + bx_2 + \mu + \beta},$$
$$p_{20} = \frac{\alpha}{ax_1 + bx_2 + \alpha}, p_{27} = \frac{bx_2}{ax_1 + bx_2 + \alpha}, p_{31} = \frac{\gamma}{ax_1 + bx_2 + \gamma}, p_{3,10} = \frac{ax_1}{ax_1 + bx_2 + \gamma}\ p_{48} = \frac{\mu}{\mu + \beta},$$
$$\mu_0 = \frac{1}{ax_1 + bx_2}, \mu_1 = \frac{1}{ax_1 + bx_2 + \mu + \beta}, \mu_2 = \frac{1}{ax_1 + bx_2 + \alpha}, \mu_3 = \frac{1}{ax_1 + bx_2 + \gamma},$$
$$\mu_1' = \frac{\gamma\beta + (ax_1 + bx_2)(\mu + \gamma)}{\gamma\beta(ax_1 + bx_2 + \mu + \beta)}, \mu_2' = \frac{ax_1 + \alpha}{\alpha(ax_1 + bx_2 + \alpha)} = W_2^{H^*}(0), \mu_7 = \frac{1}{\beta} = W_7^{S^*}(0),$$
$$\mu_3' = \frac{\gamma\beta + (ax_1 + bx_2)(\mu + \gamma + \beta)}{\gamma\beta(ax_1 + bx_2 + \gamma)}, W_1^{S^*}(0) = \frac{(ax_1 + bx_2 + \beta)(\mu + \beta)(\gamma + \beta) + \mu\gamma(ax_1 + bx_2)}{\beta(\mu + \beta)(\gamma + \beta)(ax_1 + bx_2 + \mu + \beta)}$$

We can obtain the following results

$$MTSF = \frac{\begin{array}{c}(ax_1 + 2bx_2 + \alpha)\{(ax_1 + bx_2 + \mu + \beta)(ax_1 + bx_2 + \gamma) - \mu\gamma\} + bx_2 \\ (ax_1 + bx_2 + \alpha)(ax_1 + bx_2 + \mu + \gamma)\end{array}}{\{(ax_1 + bx_2 + \mu + \beta)(ax_1 + bx_2 + \gamma) - \mu\gamma\}\{(ax_1 + bx_2)(ax_1 + bx_2 + \alpha) - \alpha bx_2\} -}$$

$$A_0(\infty) = \frac{\begin{array}{c}\beta bx_2(ax_1 + bx_2 + \gamma)(ax_1 + bx_2 + \alpha)\alpha\gamma\beta(ax_1 + \alpha)\{(ax_1 + bx_2) \\ (ax_1 + bx_2 + \gamma + \mu) + \beta(ax_1 + bx_2 + \gamma)\}\end{array}}{\begin{array}{l}\alpha^2[(ax_1 + bx_2 + \gamma + \mu)(ax_1 + bx_2)\{\gamma\beta + bx_2(\mu + \gamma)\} + \beta\{(ax_1 + bx_2)(\gamma\beta + \mu bx_2) + \\ \beta\gamma^2] + \gamma\{\beta(ax_1 + \alpha) - \alpha bx_2\}\{ax_1(ax_1 + bx_2)(ax_1 + bx_2 + \mu + \gamma) - \gamma\beta(ax_1 + bx_2 + \gamma)\}\end{array}}$$

$$R_0(\infty) = \frac{ax_1\alpha\gamma\beta(ax_1 + \alpha)\{(ax_1 + bx_2)(ax_1 + bx_2 + \gamma + \mu) + \beta(ax_1 + bx_2 + \gamma)\}}{\begin{array}{l}\alpha^2[(ax_1 + bx_2 + \gamma + \mu)(ax_1 + bx_2)\{\gamma\beta + bx_2(\mu + \gamma)\} + \beta\{(ax_1 + bx_2)(\gamma\beta + \mu bx_2) + \\ \beta\gamma^2] + \gamma\{\beta(ax_1 + \alpha) - \alpha bx_2\}\{ax_1(ax_1 + bx_2)(ax_1 + bx_2 + \mu + \gamma) - \gamma\beta(ax_1 + bx_2 + \gamma)\}\end{array}}$$

$$T_0(\infty) = \frac{bx_2\alpha^2\gamma\beta(ax_1 + bx_2 + \gamma)(ax_1 + bx_2 + \beta + \mu)}{\begin{array}{l}(\beta + \mu)[\alpha^2[(ax_1 + bx_2 + \gamma + \mu)(ax_1 + bx_2)\{\gamma\beta + bx_2(\mu + \gamma)\} + \beta\{(ax_1 + bx_2) \\ (\gamma\beta + \mu bx_2) + \beta\gamma^2] + \gamma\{\beta(ax_1 + \alpha) - \alpha bx_2\}(ax_1(ax_1 + bx_2)(ax_1 + bx_2 + \mu + \gamma) - \\ \gamma\beta(ax_1 + bx_2 + \gamma))]\end{array}}$$

$$B_0^H(\infty) = \frac{ax_1\gamma\beta(ax_1 + \alpha)\{(ax_1 + bx_2 + \beta)(ax_1 + bx_2 + \gamma + \mu) + \mu(ax_1 + bx_2)\}}{\begin{array}{l}\alpha^2[(ax_1 + bx_2 + \gamma + \mu)(ax_1 + bx_2)\{\gamma\beta + bx_2(\mu + \gamma)\} + \beta\{\{ax_1 + bx_2\}(\gamma\beta + \mu bx_2) + \beta\gamma^2] + \\ \gamma\{\beta(ax_1 + \alpha) - \alpha bx_2\}\{ax_1(ax_1 + bx_2)(ax_1 + bx_2 + \mu + \gamma) - \gamma\beta(ax_1 + bx_2 + \gamma)\}\end{array}}$$

$$B_0^S(\infty) = \frac{\begin{array}{c}bx_2\alpha\gamma[(\beta + \gamma)(\beta + \mu)\{(ax_1 + \alpha)\{(ax_1 + bx_2 + \beta)(ax_1 + bx_2 + \gamma) + \\ \mu ax_1(ax_1 + bx_2)\} + \alpha(ax_1 + bx_2 + \gamma)\mu\gamma(ax_1 + bx_2)]\end{array}}{\begin{array}{l}(\beta + \gamma)(\beta + \mu)[\alpha^2[(ax_1 + bx_2 + \gamma + \mu)(ax_1 + bx_2)\{\gamma\beta + bx_2(\mu + \gamma)\} + \beta\{(ax_1 + bx_2)(\gamma\beta + \mu bx_2) + \\ \beta\gamma^2] + \gamma\{\beta(ax_1 + \alpha) - \alpha bx_2\}\{ax_1(ax_1 + bx_2)(ax_1 + bx_2 + \mu + \gamma) - \gamma\beta(ax_1 + bx_2 + \gamma)\}]\end{array}}$$

1.8 Graphical Presentation of Reliability Indices

The reliability indices are presented graphically in following Figures 1.2–1.4.

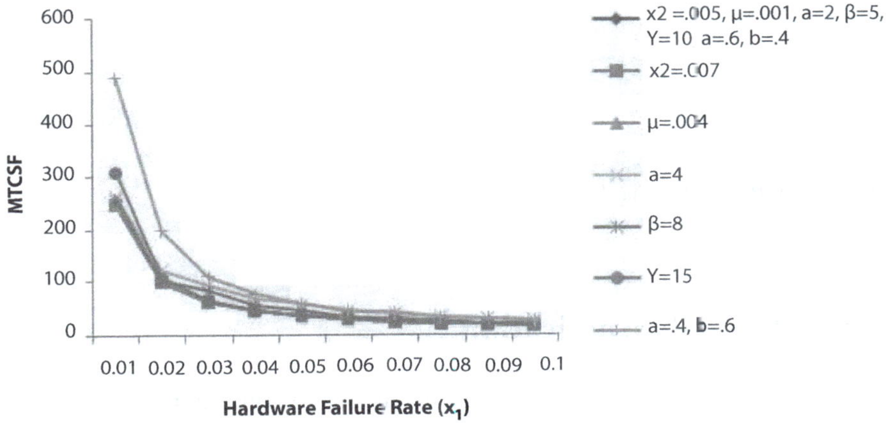

Figure 1.2 MTSF vs hardware failure rate (x_1).

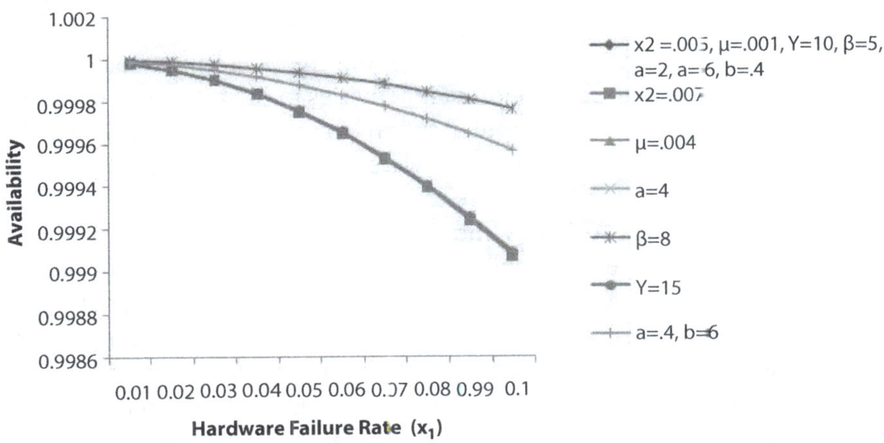

Figure 1.3 Availability vs hardware failure rate (x_1).

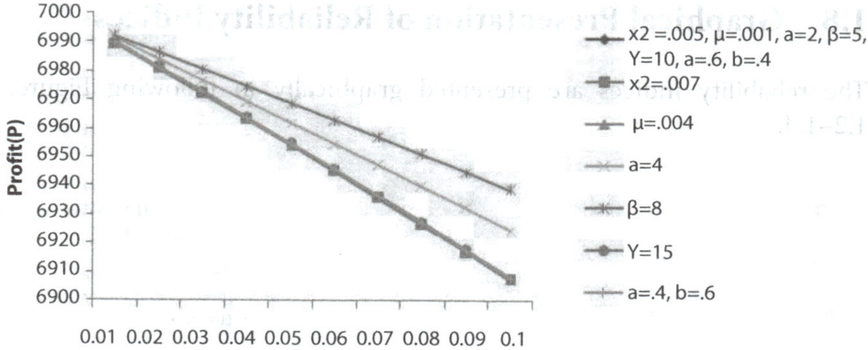

Figure 1.4 Profit (P) vs hardware failure rate (x_1).

1.9 Real-Life Application

The example of the present study can be taken as a computer system operating in a shopping mall. In the mall, a computer system is used for billing of luggage which is handled by an operator (manpower). Another similar unit (computer system) is also taken as a spare in cold standby which is used after the failure of the operative unit and this provision is called unit wise redundancy for a computer system. An operator (service facility) is hired to take the responsibility for observing the operating situations of the systems and also to carry out repair activities in the form of h/w repair and s/w upgradation. The computer system may fail because of its hardware failure or due to the need of software upgradation. Generally, there is a need of repair for the hardware at its failure and upgradation for the software when it fails to follow the instructions or programming as per requirement. Therefore, the single service facility undertakes hardware for repair and upgrades the software. The service facility is subjected to failure during hardware repair and thus the provision of treatment of the service facility has been made so that the repair activities may be resumed as soon as possible. On the other hand, priority may be given to the repair discipline and so here priority to s/w upgradation in one unit (operating unit) is given over the h/w repair of the other unit (already failed due to nonfunctioning of the h/w) in view of the fact that h/w component has already been available for its use in the operating unit.

1.10 Conclusion

The system model has been analyzed stochastically by taking one more unit (computer system) in cold standby and so-called unit wise cold standby redundancy. There is a single server which undertakes hardware under repair when it fails and software for upgradation whenever needed. The upgradation of s/w has been given priority over the h/w repair. The trend of reliability indices MTCSF, availability and profit has been noticed w.r.t h/w failure rate (x_1) and for the fixed values of parameters associated with h/w repair rate, s/w upgradation rate and treatment rate of server as shown in Figures 1.2–1.4, respectively. It is observed that MTCSF, availability, and profit function go on decreasing with increase in components failure rates and server failure rate (μ) while they increase with the increase of h/w repair rate (α), s/w upgradation rate (β), and treatment rate (ϒ). Hence, it is suggested that a computer system can be made more useful and profitable by providing cold standby redundancy and proper repair facilities for hardware repair and software upgradation.

References

1. Bao, X. and Cui, L., A study on reliability for a two-item cold standby markov repairable system with neglected failures. *Commun. Stat Theory Methods*, 41, 21, 3988–3999, 2012.
2. Bhardwaj, R.K. and Singh, R., Steady state behavior of a cold-standby system with server failure and arbitrary repair, replacement & treatment. *Int. J. Appl. Eng. Res.*, 9, 24, 26563–26578, 2014.
3. El-Said, K.M. and El-Sherbeny, M.S., Stochastic analysis of a two-unit cold standby system with two-stage repair and waiting time. *Sankhya: Indian J. Stat. B*, 72, 1, 1–10, 2010.
4. Friedman, M.A. and Tran, P., Reliability techniques for combined hardware/software systems, in: *Conference: Reliability and Maintainability Symposium, Proceedings Annual*, pp. 290–293, IEEE, Las Vegas, NV, USA, 1992.
5. Kumar, A., Baweja, S., Barak, M.S., Stochastic behavior of a cold standby system with maximum repair time. *Decis. Sci. Lett.*, 4, 4, 569–578, 2015.
6. Kumar, A. and Malik, S.C., Profit analysis of a computer system with priority to software replacement over hardware repair subject to maximum operation and repair times. *Int. J. Eng. Sci. Technol.*, 3, 10, 7452–7468, 2011.
7. Kumar, A. and Saini, M., Analysis of a single-unit system with weibull failure and repair densities subject to server failure. *Malaysian J. Sci.*, 35, 1, 15–22, 2018.

8. Kumar, A. and Saini, M., Profit analysis of a computer system with preventive maintenance and priority subject to maximum operation and repair times. *Iran J. Comput. Sci.*, 1, 3, 147–153, 2018.
9. Kumar, A. and Yadav, R.K., Stochastic analysis of a computer system with hardware redundancy and priority to software up-gradation subject to failure of service facility. *Int. J. Stat Reliability Eng.*, 7, 1, 160–167, 2020.
10. Kumar, J. and Goel, M., Availability and profit analysis of a two unit cold standby system for general distribution. *Cogent Math.*, 3, 1, 1–30, 2016.
11. Kumar, J., Malik, S.C., Anand, J., Cost-benefit analysis of a computer system with priority to s/w replacement over h/w repair. *Appl. Math. Sci.*, 6, 75, 3723–3734, 2012.
12. Lai, C.D., Xie, M., Poh, K.L., Dai, Y.S., Yang, P., A model for availability analysis of distributed software/hardware systems. *Inf. Softw. Technol.*, 44, 343–350, 2002.
13. Malik, S.C. and Anand, J., Reliability and economic analysis of a computer system with independent hardware and software failures. *Bull. Pure Appl. Sci.*, 29E, 1, 141–153, 2010.
14. Malik, S.C., Kumar, A., Yadav, R.K., Stochastic analysis of a computer system with software redundancy and failure of service facility. *Int. J. Adv. Sci. Technol.*, 29, 04, 7279–7288, 2020.
15. Meng, X., Yuan, L., Yin, R., The reliability analysis of a two-unit cold standby system with failable switch and maintenance equipment, in: *International Conference on Computational Intelligence and Security*, vol. 2, pp. 941–944, IEEE, Guangzhou, China, 2006.
16. Munday, V.J. and Malik, S.C., Reliability measures of a computer system with different repair and redundant policies for components. *Int. J. Stat Reliability Eng.*, 1, 1, 26–35, 2014.
17. Nandal, J. and Rathee, R., Stochastic analysis of a redundant system with server failure and conditional arrival time. *Int. J. Stat Reliability Eng.*, 2, 1, 94–102, 2015.
18. Sridharan, V. and Mohanavadivu, P., Stochastic behavior of two-unit standby system with two types of repairmen and patience time. *Math. Comput. Model.*, 28, 9, 63–71, 1998.
19. Welke, S.R., Labib, S.W., Ahmed, A.M., Reliability modelling of hardware/software system. *IEEE Trans. Reliab.*, 44, 3, 413–418, 1995.
20. Yadav, R.K. and Malik, S.C., Stochastic analysis of a computer system with unit wise cold standby redundancy and failure of service facility. *Int. J. Math. Eng. Manage. Sci.*, 5, 3, 529–543, 2020.

2

Mathematical Modeling and Availability Optimization of Turbine Using Genetic Algorithm

Monika Saini, Nivedita Gupta and Ashish Kumar*

Department of Mathematics & Statistics, Manipal University Jaipur, Jaipur, Rajasthan, India

Abstract

The main aim of present study is to optimize the availability of a turbine unit (TU) of steam turbine power plant (STPP). For this purpose, a mathematical model is developed using Markovian Birth-Death Process (MBDP) and Chapman-Kolmogorov differential equations derived for a proposed model. The analytical solution of mathematical model is derived for a particular case by considering exponential distribution for random variables associated with failure and repair rates. By using nature-inspired algorithm (NIA), namely genetic algorithm (GA), an effort is made to attain the global solution of the TU. The derived results are presented in tabular and graphical forms.

Keywords: Availability, optimization, genetic algorithm, Markov birth-death process

2.1 Introduction

In spite of advancement of renewable energy resources, steam turbine power plants (STPP) are the primary source of electricity generation in the world. In most of the STPPs, coal is used as a fuel. Though on earth, sufficient coal reserves are available for next two centuries, coal causes severe negative impact on environment. During the last few decades, other

**Corresponding author*: ashishbarak2020@gmail.com

S. C. Malik, Deepak Sinwar, Ashish Kumar, S. R. Gadde, Prasenjit Chatterjee and Bui Thanh Hung (eds.) Computational Intelligence in Sustainable Reliability Engineering, (23–46) © 2023 Scrivener Publishing LLC

sources like oil, natural gases and uranium were also used as fuel. But now, most of the energy companies focusing on the process improvement of plant by increasing the efficiency of plant and simultaneously the concentration is also given on decreasing the diverse environmental effects. The performance of any plant can be enhanced by increasing the reliability and availability of its components as these are the prominent measures of system effectiveness. STPPs are very complex systems having several components, like turbine, generator, cooling tower, heat recovery steam generator, etc. But among these turbines play a crucial role. Many researchers tried to investigate the reliability and availability of complex systems like industrial systems, power plants, fertilizer manufacturing units through various reliability evaluation techniques. Though, steam turbine power plant and its components do not explore so extensively as far as reliability evaluation is concerned. Sabouhi *et al.* [1] developed a reliability model for combined cycle power plants and investigated the availability of it. Carazas and Desouza [2] conducted the availability investigation of gas turbines of power plants. Gupta *et al.* [3] investigated the operational availability of generators used in STPPs using Markov birth-death process and supplementary variable technique. Erdem *et al.* [4] comparatively measured the energetic and exegetic performance of coal fired thermal power plants situated in Turkey. Carazas *et al.* [5] analyzed the availability of heat recovery steam generators of thermal power plants. Shopeju and Oyedepo [6] conducted a broad literature review on the use of stochastic models in power plants. Gupta *et al.* [7] adopted the RAMD technique for reliability investigation of generator of STTPs. Nasiriani *et al.* [8] conducted the reliability evaluation of ocean thermal energy conversion power plants. Luo *et al.* [9] performed the operational availability optimization of petrochemical complex under consideration of equipment failure. Radin [10] developed the mathematical model for increasing the flexibility and reliability of power units in thermal power plants based on steam turbines. Sultanov *et al.* [11] proposed the methodology to assess the actual performance of steam turbine plant-based technical conditions of system using reliability indices. Sunil *et al.* [12] proposed a mathematical model for boiler drum to assess its reliability. Prock [13] suggested a mathematical model for steam generator to detect the sensor fault. Arora and Kumar [14] analyzed the availability of steam and power generation systems using semi-Markovian approach. Bhangu *et al.* [15] analyzed the availability of thermal power plants. Gupta [16] used stochastic models and derived expression for availability of critical engineering systems. De Souza [17] proposed several strategies to assess the performance of thermal power plants. Bose *et al.* [18] investigated reliability and maintainability of thermal power plants using Markovian approach. Tanuma [19, 20] described in detail the configurations and working methodology of steam turbine power plants. De Souza *et al.* [21] assess the reliability of combined cycle steam turbine power plants. Kumar *et al.* [22] suggested a Markov model to evaluate the availability of power generation system. Houdova *et al.* [23] developed an approach for solving availability problems and to optimize the cost in power plants. Kim [24] conducted the failure analysis for a last-stage blade in low-pressure power plants. Rajesh and Prasad [25] derived various system effectiveness measures for a three-unit gas turbine power generating unit with seasonal effect. Ueasin *et al.* [26] carried out the performance assessment of the biomass steam turbine power plant and optimized the results. Okafor and Atikpakpa [27] used Markovian approach for assessing the reliability of a steam and gas turbine unit of power station. Oyedepo and Fagbenle [28] carried out a study to implement maintenance strategies in thermal power plants.

The whole manuscript is divided into six sections, including present introductory section, the system description, notations are given in section 2.2, section 2.3 developed a mathematical model, description of optimization technique is appended in section 2.4, section 2.5 described the results in detail, and the last section included the results.

2.2 System Description, Notations, and Assumptions

2.2.1 System Description

The turbine subsystem is a configuration of the following components as given in Figure 2.1.

- Subsystem A (high-pressure turbine): the high-pressure (HP) turbine is the initial turbine in main engine, which receive steam from the main steam system. The high-pressure turbines are the smallest of the set, with the shortest blades, and they receive steam from the boiler at the highest temperature and pressure. It is designed in such a way that it extracts work out of high-pressure steam. This subsystem consists of one unit which is connected in series with subsequent subsystems.
- Subsystem B (Intermediate pressure turbine): Intermediate-pressure turbine is just next after the high-pressure turbine. After high-pressure turbine steam enters intermediate

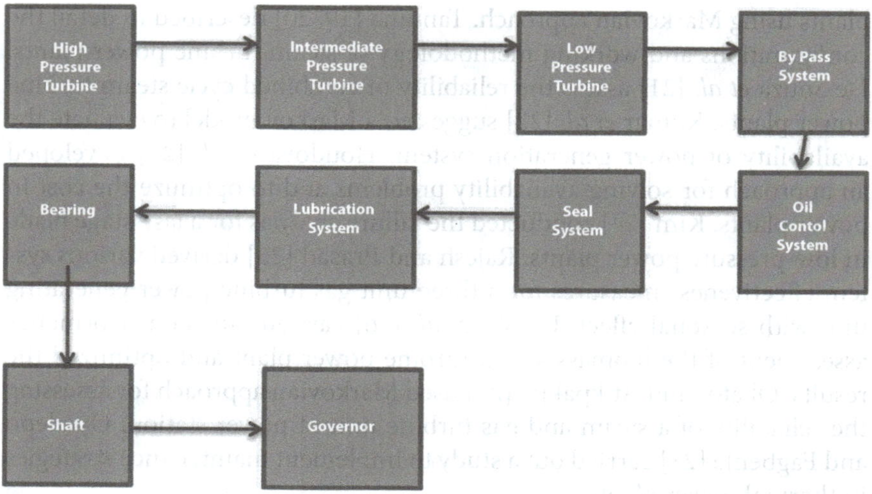

Figure 2.1 Configuration diagram of turbine.

pressure turbine and generate energy. Intermediate-pressure turbine has longer bladed than high-pressure turbine and is optimized for steam at a lower-entry temperature and pressure. This subsystem consists of one unit which is connected in series with subsequent subsystems.

- Subsystem C (low-pressure turbine): In turbine, the high-pressure steam is used to drive the turbine blades. In this process, it loses its energy and gets converted to low-pressure steam. In case of reheating, since the steam which is to be further expanded is a low-pressure steam, therefore, we require a low-pressure turbine to expand it. It has bigger blades than intermediate pressure turbine. This subsystem consists of one unit which is connected in series with subsequent subsystems.
- Subsystem D (bypass system): The main function of turbine bypass system is to isolate the bypass loop during operation. Failure of it damaged seats and valves, which result in lost energy and loss of control during start-up and turbine trips. This subsystem consists of one unit, which is connected in series with subsequent subsystems.
- Subsystem E (oil control system): Oil control system is mainly used to make hydraulic part of steam turbine functional.

This subsystem consists of one unit, which is connected in series with subsequent subsystems.

- Subsystem F (seal system): A seal oil systemprovides cool and clean seal oil to the seals at the proper pressure. It also prevents gas from leaking out of the compressor. This subsystem consists of one unit which is connected in series with subsequent subsystems.
- Subsystem G (lubrication system): Steam turbine lubrication system is normally needed to supply oil for trip-and-throttle valve, power cylinder, or similar unit such as combined pressure lubrication, control oil unit, etc. This subsystem consists of one unit which is connected in series with subsequent subsystems.
- Subsystem H (bearing): Bearing plays a supportive role in turbine rotor. It also projects the shaft from any tear and wear during the running condition. Set of bearings jointly make a subsystem, which is connected in series with subsequent subsystems.
- Subsystem I (shaft): The shaft connects the turbine to the generator, turning at the same speed as the turbine. This subsystem consists of one unit, which is connected in series with subsequent subsystems.
- Subsystem J (governor): A steam turbine governor provides startup and on-line control for steam turbine driven generators and mechanical drives such as compressors and pumps. This subsystem consists of one unit which is connected in series with subsequent subsystems.

2.2.2 Notations

: Indicates system is in full capacity

: Indicates system is in failed states

A, B, C, D, E, F, G, H, I, J: Represent units working with full capacity

a, b, c, d, e, f, g, h, i, j: Represent failed units

$S_i (0 \le i \le 10)$: Represent the different states

$\alpha_i (0 \le i \le 10)$: The constant failure rates of the subsystems *A, B, C, D, E, F, G, H, I, J* respectively

$\beta_i (0 \le i \le 10)$: Represent the repair rates of the subsystem *A, B, C, D, E, F, G, H, I, J*, respectively

$P_0(t)$: The probability that system is in full capacity at initial time t
$P_i(x, t)$ $(0 \leq i \leq 10)$: The probability that the system is in i^{th} state at time t has an elapsed repair timex

2.2.3 Assumptions

- The failure rates follow exponential distribution and repair rates follows arbitrary distribution and are independent to each other.
- There are no simultaneous failures among the subsystems.
- There are sufficient repair and replacement facilities. Repairmen always present in plant and performance wise repaired system is as good as new.

2.3 Mathematical Modeling of the System

Here we are using supplementary variable technique to obtain the set of differential-difference equations associated with the model as given in Figure 2.2.

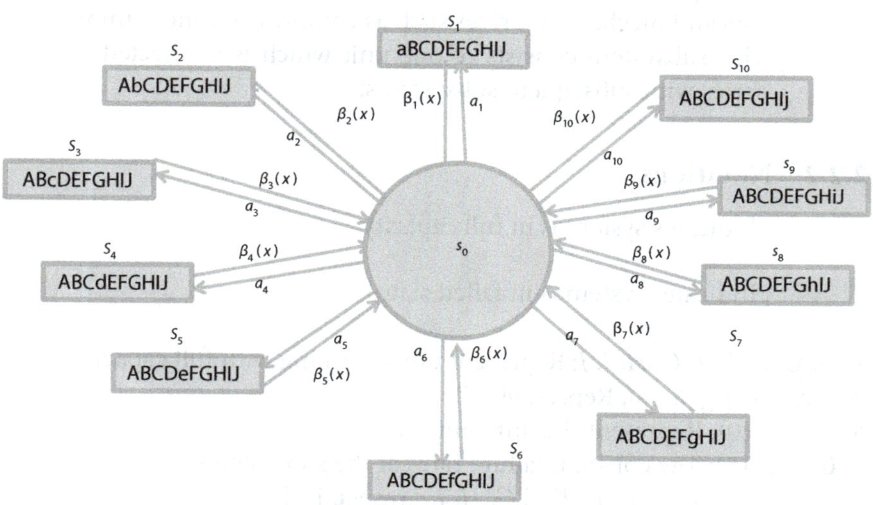

Figure 2.2 State changeover diagram of a turbine subsystem in steam turbine power plant.

$$P_0(t+\Delta t)=(1-\alpha_1\Delta t-\alpha_2\Delta t-\alpha_3\Delta t-\alpha_4\Delta t-\alpha_5\Delta t-\alpha_6\Delta t-\alpha_7\Delta t-\alpha_8\Delta t-\alpha_9\Delta t-\alpha_{10}\Delta t)P_0(t)+$$
$$\int_0^\infty \beta_1(x)P_1(x,t)\Delta t dx+\int_0^\infty \beta_2(x)P_2(x,t)\Delta t dx+\int_0^\infty \beta_3(x)P_3(x,t)\Delta t dx+\int_0^\infty \beta_4(x)P_4(x,t)\Delta t dx+$$
$$\int_0^\infty \beta_5(x)P_5(x,t)\Delta t dx+\int_0^\infty \beta_6(x)P_6(x,t)\Delta t dx+\int_0^\infty \beta_7(x)P_7(x,t)\Delta t dx+\int_0^\infty \beta_8(x)P_8(x,t)\Delta t dx+$$
$$\int_0^\infty \beta_9(x)P_9(x,t)\Delta t dx+\int_0^\infty \beta_{10}(x)P_{10}(x,t)\Delta t dx$$

Dividing both the sides of above equation by Δt, we get

$$\frac{P_o(t+\Delta t)-P_0(t)}{\Delta t}=(-\alpha_1-\alpha_2-\alpha_3-\alpha_4-\alpha_5-\alpha_6-\alpha_7-\alpha_8-\alpha_9-\alpha_{10})P_0(t)+\int_0^\infty \beta_1(x)P_1(x,t)dx+$$
$$\int_0^\infty \beta_2(x)P_2(x,t)dx+\int_0^\infty \beta_3(x)P_3(x,t)dx+\int_0^\infty \beta_4(x)P_4(x,t)dx+\int_0^\infty \beta_5(x)P_5(x,t)dx+$$
$$\int_0^\infty \beta_6(x)P_6(x,t)dx+\int_0^\infty \beta_7(x)P_7(x,t)dx+\int_0^\infty \beta_8(x)P_8(x,t)dx+\int_0^\infty \beta_9(x)P_9(x,t)dx+$$
$$\int_0^\infty \beta_{10}(x)P_{10}(x,t)dx$$

As $\Delta t \to 0$

$$\lim_{\Delta t\to 0}\frac{P_o(t+\Delta t)-P_0(t)}{\Delta t}=(-\alpha_1-\alpha_2-\alpha_3-\alpha_4-\alpha_5-\alpha_6-\alpha_7-\alpha_8-\alpha_9-\alpha_{10})P_0(t)+\int_0^\infty \beta_1(x)P_1(x,t)dx+$$
$$\int_0^\infty \beta_2(x)P_2(x,t)dx+\int_0^\infty \beta_3(x)P_3(x,t)dx+\int_0^\infty \beta_4(x)P_4(x,t)dx+\int_0^\infty \beta_5(x)P_5(x,t)dx+$$
$$\int_0^\infty \beta_6(x)P_6(x,t)dx+\int_0^\infty \beta_7(x)P_7(x,t)dx+\int_0^\infty \beta_8(x)P_8(x,t)dx+\int_0^\infty \beta_9(x)P_9(x,t)dx+$$
$$\int_0^\infty \beta_{10}(x)P_{10}(x,t)dx$$

$$\Rightarrow \frac{dP_o}{dt} + (\alpha_1 + \alpha_2 + \alpha_3 + \alpha_4 + \alpha_5 + \alpha_6 + \alpha_7 + \alpha_8 + \alpha_9 + \alpha_{10})P_0(t) = \int_0^\infty \beta_1(x)P_1(x,t)dx + \int_0^\infty \beta_2(x)P_2(x,t)dx +$$

$$\int_0^\infty \beta_3(x)P_3(x,t)dx + \int_0^\infty \beta_4(x)P_4(x,t)dx + \int_0^\infty \beta_5(x)P_5(x,t)dx + \int_0^\infty \beta_6(x)P_6(x,t)dx +$$

$$\int_0^\infty \beta_7(x)P_7(x,t)dx + \int_0^\infty \beta_8(x)P_8(x,t)dx + \int_0^\infty \beta_9(x)P_9(x,t)dx + \int_0^\infty \beta_{10}(x)P_{10}(x,t)dx$$

$$\Rightarrow \frac{dP_o}{dt} + \sum_{i=1}^{10} \alpha_i P_o(t) = \sum_{i=1}^{10} \int_0^\infty \beta_i(x)P_i(x,t)dx$$

$$\Rightarrow \left(\frac{d}{dt} + u_0\right)P_o(t) = f_0 \tag{2.1}$$

where

$$u_0 = \sum_{i=1}^{10} \alpha_i ; f_0 = \sum_{i=1}^{10} \int_0^\infty \beta_i(x)P_i(x,t)dx$$

Now

$$P_1(x + \Delta x, t + \Delta t) = \alpha_1 \Delta t P_0(t) + (1 - \beta_1(x)\Delta x) P_1(x, t) \tag{2.2}$$

Differentiating equation (2.2) with respect to x and t partially, we get

$$\Rightarrow \left(\frac{\partial}{\partial x} + \frac{\partial}{\partial t} + \beta_1(x)\right)P_1(x,t) = \alpha_1 P_0(t) \tag{2.3}$$

Similarly, we can obtain

$$\left(\frac{\partial}{\partial x} + \frac{\partial}{\partial t} + \beta_i(x)\right)P_i(x,t) = \alpha_i P_0(t); \qquad i = 2,3,4,5,6,7,8,9,10 \tag{2.4}$$

The boundary conditions are given as

$$P_1(0, t) = \alpha_1 P_0(t) \quad P_2(0, t) = \alpha_2 P_0(t)$$

$$P_3(0,t)=\alpha_3 P_0(t) \quad P_4(0,t)=\alpha_4 P_0(t)$$

$$P_5(0,t)=\alpha_5 P_0(t) \quad P_6(0,t)=\alpha_6 P_0(t)$$

$$P_7(0,t)=\alpha_7 P_0(t) \quad P_8(0,t)=\alpha_8 P_0(t)$$

$$P_9(0,t)=\alpha_9 P_0(t) \quad P_{10}(0,t)=\alpha_{10} P_0(t) \tag{2.5}$$

and initial conditions are given as

$$P_0(0)=1$$

$$P_i(0)=0; \qquad i=1 \ to \ 10 \tag{2.6}$$

Equations (2.1), (2.3), and (2.4) together with boundary and initial conditions constitutes Chapman-Kolmogorov differential difference equation.

In particular case, to show the importance of results based on availability of the system and profit analysis, we assume repair rates to follow exponential distribution. Then the above set of equations, boundary, and initial conditions reduces as follows:

$$\left(\frac{d}{dt}+u_0\right)P_o(t)=\beta_1P_1(t)+\beta_2P_2(t)+\beta_3P_3(t)+\beta_4P_4(t)+\beta_5P_5(t)+\beta_6P_6(t)+ \beta_7P_7(t)+\beta_8P_8(t)+\beta_9P_9(t)+\beta_{10}P_{10}(t) \tag{2.7}$$

$$\left(\frac{d}{dt}+\beta_i\right)P_i(t)=\alpha_iP_0(t); \qquad i=1,2,3,4,5,6,7,8,9.10 \tag{2.8}$$

and initial conditions are given as

$$P_0(0)=1$$

$$P_i(0)=0; \qquad i=1 \ to \ 10 \tag{2.9}$$

In industries and large plants, the steady state probabilities and long-time availability of the system is required and here this is calculated by taking $\frac{d}{dt}=0$ as $t \to \infty$, $P_i(t) = P_i$ in equations (2.7) and (2.8), we get

$$\mathrm{B}_1 P_1 = \alpha_1 P_0 \qquad \mathrm{B}_2 P_2 = \alpha_2 P_0$$

$$\mathrm{B}_3 P_3 = \alpha_3 P_0 \qquad \mathrm{B}_4 P_4 = \alpha_4 P_0$$

$$\mathrm{B}_5 P_5 = \alpha_5 P_0 \qquad \mathrm{B}_6 P_6 = \alpha_6 P_0$$

$$\mathrm{B}_7 P_7 = \alpha_7 P_0 \qquad \mathrm{B}_8 P_8 = \alpha_8 P_0$$

$$\mathrm{B}_9 P_9 = \alpha_9 P_0 \qquad \mathrm{B}_{10} P_{10} = \alpha_{10} P_0$$

that is

$$P_j = \frac{\alpha_j}{\beta_j} P_0; \quad j = 1,2,3,4,5,6,7,8,9,10 \tag{2.10}$$

Solving set of equations (2.7) and (2.8) together, we get the probability of being in states S_1 *to* S_{10} in terms of P_0 as following

Using normalizing condition, $\sum_{i=0}^{10} P_i = 1$, we have

$$\sum_{i=0}^{10} P_i = 1$$

$$\Rightarrow P_0 + P_1 + P_2 + P_3 + P_4 + P_5 + P_6 + P_7 + P_8 + P_9 + P_{10} = 1$$

$$\Rightarrow P_0 + \frac{\alpha_1}{\beta_1}P_0 + \frac{\alpha_2}{\beta_2}P_0 + \frac{\alpha_3}{\beta_3}P_0 + \frac{\alpha_4}{\beta_4}P_0 + \frac{\alpha_5}{\beta_5}P_0 + \frac{\alpha_6}{\beta_6}P_0 + \frac{\alpha_7}{\beta_7}P_0 + \frac{\alpha_8}{\beta_8}P_0 + \frac{\alpha_9}{\beta_9}P_0 + \frac{\alpha_{10}}{\beta_{10}}P_0 = 1$$

$$\Rightarrow P_0 = \frac{1}{\left(1 + \frac{\alpha_1}{\beta_1} + \frac{\alpha_2}{\beta_2} + \frac{\alpha_3}{\beta_3} + \frac{\alpha_4}{\beta_4} + \frac{\alpha_5}{\beta_5} + \frac{\alpha_6}{\beta_6} + \frac{\alpha_7}{\beta_7} + \frac{\alpha_8}{\beta_8} + \frac{\alpha_9}{\beta_9} + \frac{\alpha_{10}}{\beta_{10}}\right)} \tag{2.11}$$

The steady state availability of the system is given as

$$A_v = P_0$$

$$\Rightarrow A_v = \frac{1}{\left(1+\frac{\alpha_1}{\beta_1}+\frac{\alpha_2}{\beta_2}+\frac{\alpha_3}{\beta_3}+\frac{\alpha_4}{\beta_4}+\frac{\alpha_5}{\beta_5}+\frac{\alpha_6}{\beta_6}+\frac{\alpha_7}{\beta_7}+\frac{\alpha_8}{\beta_8}-\frac{\alpha_9}{\beta_9}+\frac{\alpha_{10}}{\beta_{10}}\right)} \tag{2.12}$$

2.4 Optimization

Computational intelligence-based algorithms are extensively used to obtain the optimal solution of real-world problems. Computational intelligence strategies are basically classified as population based and trajectory-based techniques. Population-based approaches are inspired by the natural evolution process of selection, fitness, and reproduction, thereby providing optimum solutions, and gaining more popularity nowadays. In present study, optimization of turbine unit is performed using well-known population-based technique, namely genetic algorithm (GA). For simulating the experiments, we have used MATLAB R2019a on Windows 10 64-bit operating system having 8 GB of RAM and Intel Core i5 8th generation CPU.

2.4.1 Genetic Algorithm

Genetic algorithm is a popular evolutionary computation that is inspired by biological evolution process. Holland [29] developed the genetic algorithms. This process is based on the fitness function of individuals. The fitness function is the objective function decided with the help of the decision parameters. The algorithm searches optimum solution of problem in a search space by evaluating the value of fitness function. In this algorithm, by assigning initialize values to decision parameters initial solution of objective function is derived. The following steps involved in searching process of GA:

- Chromosomes encoding and initial random population generation
- Fitness value calculation for optimization
- The algorithm converges upon satisfaction of stopping criteria (meeting required constraints/conditions), otherwise moves towards next step.
- Parent selection and new population generation by genetic operators' crossover and mutation
- Go to step 2.

2.5 Results and Discussion

In this section, the effect of various failure and repair rates on availability is investigated for a particular case. The availability of turbine unit is obtained w.r.t to failure rate (α_1) of subsystem-high-pressure turbine by taking variation in failure rate values. The values of α_1 is α_1 = 0.0055, 0.0065, 0.0075, 0.0085, 0.0095, 0.0105, 0.0115, 0.0125, 0.0135, 0.0145 and repair rate of high-pressure turbine is taken constant as $\beta_1 = 0.45$. The failure and repair rates of other subsystems is taken as follows:

$\alpha_2 = 0.0072$, $\alpha_3 = 0.0043$, $\alpha_4 = 0.0092$, $\alpha_5 = 0.00145$, $\alpha_6 = 0.007$, $\alpha_7 = 0.00214$, $\alpha_8 = 0.00427$, $\alpha_9 = 0.00821$, $\alpha_{10} = 0.009$, and $\beta_2 = 0.92$, $\beta_3 = 0.28$, $\beta_4 = 0.59$, $\beta_5 = 0.88$, $\beta_6 = 1.004$, $\beta_6 = 0.154$, $\beta_7 = 0.218$, $\beta_8 = 0.94$, $\beta_{10} = 0.75$

The availability of the system is calculated by above parameter values in equation (2.12) and results are highlighted in Figure 2.3.

However, availability of subsystem-A, i.e., high-#pressure turbine is also obtained w.r.t to repair rate (β_1) by varying its values as given below β_1 = 0.45, 0.55, 0.65, 0.75, 0.85, 0.95, 1.05, 1.15, 1.25, 1.35 and constant value of failure rate is considered as α_1 0.0055.The failure and repair rates of other subsystems have been taken as; $\alpha_2 = 0.0072$, $\alpha_3 = 0.0043$, $\alpha_4 = 0.0092$, $\alpha_5 = 0.00145$, $\alpha_6 = 0.007$, $\alpha_7 = 0.00214$, $\alpha_2 = 0.00427$, $\alpha_2 = 0.00821$, $\alpha_2 = 0.009$ and $\beta_2 = 0.92$, $\beta_3 = 0.28$, $\beta_4 = 0.59$, $\beta_5 = 0.88$, $\beta_6 = 1.004$, $\beta_7 = 0.154$, $\beta_8 = 0.218$, $\beta_9 = 0.94$, $\beta_{10} = 0.75$

Using the parametric values given in Table 2.1, the availability of the system is obtained using genetic algorithm above values of parameters and results are highlighted in Figure 2.4.

From Figure 2.2, it is revealed that availability of the turbine unit decreases with respect to the variation in failure rates. Figure 2.3 revealed that variation in repair rate immediately enhance the availability of unit. Though it shows the local behavior of plant. From Tables 2.2 to 2.5, it is

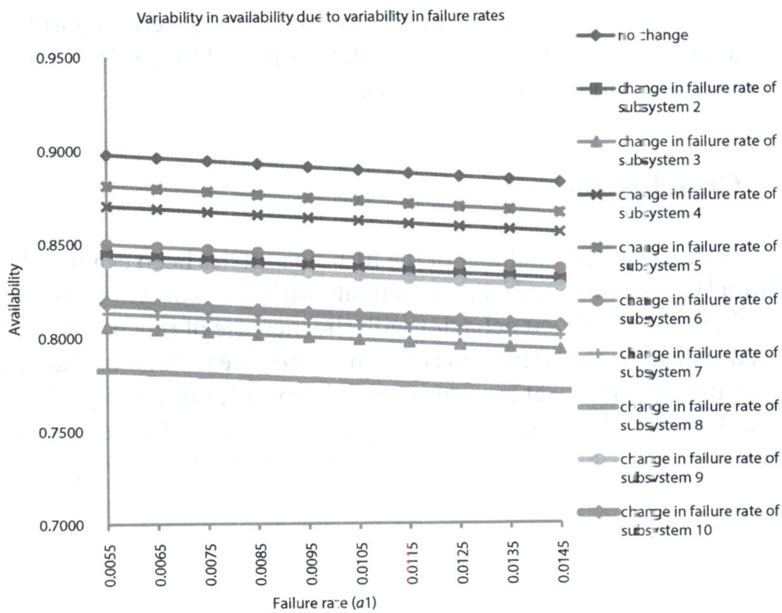

Figure 2.3 Effect of variation in failure rates on system's availability w.r.t failure rate α_1.

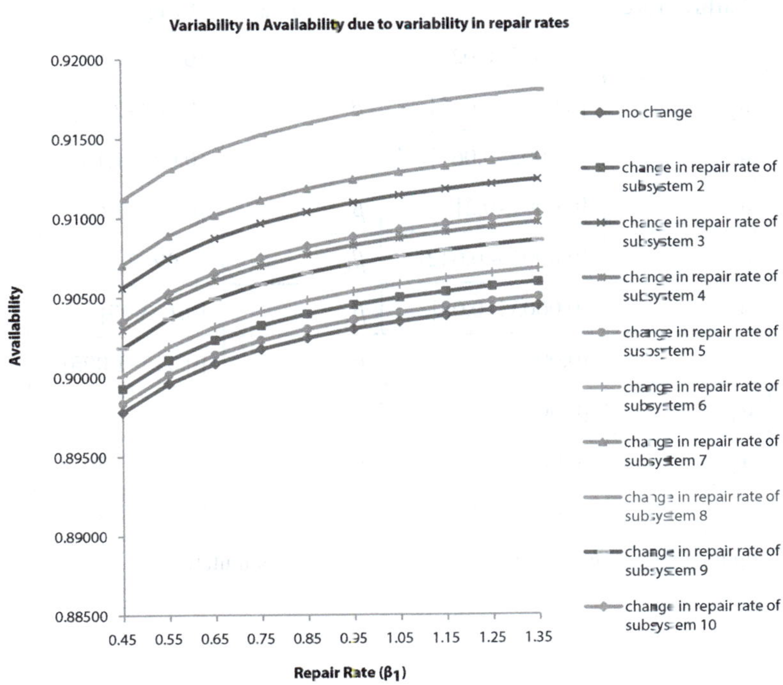

Figure 2.4 Effect of variation in repair rates on system's availability w.r.t. repair rate β_1.

depicted that global value of availability by using genetic algorithm is 0.9936 at crossover probability, 0.7; mutation probability, 0.5; number of evolutions, 450; and population size, 65.

2.6 Conclusion

For a particular case, availability of the turbine unit has been evaluated, and its optimal value is obtained using genetic algorithm. It is observed that availability showed an inclined behavior with the increase of repair rate while it declined sharply with the increase of failure rate. The turbine unit achieved its optimum availability at crossover probability, 0.7; mutation probability, 0.5; number of evolutions, 450; and population size, 65. The results can be used by system designers during the designing stage of turbine systems.

Table 2.1 Search space for the parameters of turbine unit.

Failure rate	Range	Repair rate	Range
α_1	[0.0005–0.02]	β_1	[0.015–1.5]
α_2	[0.0003–0.075]	β_2	[0.063–2.35]
α_3	[0.0002–0.064]	β_3	[0.025–1.15]
α_4	[0.0001–0.2]	β_4	[0.012–1.65]
α_5	[0.00025–0.042]	β_5	[0.09–2.15]
α_6	[0.00035–0.08]	β_6	[0.1–3.05]
α_7	[0.00025–0.03]	β_7	[0.045–1.005]
α_8	[0.00052–0.055]	β_8	[0.052–1.1]
α_9	[0.00015–0.15]	β_9	[0.06–2.35]
α_{10}	[0.00022–0.25]	β_{10}	[0.082–1.85]

Population size = 65, evolution = 450, crossover = 0.8, mutation = 0.5.

Table 2.2 Values of estimated parameters and availability with respect to population size having number of evolutions = 450, crossover probability = 0.8 and mutation probability = 0.5.

Population	**12**	**24**	**36**	**48**	**60**	**72**	**84**	**96**	**108**
Availability	0.9752	0.9784	0.9631	**0.9852**	0.9683	0.9664	0.9724	0.981	0.9702
α_1	0.0295	0.0113	0.0034	**0.0038**	0.0398	0.0322	0.0001	0.0013	0.0273
α_2	0.0474	0.0677	0.1086	**0.0536**	0.0207	0.0987	0.0357	0.0137	0.1502
α_3	0.0455	0.0434	0.0358	**0.0065**	0.0016	0.0206	0.0089	0.001	0.0086
α_4	0.0215	0.0048	0.0146	**0.0186**	0.0957	0.0007	0.0119	0.0073	0.0041
α_5	0.004	0.0178	0.0075	**0.0041**	0.0028	0.0036	0.0021	0.0018	0.0009
α_6	0.024	0.0084	0.0005	**0.0433**	0.0213	0.0052	0.0557	0.0324	0.0278
α_7	0.0025	0.0078	0.0264	**0.0073**	0.006	0.0013	0.0473	0.0269	0.0079
α_8	0.0294	0.0189	0.0283	**0.0034**	0.0041	0.0117	0.0325	0.0702	0.0341
α_9	0.0167	0.0704	0.1471	**0.0029**	0.0243	0.0624	0.0054	0.0981	0.0696
α_{10}	0.0243	0.0733	0.0084	**0.0371**	0.0122	0.0287	0.0068	0.0493	0.0382
β_1	1.5501	5.3742	0.1487	**0.8495**	1.0455	0.4802	1.7892	2.3546	3.6251
β_2	0.2273	1.0054	2.2172	**2.2104**	1.3411	1.382	1.8295	0.3663	1.2622

(Continued)

Table 2.2 Values of estimated parameters and availability with respect to population size having number of evolutions = 450, crossover probability = 0.8 and mutation probability = 0.5. *(Continued)*

Population	**12**	**24**	**36**	**48**	**60**	**72**	**84**	**96**	**108**
Availability	0.9752	0.9784	0.9631	**0.9852**	0.9683	0.9664	0.9724	0.981	0.9702
β_3	0.7803	0.7841	1.3994	**4.2041**	0.0916	0.4858	1.8517	2.1029	0.5417
β_4	0.7841	0.9941	1.0245	**0.7975**	3.5132	1.3549	1.3337	0.8931	3.1008
β_5	1.3453	2.5268	0.873	**0.256**	0.143	0.3384	2.8882	0.3252	2.2543
β_6	0.1544	0.7022	5.9191	**0.9889**	1.5857	1.6599	3.7473	4.3303	0.9721
β_7	0.2953	0.6201	0.316	**0.2773**	1.3614	0.3603	0.4963	2.2063	0.269
β_8	0.561	0.1486	0.6748	**1.5103**	0.8741	1.1475	2.0015	0.8796	4.455
β_9	4.3585	1.3264	1.8414	**1.4119**	1.8879	1.9255	1.505	1.5426	3.6756
β_{10}	1.2624	2.9952	0.572	**1.2618**	1.1091	1.5502	2.6757	5.4741	1.248

Table 2.3 Values of estimated parameters and availability with respect to number of evolutions having population size = 65, crossover probability = 0.8 and mutation probability = 0.5.

Evolution	**50**	**100**	**150**	**200**	**250**	**300**	**350**	**400**	**450**
Availability	0.9648	0.9517	0.9742	0.9782	0.9726	**0.9846**	0.9809	0.9806	0.9648
α_1	0.0288	0.0071	0.0037	0.0526	0.1128	**0.0549**	0.0299	0.0031	0.0204
α_2	0.0078	0.0706	0.1491	0.2296	0.0944	**0.0319**	0.1267	0.1651	0.0129
α_3	0.0028	0.0248	0.004	0.0383	0.0082	**0.0173**	0.0174	0.0084	0.0164
α_4	0.0657	0.0192	0.0019	0.0699	0.035	**0.0581**	0.0679	0.0013	0.0097
α_5	0.0021	0.0157	0.0537	0.0303	0.0107	**0.0056**	0.0075	0.0332	0.0041
α_6	0.0238	0.0095	0.0279	0.0549	0.0019	**0.0558**	0.0076	0.0046	0.0176
α_7	0.0225	0.0029	0.0099	0.0069	0.0013	**0.0061**	0.0132	0.0459	0.0065
α_8	0.0037	0.0362	0.0096	0.0147	0.0256	**0.0023**	0.0258	0.0096	0.0089
α_9	0.1509	0.0044	0.0101	0.0202	0.0155	**0.0027**	0.0904	0.2033	0.1374
α_{10}	0.0184	0.0169	0.1228	0.0896	0.0521	**0.0411**	0.0071	0.0331	0.0444
β_1	0.217	1.1124	2.7569	2.0613	0.8007	**2.2667**	1.6487	0.9131	0.8606
β_2	7.1686	0.2902	0.8205	1.6351	3.5558	**1.0675**	2.0798	1.1179	2.6974

(Continued)

Table 2.3 Values of estimated parameters and availability with respect to number of evolutions having population size = 65, crossover probability = 0.8 and mutation probability = 0.5. (*Continued*)

Evolution	**50**	**100**	**150**	**200**	**250**	**300**	**350**	**400**	**450**
Availability	0.9648	0.9517	0.9742	0.9782	0.9726	**0.9846**	0.9809	0.9806	0.9648
β_3	0.7142	0.6127	1.479	1.0031	0.2983	**1.4905**	0.4364	1.3455	0.8442
β_4	6.0944	2.441	0.8754	3.8657	1.3173	**2.1648**	3.2976	3.4957	1.0166
β_5	3.6976	1.2403	2.2572	2.7278	0.5085	**2.2652**	0.5453	1.2309	1.9756
β_6	3.8651	2.3888	1.3245	2.9487	0.5119	**8.0135**	2.2607	2.1054	3.8999
β_7	4.4934	1.2173	0.1742	2.0863	1.5147	**0.2166**	0.3184	0.5621	1.5098
β_8	1.3557	2.087	3.8214	1.0014	0.5174	**0.2087**	1.3004	0.2452	0.8078
β_9	3.2834	2.2095	2.1977	0.9779	2.4455	**1.0933**	4.9233	5.1242	4.3554
β_{10}	0.4287	1.7995	6.6863	2.2679	9.7599	**1.3966**	2.047	2.4583	0.6992

Table 2.4 Values of estimated parameters and availability with respect to crossover probability having number of evolutions = 450, population size = 65 and mutation probability = 0.5.

Crossover	**0.1**	**0.2**	**0.3**	**0.4**	**0.5**	**0.6**	**0.7**	**0.8**	**0.9**
Availability	0.9697	0.97	0.9799	0.9628	0.9617	0.9797	**0.9936**	0.9648	0.9697
α_1	0.0239	0.0268	0.0034	0.0182	0.0016	0.0614	**0.0456**	0.0204	0.0657
α_2	0.018	0.0932	0.132	0.0419	0.1001	0.0241	**0.0255**	0.0129	0.0348
α_3	0.0027	0.0457	0.0058	0.01	0.0129	0.0005	**0.0114**	0.0164	0.0003
α_4	0.0344	0.0741	0.0153	0.0081	0.0066	0.071	**0.0006**	0.0097	0.0736
α_5	0.0013	0.0019	0.0022	0.0017	0.0016	0.0167	**0.0002**	0.0041	0.001
α_6	0.0701	0.085	0.0216	0.0124	0.0742	0.1032	**0.0189**	0.0176	0.2523
α_7	0.0152	0.014	0.0087	0.0098	0.0025	0.015	**0.004**	0.0065	0.0056
α_8	0.0101	0.0411	0.0042	0.1099	0.002	0.0045	**0.023**	0.0089	0.0115
α_9	0.0427	0.0722	0.0177	0.0044	0.0121	0.0553	**0.0188**	0.1374	0.0169
α_{10}	0.0025	0.0142	0.0084	0.0467	0.0285	0.0072	**0.0669**	0.0444	0.0673
β_1	2.0731	1.9394	0.3266	2.1901	0.674	0.9926	**1.153**	0.8606	1.9083
β_2	0.6078	4.3047	1.9496	6.7629	0.5608	0.6797	**2.089**	2.6974	0.9649

(Continued)

Table 2.4 Values of estimated parameters and availability with respect to crossover probability having number of evolutions = 450, population size = 65 and mutation probability = 0.5. (*Continued*)

Crossover	**0.1**	**0.2**	**0.3**	**0.4**	**0.5**	**0.6**	**0.7**	**0.8**	**0.9**
Availability	0.9697	0.97	0.9799	0.9628	0.9617	0.9797	**0.9936**	0.9648	0.9697
β_3	4.7153	1.3292	0.4146	0.8951	0.2145	0.5451	**2.336**	0.8442	0.9226
β_4	0.5559	1.0253	3.1408	1.4124	0.7288	2.2738	**0.3124**	1.0166	1.7858
β_5	2.1463	2.7983	1.2111	1.62	0.4519	3.2983	**2.6912**	1.9756	0.2774
β_6	7.7783	3.6947	1.1591	0.1821	0.4285	5.1629	**5.118**	3.8999	1.8649
β_7	0.4748	1.1916	0.1011	0.271	0.6266	0.2683	**1.0503**	1.5098	0.2918
β_8	3.4365	0.9239	0.6074	3.223	0.2877	1.1277	**1.5984**	0.8078	1.6126
β_9	1.47	1.1009	0.7446	0.3487	2.1008	2.774	**2.0541**	4.3554	0.4922
β_{10}	1.9323	2.2301	0.3054	2.4415	1.4615	1.4653	**4.1582**	0.6992	1.2365

Table 2.5 Values of estimated parameters and availability with respect to mutation probability having number of evolutions = 450, crossover probability = 0.8 and population size = 65.

Mutation	**0.06**	**0.12**	**0.18**	**0.24**	**0.3**	**0.36**	**0.42**	**0.48**	**0.54**
Availability	0.9588	0.9753	0.9521	0.9624	0.9571	0.9563	0.9675	0.9687	**0.9846**
α_1	0.0017	0.0303	0.0766	0.1036	0.0101	0.0712	0.0257	0.0163	**0.0049**
α_2	0.0183	0.0092	0.0997	0.0312	0.0182	0.0231	0.0064	0.0366	**0.0221**
α_3	0.0034	0.0102	0.0038	0.047	0.0015	0.0176	0.0054	0.0093	**0.007**
α_4	0.0314	0.0022	0.0259	0.0255	0.0173	0.0124	0.014	0.0902	**0.003**
α_5	0.0064	0.0011	0.0303	0.0236	0.0019	0.0055	0.0081	0.0056	**0.0138**
α_6	0.0814	0.0584	0.1661	0.009	0.0035	0.0293	0.0534	0.006	**0.0032**
α_7	0.0215	0.0034	0.0047	0.0041	0.0068	0.0149	0.0042	0.0468	**0.009**
α_8	0.0615	0.0013	0.0024	0.0104	0.0255	0.0385	0.0015	0.0064	**0.013**
α_9	0.0349	0.0076	0.0005	0.0386	0.0253	0.0073	0.0814	0.0692	**0.033**
α_{10}	0.0498	0.0128	0.0025	0.0105	0.001	0.0306	0.0674	0.0085	**0.0722**
β_1	3.4178	0.0619	1.0545	2.8777	0.9142	1.2139	0.3421	0.8169	**0.8325**
β_2	2.2941	2.7479	0.4828	1.2594	0.4245	0.5185	1.1986	0.7637	**2.5013**

(Continued)

Table 2.5 Values of estimated parameters and availability with respect to mutation probability having number of evolutions = 450, crossover probability = 0.8 and population size = 65. *(Continued)*

Mutation	**0.06**	**0.12**	**0.18**	**0.24**	**0.3**	**0.36**	**0.42**	**0.48**	**0.54**
Availability	0.9588	0.9753	0.9521	0.9624	0.9571	0.9563	0.9675	0.9687	**0.9846**
β_3	0.6115	0.5575	0.2129	1.441	0.0877	0.1396	0.5261	0.6624	**2.5122**
β_4	1.7034	1.4939	1.297	3.3144	1.2377	0.7691	1.1895	2.8833	**0.3471**
β_5	0.469	2.322	0.6383	1.8374	2.3194	1.2681	1.092	0.9395	**0.9928**
β_6	4.5365	0.6498	2.9172	0.6451	0.3864	6.5168	2.9511	5.0429	**0.459**
β_7	1.5771	1.7065	0.1404	2.9346	0.2586	1.5124	0.7601	0.8023	**0.5308**
β_8	1.307	2.6607	1.0379	0.2699	0.5679	4.7616	0.932	0.6751	**0.3633**
β_9	0.7555	0.4754	0.7849	1.7777	4.7958	1.4518	2.2577	2.5339	**2.0866**
β_{10}	1.4687	0.7779	0.2704	1.4606	0.3059	1.8541	1.3761	2.683	**1.2766**

References

1. Sabouhi, H., Abbaspour, A., Fotuhi-Firuzabad, M., Dehghanian, P., Reliability modeling and availability analysis of combined cycle power plants. *Int. J. Electr. Power Energy Syst.*, 79, 108–119, 2016.
2. Carazas, F. and Desouza, G., Availability analysis of gas turbines used in power plants. *Int. J. Thermodyn.*, 12, 1, 28–37, 2009.
3. Gupta, N., Saini, M., Kumar, A., Operational availability analysis of generators in steam turbine power plants. *SN Appl. Sci.*, 2, 4, 1–11, 2020.
4. Erdem, H.H., Akkaya, A.V., Cetin, B., Dagdas, A., Sevilgen, S.H., Sahin, B., Atas, S., Comparative energetic and exergetic performance analyses for coal-fired thermal power plants in Turkey. *Int. J. Therm. Sci.*, 48, 11, 2179–2186, 2009.
5. Carazas, F.J.G., Salazar, C.H., De Souza, G.F.M., Availability analysis of heat recovery steam generators used in thermal power plants. *Energy*, 36, 6, 3855–3870, 2011.
6. Shopeju, O.O. and Oyedepo, S.O., A comprehensive review of thermal power plants reliability using stochastic methods. *IOP Conf. Ser.: Mater. Sci. Eng.* IOP Publishing, 1107, 1, 012151, April 2021.
7. Gupta, N., Kumar, A., Saini, M., Reliability and maintainability investigation of generator in steam turbine power plant using RAMD analysis. *IOP Conf. Ser.: Mater. Sci. Eng.* IOP Publishing, 1714, 1, 012009, 2021.
8. Nasiriani, K., Ghaedi, A., Nafar, M., Reliability evaluation of power systems containing ocean thermal energy conversion power plants. *Sci. Iran.*, 29, 4, 1957–1974, 2020.
9. Luo, X., Zhang, B., Chen, Y., Mo, S., Operational planning optimization of steam power plants considering equipment failure in petrochemical complex. *Appl. Energy*, 112, 1247–1264, 2013.
10. Radin, Y.A., Improving the flexibility and reliability of steam power units at thermal power plants. *Therm. Eng.*, 68, 6, 481–489, 2021.
11. Sultanov, M.M., Ivanitckii, M.S., Lunenko, V.S., Development of approaches to assessing the actual technical condition of steam turbines based on reliability indicators, in: *2021 3rd International Youth Conference on Radio Electronics, Electrical and Power Engineering (REEPE)*, IEEE, pp. 1–5, March 2021.
12. Sunil, P.U., Barve, J., Nataraj, P.S.V., Mathematical modeling, simulation and validation of a boiler drum: Some investigations. *Energy*, 126, 312–325, 2017.
13. Prock, J., Mathematical modeling of a steam generator for sensor fault detection. *Appl. Math. Model.*, 12, 6, 581–609, 1988.
14. Arora, N. and Kumar, D., Availability analysis of steam and power generation systems in the thermal power plant. *Microelectron. Reliab.*, 37, 5, 795–799, 1997.
15. Bhangu, N.S., Singh, R., Pahuja, G.L., Availability performance analysis of thermal power plants. *J. Inst. Eng. (India): C*, 100, 3, 439–448, 2019.

16. Gupta, S., Stochastic modelling and availability analysis of a critical engineering system. *Int. J. Qual. Reliabil. Manage.*, 36, 5, 782–796, 2019.
17. De Souza, G.F.M., *Thermal Power Plant Performance Analysis*, Springer, London, 2012.
18. Bose, D., Chattopadhyay, S., Bose, G., Adhikary, D., Mitra, S., RAM investigation of coal-fired thermal power plants: A case study. *Int. J. Ind. Eng. Comput.*, 3, 3, 423–434, 2012.
19. Tanuma, T., Introduction to steam turbines for power plants, in: *Advances in Steam Turbines for Modern Power Plants*, pp. 3–9, Woodhead Publishing, India, 2017.
20. Tanuma, T. (Ed.), *Advances in Steam Turbines for Modern Power Plants*, Woodhead Publishing, 2017.
21. De Souza, G.F.M., Carazas, F.J.G., dos Santos Guimarães, L., Rodriguez, C.E.P., Combined-cycle gas and steam turbine power plant reliability analysis, in: *Thermal Power Plant Performance Analysis*, pp. 221–247, Springer, London, 2012.
22. Kumar, R., Sharma, A., Tewari, P., Markov approach to evaluate the availability simulation model for power generation system in a thermal power plant. *Int. J. Ind. Eng. Comput.*, 3, 5, 743–750, 2012.
23. Houdova, L., Houdova, L., Jelinek, L., Janecek, E., Approach to solving the task of availability prediction and cost optimization of a steam turbine, in: *Proceedings of the ITI 2010, 32nd International Conference on Information Technology Interfaces*, IEEE, pp. 629–634, June 2010.
24. Kim, H.J., Fatigue failure analysis of last stage blade in a low pressure steam turbine. *Eng. Fail. Anal.*, 6, 2, 93–100, 1999.
25. Rajesh, G.T. and Prasad, J., Reliability and availability analysis for a three-unit gas turbine power generating system with seasonal effect and FCFS repair pattern. *Int. J. Appl. Eng. Res.*, 13, 12, 10948–10964, 2018.
26. Ueasin, N., Wongchai, A., Nonthapot, S., Performance assessment and optimization of biomass steam turbine power plants by data envelopment analysis. *Int. J. Energy Econ. Policy*, 5, 3, 668–672, 2015.
27. Okafor, C.E. and Atikpakpa, A., Availability assessment of steam and gas turbine units of a thermal power station using Markovian approach. *Arch. Curr. Res. Int.*, 6, 4, 1–17, 2016.
28. Oyedepo, S.O. and Fagbenle, R.O., A study of implementation of preventive maintenance programme in Nigeria power industry—Egbin thermal power plant, case study. *Energy Power Eng.*, 3, 207–220, 2011.
29. Holland, J., Genetic algorithms. *Sci. Am.*, 267, 1, 66–73, 1992.

3

Development of Laplacian Artificial Bee Colony Algorithm for Effective Harmonic Estimator Design

Aishwarya Mehta[1], Jitesh Jangid[1], Akash Saxena[2], Shalini Shekhawat[3*] and Rajesh Kumar[4]

[1]Department of Electrical Engineering, Swami Keshvanand Institute of Technology, Management and Gramothan, Jaipur, India
[2]School of Computing Science and Engineering, VIT Bhopal University, Bhopal, India
[3]Department of Mathematics, SKIT, Jaipur, India
[4]Department of Electrical Engineering, MNIT, Jaipur, India

Abstract

Adverse effects of harmonic with ever increasing nonlinear loads is major concern these days. Harmonic estimation problem is an important problem as accurate estimation of harmonic signals can be used for designing the harmonic filters. The estimator design problem is considered as a challenging problem and still a potential area of research due to the fact that diverse operating conditions are having impact on the performance of the estimator. For designing the estimator a hybrid approach based on least square error minimization with the help of a new version of Artificial Bee Colony algorithm is proposed. The proposed version employs a Laplacian factor based update equation in scout bee phase. For proving the modification meaningful, first the proposed algorithm is tested on several standard benchmark problems and then it is applied on estimator design problem. Results reported in both parts indicate that proposed modification is meaningful and performance of LABC is comparable with many state of the art algorithms.

Keywords: Fast Fourier transform, artificial neural network, wavelet transform, artificial bee colony

**Corresponding author*: shekhawatshalini17@gmail.com

S. C. Malik, Deepak Sinwar, Ashish Kumar, S. R. Gadde, Prasenjit Chatterjee and Bui Thanh Hung (eds.) Computational Intelligence in Sustainable Reliability Engineering, (47–96) © 2023 Scrivener Publishing LLC

3.1 Introduction

In the current scenario, demand for electrical power is rapidly increasing day by day due to escalation in population. Our power system network is based upon the consumer-centric policies. These policies describe that consumer satisfaction is a priority of the service providers. For consumers' satisfaction, engineers try to provide a reliable electrical power supply to the distribution network. The reliability of electrical supply is maintained by the semiconductor devices. Semiconductor devices are compact in size and a large amount of electrical power can be transmitted without disturbing reliability. Due to the nonlinear behavior of these devices, network parameters like power factor, voltage and current signals are badly affected. Solid power state devices generate harmonics and these are primary cause of low overloading capability, etc. Harmonics have attracted the attention of researchers over the last few decades due to the complexity associated with them. Harmonics are pollutants of current and voltage signals. In other words, harmonics are defined as multiples of the fundamental frequency (50 Hz). According to its definition harmonics are categorized into four different categories. First one is even harmonics that is even multiple fundamental harmonics. The second one is odd harmonics which is an odd multiple of fundamental harmonics. The third one is sub harmonics, these harmonics are noninteger multiple of the fundamental frequency and also less than the fundamental frequency. The fourth one is inter harmonics which are noninteger multiple of fundamental harmonics and greater than the fundamental frequency.

Current and voltage signals are polluted by the power converters, lighting devices, etc. These polluted signals disturb the performance of the power system network. Losses namely eddy current losses, skin effect along with effects namely improper functioning of power network equipment, reduction of the life span of power equipment, etc are associated with harmonic pollution. To protect the system from disturbances, mitigation filters are designed. Mitigation filter helps to reduce harmonics. To design an efficient mitigation filter, it is required that the exact value of phase and amplitude of power harmonics are known. The exact values of phase and amplitude components of power harmonics are measured by the harmonic estimators.

From the above discussion, it is concluded that harmonic estimation is necessary for designing a mitigation filter. In this consequence, two steps are taken. In the first step, amplitude and phase components are estimated for power harmonics, after this estimation process, the second step is taken in which harmonic mitigation filters are designed to mitigate harmonics.

In the past decades, harmonic estimation is done by different techniques, in these techniques, the most preferred technique is the fast Fourier transform (FTT). FTT provides optimal solutions in the absence of interharmonics. When interharmonics present in the network, it suffers from the picket fence effect. To overcome the shortcomings of FFT, Wavelet Transform (WT) is used. WT uses mother wavelets, which are the decomposition form of frequency. WT accuracy depends on the mother wavelets. Hence, the selection of mother wavelets is a crucial part of these researches.

Artificial Neural Network (ANN) is also used to estimate components of power harmonics. ANN uses direction links to connect neurons. In ANN, the order of power harmonics is proportional to direction links. When the order of power harmonics increases then ANN does not provide precise solutions. From the above discussion, it is validated that traditional methods do not have the capability to estimate higher-order power harmonic components. Further, researchers adopt the metaheuristic techniques for accurate estimation of multiple harmonics in a signal.

In the past several years, many harmonics estimation techniques are used. These techniques are classified into two categories: first category belongs to conventional techniques shown in Table 3.1 and second one belongs to meta-heuristic techniques.

To improve efficiency of estimation, meta-heuristic methods are used. These methods can escape from local optima and reach global optima as soon as possible without any delay. Meta-heuristic techniques are further classified into some categories. These categories are described as

- Algorithms based on swarm intelligence: These algorithms depend on the population of swarms and inspired by environment activities. In these algorithms, swarms live in groups, sharing each and every information in the group. The sharing of information is useful to find an optimum result of a real-world problem. Ant-Lion Optimization Algorithm (ALO) [1], Particle Swarm Optimization (PSO) [2], Grey Wolf Optimization Algorithm (GWO) [3], Ant Colony Optimization (ACO) [4] are examples of these techniques.
- Algorithms based on human behavior: Algorithms based on human activities lie in this category. The human-based algorithm includes Teaching Based Optimization Algorithm (TLBO) [5], Harmony Search (HS) [6], Exchange Market Algorithm (EMA) [7], Coronavirus Herd Immunity Optimizer (CHIO) [8], and so on.

Table 3.1 Comparison of conventional harmonics estimation techniques.

Technique	Merits	Demerits
DWT	*Time-frequency domain *Multiresolution ability *Easy implementation with filter bank	*Spectrum is terms of frequency bands *Mother wavelet selection is arbitrary and affects the accuracy
HHT	*Posteriori data processing Basis function (IMF) is adaptive	*Additional tools needed for interpretation of transformed parameters
CZT	*Flexibility on data sampling and number of samples	*Computationally more demanding *Additional data storage requirement
Prony	*Can detect all attributes of signal: frequency, amplitude, phase and damping factor *Very high resolution and no side lobes	*Computationally inefficient *Highly prone to noise and model mismatch
MUSIC	*Good accuracy with short data length	*Priori knowledge of frequency search range is needed *High computational time
ANN	*Self-adaptive *Fairly good accuracy with noisy data samples	*Multilayered complex structure *Greedy nature *May be trapped in the local minima

- Algorithms based on physics: This type of algorithms is made by the principle and theorem of physics which are used to solve multidimension problems. Some examples of this category are Big Bang-Big Crunch Algorithm (BB-BC) [9], Space Gravitational Algorithm (SGA) [10], Central Force Optimization (CFO) [11], Magnetic Optimization Algorithm (MOA) [12], etc.

- Algorithms based on evolution theory: These algorithms are developed by evolution theory which is given by Charles Darwin. Inheritance property helps to improve results from the previous one result. Differential algorithms and genetic algorithms are examples of evolution based algorithms.
- Hybrid algorithms: Hybrid algorithms are formed by the combination of two or more algorithms. By this combination, results are better than the previous results. Some hybrid algorithms are Grey Wolf Optimization (GWO) with particle swarm optimization (PSO) [13], ant colony optimization (ACO) with PSO [14] and adaptive neuro-fuzzy inference system (ANFIS) and artificial bee colony (ABC) algorithm [15] etc.

In the literature, many techniques like fast Fourier transform (FFT) [16], Discrete Fourier transform (DFT) [17], adaptive notch filtering (ANF) [18], Kalman Filtering (KF) [19] exists. Hybrid based techniques include artificial neural network (ANN) [20], ADALINE [21], etc. are used to reduce the adverse effects of harmonics. In these techniques, researcher focused on calculation, mapping, reduction and estimation of electrical harmonics problems and minimize their side effects [22, 23].

In the field of optimization, a single algorithm is not able to solve all kinds of scientific problems which is also proven through no free lunch theorem. To increase the capabilities of algorithms, researchers try to develop new algorithms and variants of existing algorithms. These newly developed algorithms and variants help to improve the results and also increase the efficacy of the existing algorithm.

In 2005, Karaboga invented the artificial bee colony algorithm (ABC) [24] which is inspired by nature and uses foraging activity of honey bees. The main motivation to use ABC is that ABC algorithm involves a few control parameters only which enables the ABC algorithm to compete with other population-based algorithms. ABC gained attention due to its simplicity and ease of implementation process and has been applied to solve many practical optimization problems. While this algorithm is robust, it also has a poor convergence rate and sometimes also stuck into local minima. To remove these problems of ABC, researchers are trying to develop variants and also hybrid techniques to improve the results.

Recently, some updated versions of ABC algorithm are developed which are tested on engineering problems, these variant of ABC are as follows: Gbest-guided artificial bee colony [25] introduced by Zhu *et al.* Enhanced artificial bee colony optimization [26] given by Tsai *et al.* Modified artificial bee colony [27] introduced by Nirmala Sharma *et al.* Island artificial bee colony for global optimization (IABC) [28], Natural selection methods for artificial bee colony with new versions of onlooker bee [29] and hybrid artificial bee colony [30] are introduced by Mohammed *et al.* After so many developments, ABC algorithm still suffered with some drawbacks, which are as follows

1. The convergence rate of ABC is slow in comparison to other population-based algorithms.
2. During multimodal complex problems, the ABC algorithm is stuck into local minima's and unable to find global optima.

To amend these drawbacks a new version of ABC is proposed in this work named as Laplacian Factor based artificial bee colony (LABC). The major contributions of this paper are as follows

1. We have developed an advanced version of the ABC algorithm, which is not only able to handle unimodal, multimodal, and fixed dimension multimodal test functions for optimization purpose but also applied on real-time applications.
2. Validation of LABC is done with the help of conventional benchmark test functions and CEC-2017 test functions. Various analyses are conducted to evaluate performance of LABC.
3. Two classical waveforms for designing a harmonic estimator are considered to validate the efficacy of the harmonic estimator.

The reminder of the paper is structured as follows: section 3.2 describes the problem formulation of harmonic estimation. Section 3.3 represents original ABC and proposed variant. Sections 3.4 and 3.5 help to validate the modified version of ABC with numerical and analytical processes respectively. In section 3.6 design analysis of harmonic estimator is presented. Last section 3.7 provides a summarized conclusion of the work done in paper.

3.2 Problem Formulation of Harmonics

In the introduction section, it is seen that harmonics create many problems and also disturb the voltage and current waveforms. In this consequence, it is necessary to get an exact estimation of current and voltage signals of power harmonics. Generally, current and voltage signals are mathematically expressed as

$$P(t) = \sum_{r=1}^{R} M_r \sin(w_r t + \phi_r) + M_{DC} exp(-\theta_{DC} t) + U(t)w \quad (3.1)$$

where R represents order of power harmonics, w_r is angular frequency which is measured through $w_r = r2\pi f_{fun}$, f_{fun} represents component fundamental frequency for power harmonics. Additive White Gaussian Noise (AWGN) [31] is denoted by $U(t)$. $M_{DC} exp(-\theta_{DC} t)$ is used to represent exponential decaying term of DC component.

Now, the signal is sampled in sampling time and applies Taylor series expansion mentioned in [32]. After this, implementation harmonic problem equation can be written as

$$P(a) = \sum_{r=1}^{R} [M_r \sin(w_r a T_s)\cos(\phi_r) + M_r \cos(w_r a T_s)\sin(\phi_r)] + M_{DC} - M_{DC}\, exp(-\phi_{DC} a T_s) + U(a)$$

Now, we transform the above equation in parametric expression, which is mathematically represented as

$$P(n) = D(n)\phi(n) \quad (3.2)$$

Here, D(n) is estimated by following equation

$$D(n) = [\sin(w_1 a T_s)\cos(w_1 a T_s \ldots \sin(w_r a T_s)\cos(w_r a T_s) 1 - a T_s]^T \quad (3.3)$$

Now, all the unknown parameters are expressed as below

$$\phi(n) = [M_1 \sin(\phi_1)\cos(\phi_1) \ldots M_r \sin(\phi_k)\cos(\phi_k) \ldots M_{DC} \ldots \phi_{DC}]^T \quad (3.4)$$

The objective function of harmonic estimation problem is formulated as

$$O = Min\left(\sum_{n=1}^{N} (P_n - P_{nest})^2\right) \quad (3.5)$$

Here, O is used to represent the objective function, P_n is original signal of harmonic problem, while P_{nest} is measured signal of harmonic problem.

3.3 Development of Laplacian Artificial Bee Colony Algorithm

3.3.1 Basic Concepts of ABC

ABC algorithm uses the foraging behavior of honey bees. These honey bees have three special qualities which are listed below

- They search for a new food location and share this information with other honey bees of the group.
- Find a new food source location which is rich in nectar amount.
- Neglect the previous food source location when a new location in nectar amount then previous one is found richer.

The honey bees, which are present in a group, provide collective information that is used to locate a new food source location that is richer in nectar amount. This collective information depends on behaviour of honey bees, so honey bees are divided into three groups:

1. Employed Honey Bees.
2. Onlooker Honey Bees.
3. Scout Honey Bees.

50% of honey bees are known as employed bees and the remaining 50% of bees are onlooker bees. Employed bees search for a new source of food and determine the amount of nectar. Now employed bees share the information of new food sources with onlooker bees which are waiting for food source information. Onlooker bees have the capability to take the decision of which source provides more amount of food. Employed bees share their information with the help of waggle dance. Scout bees are responsible to search new food sources.

In the ABC algorithm, location of food source is considered as the solution of the optimization problem shown in algorithm 1. These food source location are related to the amount of nectar that is known as the fitness of the solution. At the beginning of the optimization process, ABC randomly generates initial bee population in the hive is P of SN solutions. Here SN represents the number of food sources present in the environment. x_i represents each solution of optimization problem where i = 1,2,3…SN. x_i is a D-dimensional vector, here D is used for optimization parameters. After this initialization process, bees try to update their solution by repeating their process, these repeated cycles are represented as RC = 1, 2,…, MCN, here MCN denotes maximum cycle numbers which help to get an optimal solution. Employed bees, onlooker bees, and scout bees help to find an optimal solution of the optimization process in the repeated cycle. Employed bees find a new food location which consists a large amount of nectar, when the new food location has more nectar amount then the employed bees forget the previous food location. After getting a new food source location employed bees come back to the hive and share information with onlooker bees which are waiting for employed bees. After getting information, onlooker bees collect their food.

An artificial onlooker bee identifies a food source P_i with the help of probability level, which is determined using the formula below

$$P_i = \frac{fit_i}{\sum_{n=1}^{SN} fit_n} \tag{3.6}$$

New food location of artificial bee is expressed as where fit_i is the fitness value of the solution i which is directly proportional to the nectar amount of the food source at food location i and SN is the number of food sources which is equal to the number of employed bees (BN). New food location of artificial bee is expressed as

$$U_{ij} = x_{ij} + \varphi_{ij}\,(x_{ij} - x_{kj}) \tag{3.7}$$

where k ∈ 1, 2,..., SN and j ∈ 1, 2,..., D are randomly generated and also differs from "i." ϕ_{ij} is a radomly generated number which lies between −1 and 1. It regulates the production of nearby food sources around $x_{i,j}$ and presents a comparison of two food location of a bee. From (3.7), the perturbation on the position $x_{i,j}$ decreases as the difference between the parameters of the $x_{i,j}$ and $x_{k,j}$ decreases. As the search gets closer to the best solution in the search space, the step length gets shorter.

To find new food source location, scout bees are responsible. This new food location of scout bees are expressed as

$$x_i^j = x_{\min}^j + rand(0,1)[x_{\max}^j - x_{\min}^j] \tag{3.8}$$

In this process, a greedy selection mechanism is employed as the selection operation between the old and the new one.

3.3.2 The Proposed LABC Algorithm

In any algorithm, the position update is done by the mathematical approach which helps to improve the result of algorithms. These literature represents

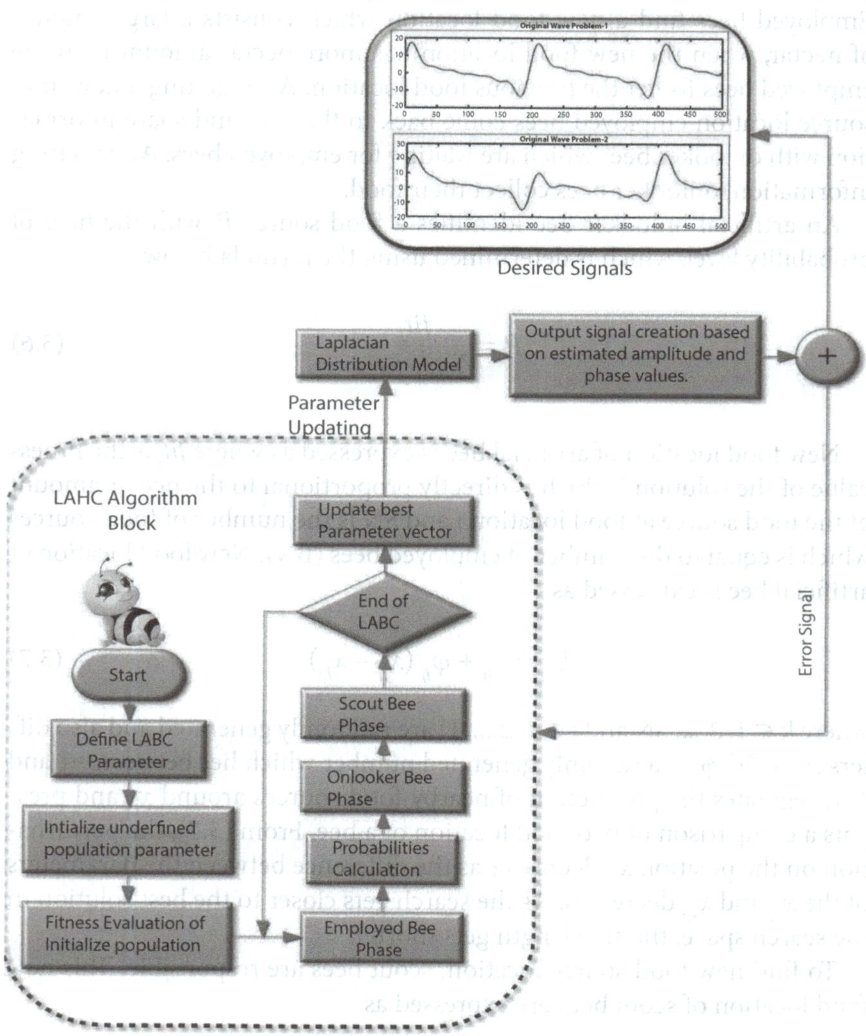

Figure 3.1 Flowchart of proposed LABC algorithm based harmonic estimator design.

effect of position update with different methods [33–35]. In this section, position of bee is updated using Laplacian operator and working mechanism of LABC shown in Figure 3.1. Laplacian operator is described as follow.

Define a number series of size (1 × Colony Size) having Laplacian distribution and take the mean (μ) and standard deviation (σ) of the series.

Define a number series Y_{prime} that is given as follows:

$$Y_{prime} = \mu + linspace(-3 \times \sigma, 3 \times \sigma, T) \tag{3.9}$$

Further, using this number series a Laplacian operator is employed here to adjust the bees position in the proposed LABC algorithm.

$$Fy = \frac{1}{\sqrt{(2*(\sigma)^2)} * exp\left(-\sqrt{\left(\frac{2}{\sigma^2}\right)}\right)^{|Y_{prime} - \mu|}} \tag{3.10}$$

The updated equation for bees with the help of equation 3.8 can be written as per following:

$$\lambda(t) = Fy \times x^j_{min} + rand(0,1)[x^j_{max} - x^j_{min}] \tag{3.11}$$

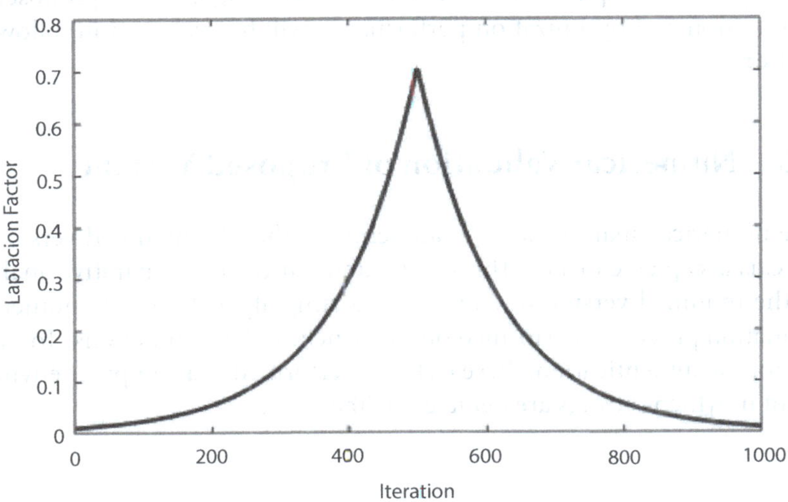

Figure 3.2 Proposed Laplacian factor.

The Laplacian factor (LF)–based mechanism will help in acceleration of the algorithm in exploration phase and it will be reduced with a larger pace in exploitation phase. As shown in Figure 3.2, it is quite obvious that during first half of iteration, the function is monotonically increasing and during later half, its monotonically decreasing. Proposed factor helps the algorithm to get redemption from stagnation in local traps and enhance the convergence of the algorithm.

3.4 Discussion

With the incorporation of this Laplacian factor, optimization performance of the ABC algorithm can be enhanced. In previous researches also, this factor has significantly enhanced the performance of optimization algorithm. Recently, Dinkar *et al.* used Laplacian number instead of random numbers in [36], Zhang *et al.* [37] introduced Laplacian Biogeography-Based Optimization. In this research, the Laplacian operator improves the search ability and the compatibility of the algorithm. Dinkar *et al.* [38] propose a modified version of freshly developed Equilibrium Optimizer (EO) with the help of Laplacian operator, which improves the macro and micro search abilities of the algorithm. These researches are strong evidences that prove that optimization capabilities can be enhanced by the incorporation of Laplacian factor. In fact, this factor can be utilized either in starting phase or decision-making phase. In other words, position update phase of any algorithm. Inspiring from these facts, LABC algorithm is proposed in this section and optimization performance will be evaluated in following section.

3.5 Numerical Validation of Proposed Variant

The numerical analysis helps to authenticate that the modified version of ABC is acceptable or not. This can be done through comparative analysis of the modified version of ABC with existing algorithms. The numerical validation process is performed on the benchmark test functions. To make an honest authentication, these tests are performed at a computer environment in which settings are depicted in Table 3.2.

Table 3.2 System configuration.

Parameter setting of system software setting	
Version of MATLAB	MATLAB R2015a
Operating system	Windows 10
Hardware setting	
Processor	INTEL CORE i5 8th generation
Hard disc drive	1 TB
RAM	8 GB

3.5.1 Comparative Analysis of LABC with Other Meta-Heuristics

In this segment, the modified version of ABC is compared with other meta-heuristic techniques which consist original ABC, Grey Wolf Optimization (GWO) [3], Sine-Cosine Algorithm (SCA) [39], Moth Flame Optimization (MFO) [40], Bat Algorithm (BAT) [41], Bio geography-Based Optimization (BBO) [42], Salp Swarm Algorithm (SSA) [43], etc. This comparison process are tested on traditional benchmark test functions, which are reported in Table 3.3, while the shape curves of the traditional test functions are illustrated in Figure 3.3. This numerical comparison helps to prove the dominance of the modified version of ABC algorithm. Traditional benchmark test functions are divided into three groups; the first group is unimodal traditional benchmark test functions, second group is multimodal benchmark test functions, and the last group is fixed-dimension multidimension traditional benchmark test functions. To make a fair comparison, it is necessary that execution time setting is maintained similar for every competitor algorithms, this setting includes,

- Number of runs: 20
- Number of iterations: 500
- Number of search agents: 30

Table 3.3 Traditional benchmark function suit.

Benchmark function suit				
Function name	**Function details**	**Dim~**	**Search space**	**Minima**
Sphere	$F^1(\vartheta)=\sum_{A=1}^{B}\vartheta_A^2$	30	[−100,100]	0
Schwefel 2.22	$F^2(\vartheta)=\sum_{A=1}^{B}\lvert\vartheta_A\rvert+\prod_{A=1}^{B}\lvert\vartheta_A\rvert$	30	[−10,10]	0
Schwefel 1.2	$F^3(\vartheta)=\sum_{A=1}^{B}\left(\sum_{j=1}^{K}\vartheta_j\right)^2$	30	[−100,100]	0
Schwefel 2.21	$F^4(\vartheta)=\max_A\{\lvert\vartheta_A\rvert \quad 1\le A\le B$	30	[−100,100]	0
Rosenbrock	$F^5(\vartheta)=\sum_{A=1}^{B-1}[100(\vartheta_{A+1}-\vartheta A^2)^2+(\vartheta_A-1)^2]$	30	[−30,30]	0
Step	$F^6(\vartheta)=\sum_{A=1}^{B-1}([\vartheta_A+0.5])^2$	30	[−100,100]	0
Quartic	$F^7(\vartheta)=\sum_{A=1}^{B-1}A\,\vartheta_A^4+rand[0,1]$	30	[−1.28,1.28]~	0
Schwefel 2.26	$F^8(\vartheta)=\sum_{A=1}^{B}-\vartheta_A\sin\left(\sqrt{\lvert\vartheta_A\rvert}\right)$	30	[−500,500]	−418.9829 × 5
Rastrigin	$F^9(\vartheta)=\sum_{A=1}^{B}\left[\vartheta_A{}^2-10\cos(2\pi\vartheta_A+10)\right]$	30	[−5.12,5.12]	0

(Continued)

Table 3.3 Traditional benchmark function suit. (*Continued*)

Benchmark function suit				
Function name	**Function details**	**Dim~**	**Search space**	**Minima**
Ackley	$F^{10}(\vartheta)=-20\exp\left(-0.2\sqrt{\frac{1}{B}\sum_{A=1}^{B}\sigma_A{}^2}\right)$ $-\exp\left(\frac{1}{B}\sum_{A=1}^{B}\cos(2\pi\vartheta_A)\right)+20+e$	30	[−32,32]	0
Griewank	$F^{11}(\vartheta)=\frac{1}{4000}\sum_{A=1}^{B}\vartheta_A{}^2-\prod_{A=1}^{B}\cos\left(\frac{\vartheta_A}{\sqrt{A}}\right)+1$	30	[−600,600]	0
Generalized Penalized	$F^{12}(\vartheta)=\frac{\pi}{p}\{10\sin(\pi i_1)+\sum_{A=1}^{B-1}(i_A-1)^2[1+10\sin^2(\pi i_{A+1})]$ $+(i_B-1)^2\}+\sum_{A=1}^{B}b(\vartheta_A,10,100,4)$ $c_e=1+\frac{\zeta_{e+1}}{4}\quad i(\vartheta_A,h,o,l)=\begin{cases} f(\vartheta_A-v) & \vartheta_A>v \\ 0 & -v<\vartheta<vht \\ f(-\vartheta_A-v)^l & \vartheta_A<-v \end{cases}$	30	[−50,50]	0

(*Continued*)

Table 3.3 Traditional benchmark function suit. (*Continued*)

Benchmark function suit				
Function name	**Function details**	**Dim~**	**Search space**	**Minima**
Penalty	$F^{13}(\vartheta)=0.1\{sin\,B^2(3\pi\vartheta_1)+\sum_{A=1}^{B}(\vartheta_A-1)^2[1+\sin^2(3\pi\vartheta_A+1)]$ $+(\vartheta_B-1)^2[1+\sin\square^2(2\pi\zeta_B)]\}+\sum_{A=1}^{B}q(\vartheta_A,5,100,4)$	30	[−50,50]	0
Dejoung (Shekel's Foxholes)	$F^{14}(\vartheta)=\left(\frac{1}{500}+\sum_{a=1}^{25}\frac{1}{a+\sum_{a=1}^{25}(\vartheta_A-h_{Aa})^6}\right)^{-1}$	2	[−65,65]~	1
Kowalik	$F^{15}(\vartheta)=\sum_{a=1}^{11}\left[h_A-\frac{\vartheta_1(d_A^2+d_A\vartheta_2)}{d_A^2+A_A\vartheta_3+\vartheta_4}\right]^2$	4	[−5,5]	0.00030
Camel Back–6 Hump	$F^{16}(\vartheta)=4\vartheta_1^2-2.1\vartheta_1^4+\frac{1}{3}\vartheta_1^6+\vartheta_1\vartheta_2-4\vartheta_2^2+4\vartheta_2^4$	2	[−5,5]	−1.0316
Branin	$F^{17}(\vartheta)=\left(\vartheta_2-\frac{5.1}{4\pi^2}\vartheta_1^2+\frac{5}{\pi}x_1-6\right)^2+10\left(1-\frac{1}{8\pi}\right)\cos\vartheta_1+10$	2	[−5,5]	0.398

(*Continued*)

Table 3.3 Traditional benchmark function suit. (*Continued*)

Benchmark function suit				
Function name	**Function details**	**Dim~**	**Search space**	**Minima**
Goldstein and Price	$F^{18}(\vartheta) = M(\vartheta) \times B(\vartheta)\backslash\backslash$ $(\vartheta)=1+(\vartheta_1+\vartheta_2+1)^2(19-14\vartheta_1+3\vartheta_1^2-14\vartheta_2+6\vartheta_1\vartheta_2+3\vartheta_2^2)$	2	[−2,2]	3
Hartman 3~Dimensional	$F^{19}(\vartheta)=-\sum_{A=1}^{4} x_A \exp\left(-\sum_{a=1}^{3} h_{Aa}(\vartheta_a-e_{Aa})^2\right)$	3	[1,3]	−3.86
Hartman 6–Dimensional	$F^{20}(\vartheta)=-\sum_{A=1}^{4} x_A \exp\left(-\sum_{a=1}^{6} h_{Aa}(\vartheta_a-e_{Aa})^2\right)$	6	[0,1]	−3.32
Shekel 1	$F^{21}(\vartheta)=-\sum_{A=1}^{5}[(Z-h_A)(Z-h_A)^b+z_A]^{-1}$	4	[0,10]	−10.1532
Shekel 2	$F^{22}(\vartheta)=-\sum_{A=1}^{7}[(\vartheta-h_A)(\vartheta-h_A)^b+z_A]^{-1}$	4	[0,10]	−10.4028

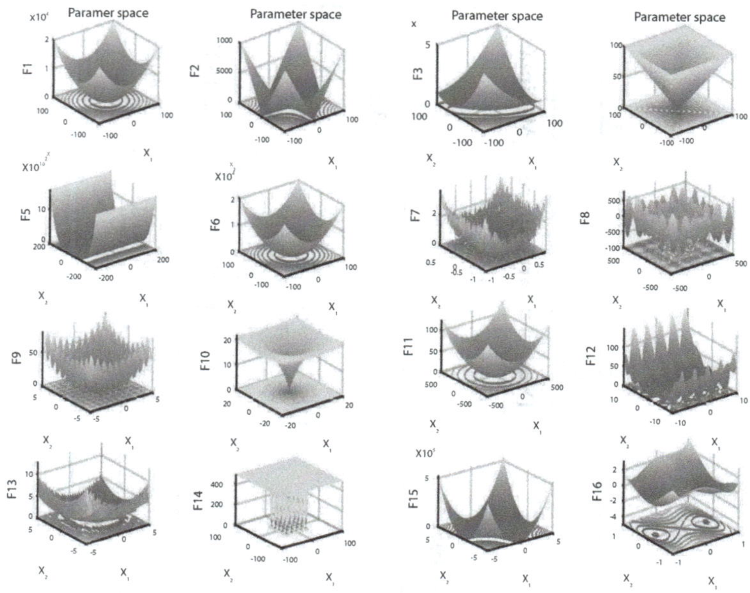

Figure 3.3 Shape curves of benchmark functions black.

Numerical results of this comparison are reported in Table 3.4. In the numerical assessment, we use following four mathematical parameters

- Average Value (Mean)
- Maximum Value (Max)
- Minimum Value (Min)
- Standard Deviation (SD)

From this comparison, the following points are noticed:

- Unimodal Benchmark Test Functions: F1–F7 test functions are known as unimodal benchmark test functions. These test functions have more exploitation ability because in these test functions local optimas are absent. From Table 3.4, it is seen that the modified version of ABC gives optimal results of the mean value for function F1, F2, F3, F4, F7 in comparison to other meta-heuristics algorithms.
- Multimodal Benchmark Test Functions: F8–F13 test functions are known as multimodal benchmark test functions. These test functions have ability to check the exploration

Table 3.4 Comparison of artificial bee colony algorithm with other algorithms.

Proposed algorithm comparison with other algorithms												
	F1				**F2**				**F3**			
Statistical values	**Mean**	**Max**	**Min**	**SD**	**Mean**	**Max**	**Min**	**SD**	**Mean**	**Max**	**Min**	**SD**
ABC	1.24E+02	2.87E+02	4.63E+01	6.91E+01	4.27E+01	1.14E+02	2.78E+00	3.19E+01	6.81E+04	1.03E+05	4.73E+04	1.40E+04
LABC	**0.00E+00**	0.00E+00	0.00E+00	0.00E+00	**4.05E−288**	5.86E−287	5.22E−299	0.00E+00	**0.00E+00**	0.00E+00	0.00E+00	0.00E+00
GWO	1.23E−27	4.95E−27	5.16E−29	1.38E−27	7.80E−17	1.63E−16	2.39E−17	4.44E−17	4.81E−06	2.34E−05	2.52E−08	6.30E−06
SCA	2.72E+01	2.29E+02	1.41E−02	6.25E+01	1.94E−02	1.11E−01	6.86E−05	3.01E−02	1.18E+04	3.16E+04	2.94E+03	7.38E+03
MFO	3.00E+03	1.00E+04	4.21E−01	4.70E+03	2.97E+01	1.00E+02	1.83E−01	2.54E+01	2.43E+04	4.27E+04	4.94E+03	1.21E+04
BAT	4.09E+04	6.01E+04	2.46E+04	9.96E+03	1.32E+07	1.76E+08	1.46E+02	4.01E+07	8.91E+04	1.57E+05	5.13E+04	2.68E+04
BBO	2.56E+00	3.64E+00	1.64E+00	6.24E−01	4.82E−01	6.02E−01	3.86E−01	6.42E−02	4.63E+02	6.60E+02	2.99E+02	1.05E+02
SSA	1.50E−07	5.31E−07	3.06E−08	1.33E−07	2.09E+00	4.43E+00	2.44E−01	1.13E+00	1.79E+03	4.44E+03	4.22E+02	1.16E+03
	F4				**F5**				**F6**			
ABC	6.38E+01	7.80E+01	5.25E+01	6.30E+00	2.53E+06	4.95E+06	1.06E+06	1.03E+06	1.07E+02	2.34E+02	3.00E+01	5.03E+01
LABC	**1.12E−237**	2.23E−236	3.01E−277	0.00E+00	2.83E+01	2.86E+01	2.80E+01	2.02E−01	2.08E+00	2.77E+00	1.50E+00	4.29E−01
GWO	8.47E−07	2.85E−06	7.37E−08	8.16E−07	**2.71E+01**	2.87E+01	2.61E+01	7.76E−01	6.84E−01	1.48E+00	1.18E−01	3.69E-01
SCA	3.78E+01	5.40E+01	2.01E+01	9.39E+00	2.09E+04	2.77E+05	3.14E+02	6.09E+04	9.33E+00	3.12E+01	5.06E+00	6.38E+00
MFO	6.95E+01	8.25E+01	5.64E+01	8.78E+00	4.02E+06	7.99E+07	1.66E+02	1.79E+07	3.02E+03	2.00E+04	5.81E−01	5.69E+03
BAT	6.99E+01	9.07E+01	5.63E+01	7.93E+00	9.73E+07	1.94E+08	3.26E+07	5.09E+07	4.13E+04	6.01E+04	2.57E+04	1.05E+04
BBO	1.49E+00	2.03E+00	1.07E+00	2.38E−01	1.71E+02	5.64E+02	5.99E+01	1.26E+02	2.41E+00	3.42E+00	1.51E+00	5.77E−01
SSA	1.15E+01	1.90E+01	5.17E+00	3.01E+00	4.91E+02	5.21E+03	2.41E+01	1.15E+03	**1.39E−07**	4.81E−07	4.95E−08	1.06E−07

(*Continued*)

Table 3.4 Comparison of artificial bee colony algorithm with other algorithms. (*Continued*)

Proposed algorithm comparison with other algorithms												
Statistical values	**Mean**	**Max**	**Min**	**SD**	**Mean**	**Max**	**Min**	**SD**	**Mean**	**Max**	**Min**	**SD**
	F7				**F8**				**F9**			
ABC	1.65E+00	4.45E+00	7.01E−01	8.72E−01	**−2.85E+60**	−1.12E+57	−4.48E+61	9.93E+60	2.45E+02	2.63E+02	2.19E+02	1.33E+01
LABC	**4.31E−05**	1.58E−04	4.90E−06	4.06E−05	−6.94E+03	−6.25E+03	−8.36E+03	4.87E+02	**0.00E+00**	0.00E+00	0.00E+00	0.00E+00
GWO	1.84E−03	4.56E−03	4.12E−04	9.80E−04	−6.03E+03	−4.03E+03	−7.90E+03	9.00E+02	2.06E+00	9.18E+00	5.68E−14	3.12E+00
SCA	7.88E−02	2.59E−01	8.14E−03	6.38E−02	−3.81E+03	−3.38E+03	−4.25E+03	2.75E+02	5.04E+01	1.31E+02	8.51E−01	3.22E+01
MFO	5.86E+00	3.24E+01	1.12E−01	8.87E+00	−8.41E+03	−6.96E+03	−9.71E+03	7.69E+02	1.60E+02	2.28E+02	1.04E+02	3.49E+01
BAT	7.23E+01	1.36E+02	1.62E+01	4.03E+01	−6.32E+03	−2.09E+03	−1.66E+04	4.32E+03	3.74E+02	4.23E+02	3.06E+02	3.60E+01
BBO	1.40E−02	2.66E−02	6.30E-03	6.35E−03	−8.26E+03	−6.85E+03	−9.45E+03	6.56E+02	5.03E+01	7.84E+01	2.59E+01	1.36E+01
SSA	1.62E−01	2.83E−01	9.03E−02	5.26E−02	−7.69E+03	−5.52E+03	−9.31E+03	9.01E+02	5.37E+01	8.66E+01	2.39E+01	1.82E+01
	F10				**F11**				**F12**			
ABC	7.70E+00	9.38E+00	5.90E+00	9.77E-01	2.22E+00	3.63E+00	1.36E+00	7.17E-01	6.13E+06	1.54E+07	8.88E+05	4.04E+06
LABC	**8.88E-16**	8.88E−16	8.88E−16	0.00E+00	**0.00E+00**	0.00E+00	0.00E+00	0.00E+00	2.75E−02	4.38E−02	1.59E−02	6.24E−03
GWO	1.03E−13	1.39E−13	7.55E−14	1.51E−14	6.52E−03	2.70E−02	0.00E+00	9.67E−03	6.84E−02	5.52E−01	9.73E−06	1.15E−01
SCA	1.24E+01	2.03E+01	4.83E−02	9.88E+00	8.15E−01	1.31E+00	5.68E−02	3.48E−01	4.46E+05	8.50E+06	8.66E−01	1.90E+06
MFO	1.70E+01	2.00E+01	2.78E+00	5.02E+00	1.90E+01	1.81E+02	7.32E−01	4.70E+01	8.94E+00	3.29E+01	3.01E+00	6.38E+00
BAT	1.99E+01	2.00E+01	1.98E+01	5.37E−02	3.45E+02	5.91E+02	2.15E+02	1.01E+02	1.44E+08	4.73E+08	2.22E+07	1.09E+08
BBO	5.80E−01	7.33E−01	3.88E−01	9.30E−02	1.00E+00	1.04E+00	9.46E−01	2.76E−02	**7.54E−03**	1.55E−02	4.00E−03	2.67E−03
SSA	2.53E+00	3.98E+00	1.50E+00	6.54E−01	1.85E−02	5.48E−02	1.03E−03	1.44E−02	7.38E+00	1.39E+01	3.61E+00	2.55E+00

(*Continued*)

Table 3.4 Comparison of artificial bee colony algorithm with other algorithms. (*Continued*)

Proposed algorithm comparison with other algorithms												
Statistical values	**Mean**	**Max**	**Min**	**SD**	**Mean**	**Max**	**Min**	**SD**	**Mean**	**Max**	**Min**	**SD**
	F13				**F14**				**F15**			
ABC	1.37E+07	4.15E+07	2.40E+06	1.02E+07	**9.99E−01**	1.01E+00	9.98E−01	2.95E−03	1.14E−03	1.26E−03	1.04E−03	6.03E−05
LABC	2.59E+00	2.98E+00	2.05E+00	3.13E−01	1.01E+00	1.22E+00	9.98E−01	5.03E−02	**3.45E−04**	3.77E−04	3.16E−04	2.01E−05
GWO	6.75E−01	1.19E+00	2.14E−01	2.30E−01	4.48E+00	1.08E+01	9.98E−01	3.83E+00	8.49E−03	2.04E−02	3.09E−04	9.95E−03
SCA	3.87E+05	2.70E+06	2.45E+00	8.39E+05	1.50E+00	2.98E+00	9.98E−01	8.78E−01	1.07E−03	1.51E−03	3.56E−04	3.78E−04
MFO	2.51E+01	8.94E+01	5.31E−01	2.25E+01	2.33E+00	5.93E+00	9.98E−01	1.82E+00	2.13E−03	2.04E−02	6.94E−04	4.33E−03
BAT	2.61E+08	5.03E+08	4.86E+07	1.32E+08	2.63E+01	1.54E+02	1.59E+00	3.68E+01	4.38E−02	1.06E−01	1.66E−03	3.23E−02
BBO	**1.07E−01**	1.54E−01	4.82E−02	2.45E−02	7.64E+00	1.64E+01	9.98E−01	4.73E+00	5.08E−03	2.04E−02	4.30E−04	7.47E−03
SSA	2.16E+01	5.53E+01	9.48E−02	1.48E+01	1.25E+00	2.98E+00	9.98E−01	5.46E−01	2.01E−03	2.04E−02	7.16E−04	4.33E−03
	F16				**F17**				**F18**			
ABC	**−1.05E+01**	−1.05E+01	−1.05E+01	3.24E−07	3.98E−01	3.98E−01	3.98E−01	3.29E−05	3.00E+00	3.00E+00	3.00E+00	1.94E−06
LABC	−5.30E+00	−5.12E+00	−7.33E+00	5.62E−01	3.98E−01	3.99E−01	3.98E−01	4.76E−04	3.02E+00	3.07E+00	3.00E+00	1.76E−02
GWO	−1.03E+00	−1.03E+00	−1.03E+00	1.98E−08	**−1.03E+01**	−5.18E+00	−10.53573	1.20E+00	7.05E+00	8.40E+01	3.00E+00	1.81E+01
SCA	−1.03E+00	−1.03E+00	−1.03E+00	6.53E−05	−3.70E+00	−9.41E−01	−5.20E+00	1.23E+00	3.00E+00	3.00E+00	3.00E+00	6.73E−05
MFO	−1.03E+00	−1.03E+00	−1.03E+00	2.28E−16	−7.16E+00	−2.42E+00	−1.05E+01	3.85E+00	**3.00E+00**	3.00E+00	3.00E+00	1.66E−15
BAT	−5.59E−01	−1.34E−01	−1.02E+00	3.37E−01	−1.29E+00	−6.61E−01	−4.72E+00	8.68E−01	1.99E+01	4.95E+01	3.02E+00	1.36E+01
BBO	−9.91E−01	−2.15E−01	−1.03E+00	1.82E−01	−5.47E+00	−1.86E+00	−1.05E+01	3.84E+00	5.70E+00	3.00E+01	3.00E+00	8.31E+00
SSA	−1.03E+00	−1.03E+00	−1.03E+00	1.43E−14	−8.81E+00	−2.43E+00	−1.05E+01	3.14E+00	3.00E+00	3.00E+00	3.00E+00	2.81E−13

(*Continued*)

Table 3.4 Comparison of artificial bee colony algorithm with other algorithms. (*Continued*)

Proposed algorithm comparison with other algorithms												
Statistical values	**Mean**	**Max**	**Min**	**SD**	**Mean**	**Max**	**Min**	**SD**	**Mean**	**Max**	**Min**	**SD**
	F19				**F20**				**F21**			
ABC	−3.86E+00	−3.86E+00	−3.86E+00	4.37E−10	**−3.32E+00**	−3.32E+00	−3.32E+00	3.98E−07	**−9.94E+00**	−7.34E+00	−1.02E+01	6.45E−01
LABC	−3.86E+00	−3.86E+00	−3.86E+00	1.69E−04	−3.32E+00	−3.31E+00	−3.32E+00	3.45E−03	−5.22E+00	−5.05E+00	−7.39E+00	5.37E−01
GWO	−3.86E+00	−3.86E+00	−3.86E+00	2.18E−03	−3.23E+00	−2.86E+00	−3.32E+00	1.09E−01	−9.15E+00	−2.63E+00	−1.02E+01	2.49E+00
SCA	−3.85E+00	−3.85E+00	−3.86E+00	1.41E−03	−2.89E+00	−1.66E+00	−3.13E+00	3.72E−01	−2.46E+00	−4.97E−01	−4.88E+00	1.68E+00
MFO	−3.86E+00	−3.86E+00	−3.86E+00	2.28E−15	−3.26E+00	−3.14E+00	−3.32E+00	6.62E−02	−5.75E+00	−2.63E+00	−1.02E+01	3.13E+00
BAT	−3.48E+00	−2.96E+00	−3.86E+00	2.76E−01	−1.80E+00	−8.57E−01	−3.23E+00	6.08E−01	−8.87E−01	−3.43E−01	−2.18E+00	5.08E−01
BBO	**−3.86E+00**	−3.86E+00	−3.86E+00	1.80E−15	−3.26E+00	−3.20E+00	−3.32E+00	6.10E−02	−5.02E+00	−2.63E+00	−1.02E+01	3.19E+00
SSA	−3.86E+00	−3.86E+00	−3.86E+00	7.30E−12	−3.24E+00	−3.18E+00	−3.32E+00	6.30E−02	−6.77E+00	−2.63E+00	−1.02E+01	3.57E+00
	F22											
ABC	**−1.04E+01**	−1.04E+01	−1.04E+01	1.39E−07								
LABC	−5.40E+00	−5.08E+00	−8.75E+00	9.88E−01								
GWO	−1.04E+01	−1.04E+01	−1.04E+01	9.36E−04								
SCA	−3.32E+00	−9.04E−01	−5.27E+00	1.65E+00								
MFO	−6.73E+00	−2.75E+00	−1.04E+01	3.46E+00								
BAT	−1.20E+00	−5.06E−01	−4.11E+00	8.30E−01								
BBO	−6.25E+00	−2.75E+00	−1.04E+01	3.55E+00								
SSA	−7.97E+00	−1.84E+00	−1.04E+01	3.48E+00								

Table 3.5 Rank analysis with competitor algorithms.

Rank analysis with competitor algorithms								
Function	**ABC**	**LABC**	**GWO**	**SCA**	**MFO**	**BAT**	**BBO**	**SSA**
F1	6	1	2	5	7	8	4	3
F2	7	1	2	3	6	8	4	5
F3	7	1	2	5	6	8	3	4
F4	6	1	2	5	7	8	3	4
F5	6	2	1	5	7	8	3	4
F6	6	3	2	5	7	8	4	1
F7	6	1	2	4	7	8	3	5
F8	1	5	7	8	2	6	3	4
F9	7	1	2	4	6	8	3	5
F10	5	1	2	6	7	8	3	4
F11	6	1	2	4	7	8	5	3
F12	7	2	3	6	5	8	1	4
F13	7	3	2	6	5	8	1	4
F14	1	2	6	4	5	8	7	3
F15	3	1	7	2	5	8	6	4
F16	1	2	5	6	3	8	7	4
F17	7	8	1	5	3	6	4	2
F18	3	5	7	4	1	8	6	2
F19	4	5	6	7	1	8	2	3
F20	1	2	6	7	4	8	3	5
F21	1	5	2	7	4	8	6	3
F22	1	6	2	7	4	8	5	3

ability of the algorithm. These test functions have more than one local optima, which increases the complexity to search global optima. The amount of local optimas depends on the dimension of the optimization problem, it means the dimension of the problem and the number of local optimas has a propositional relation. Table 3.4 represents that the exploration ability of the modified version is good in comparison to other algorithms for function F9, F10, F11.

- Fixed Dimension Multimodal Benchmark Test Functions: F14–F22 test functions are known as fixed dimension multimodal benchmark test functions. These test functions establish an inverse relation with the multimodal benchmark test functions. For fixed dimension multimodal benchmark function, only single function F15 gives better results for the modified version of ABC in comparison to other meta-heuristic algorithms.

To make a trustworthy comparison of the modified version with other meta-heuristic algorithms, we use the ranking method, results of this ranking method are reported in Table 3.5. From Table 3.5, it is observed that the modified version of ABC algorithm gets 1st rank for nine benchmark test functions in comparison to other meta-heuristic algorithms. From the numerical comparison and ranking process, it is concluded that the modified version of ABC is superior to other meta-heuristic algorithms.

3.5.2 Benchmark Test on CEC-17 Functions

To prove the supremacy of the modified version of the ABC algorithm we test this version on CEC-17 benchmark test functions. These CEC-17 functions are divided into four categories which are unimodal test functions (F1–F3), simple multimodal functions (F4–F10), hybrid functions (F11–F20), and composite functions (F21–F30) which are reported in Table 3.6. Numerical results of CEC-17 functions are reported in Table 3.7, For a fair and promising comparison of L-ABC and original ABC parameter settings (number of iteration=1000, number of run=51, dimension (Dim)=30, and number of search agents=100) are kept constant. Simulation results help to validate that L-ABC improves the convergence rate and increases the exploration area which helps to find global optima in minimum time. These results accumulated in the form of mean, SD, max, min, and p-value. These results provide a comparison between modified version of ABC and parent ABC. After inspecting these results it is clear that the modified

Table 3.6 CEC-17 benchmark functions.

Details about CEC-2017 functions	
Function name	**Minima**
Unimodal functions (Shifted and Rotated)	
Bent Cigar function (F-1)	100
Zakharov function (F-3)	300
Simple multimodal functions (Shifted and Rotated)	
Rosenbrock's function (F-4)	400
Rastrigin's function (F-5)	500
Expanded Scaffer's function (F-6)	600
Lunacek Bi Rastrigin function (F-7)	700
Non-continuous Rastrigin function (F-8)	800
Levy function (F-9)	900
Schwefel's function (F-10)	1000
Hybrid functions (HF)	
HF1 (N = 3) (F-11)	1100
HF 2 (N = 3) (F-12)	1200
HF 3 (N = 3) (F-13)	1300
HF 4 (N = 4) (F-14)	1400
HF 5 (N = 4) (F-15)	1500
HF 6 (N = 4) (F-16)	1600
HF 7 (N = 5) (F-17)	1700
HF 8 (N = 5) (F-18)	1800
HF 9 (N = 5) (F-19)	1900
HF 10 (N = 6) (F-20)	2000

(Continued)

Table 3.6 CEC-17 benchmark functions. (*Continued*)

Details about CEC-2017 functions	
Function name	**Minima**
Composite functions (CF)	
CF 1 (N = 3) (F-21)	2100
CF 2 (N = 3) (F-22)	2200
CF 3 (N= 4) (F-23)	2300
CF 4 (N = 4) (F-24)	2400
CF 5 (N = 5) (F-25)	2500
CF 6 (N = 5) (F-26)	2600
CF 7 (N = 6) (F-27)	2700
CF 8 (N = 6) (F-28)	2800
CF 9 (N = 3) (F-29)	2900
CF 10 (N = 3) (F-30)	21000

version of ABC provides better results for 24 functions for CEC-2017 test functions which are highlighted in Table 3.7. From these assessments, it is evident that the modified version is adaptable to solve the real-time optimization problems.

3.6 Analytical Validation of Proposed Variant

The analytical method is used to demonstrate the effectiveness of the modified version of ABC in comparison to other meta-heuristic algorithms. For analytical analysis, the following methods have been used here

1. Convergence Rate Test
2. Box Plot Analysis
3. Wilcoxon Rank Sum Test
4. Scalability Test.

Table 3.7 CEC-17 Benchmark analysis.

CEC-17 Benchmark analysis										
	ABC					L-ABC				
	MEAN	SD	MAX	MIN	P-value	MEAN	SD	MAX	MIN	P-value
F1	1.415E+07	5.556E+06	3.588E+07	6.149E+06	**5.652E−14**	2.380E+07	3.286E+06	3.360E+07	1.790E+07	N/A
F3	6.278E+05	2.046E+05	1.237E+06	2.900E+05	**3.304E−18**	**1.327E+05**	7.281E+03	1.480E+05	1.152E+05	N/A
F4	6.925E+02	6.211E+01	8.209E+02	5.862E+02	**3.304E−18**	**5.126E+02**	7.490E+00	5.316E+02	4.991E+02	N/A
F5	9.874E+02	1.869E+01	1.019E+03	9.293E+02	**3.304E−18**	**8.860E+02**	6.805E+00	9.002E+02	8.671E+02	N/A
F6	6.136E+02	1.676E+00	6.175E+02	6.103E+02	**3.304E−18**	6.740E+02	1.197E+00	6.762E+02	6.701E+02	N/A
F7	1.245E+03	1.790E+01	1.275E+03	1.200E+03	**3.304E−18**	1.771E+03	1.509E+01	1.787E+03	1.687E+03	N/A
F8	1.292E+03	1.783E+01	1.319E+03	1.247E+03	**3.304E−18**	**1.229E+03**	6.539E+00	1.243E+03	1.212E+03	N/A
F9	2.462E+04	4.371E+03	3.427E+04	1.589E+04	**1.555E−08**	**2.057E+04**	6.522E+02	2.186E+04	1.858E+04	N/A
F10	1.565E+04	2.891E+02	1.623E+04	1.509E+04	**3.304E−18**	**9.758E+03**	3.367E+02	1.042E+04	8.859E+03	N/A
F11	4.024E+04	7.368E+03	5.614E+04	2.515E+04	**3.304E−18**	**1.754E+03**	4.909E+01	1.863E+03	1.628E+03	N/A
F12	7.315E+09	1.278E+09	1.041E+10	3.712E+09	**3.304E−18**	**2.116E+07**	4.122E+06	3.103E+07	1.280E+07	N/A
F13	1.096E+06	6.202E+05	3.247E+06	2.533E+05	**3.304E−18**	**4.047E+04**	1.375E+04	7.164E+04	1.599E+04	N/A
F14	2.983E+06	1.001E+06	5.307E+06	1.477E+06	**3.304E−18**	**1.614E+05**	5.795E+04	3.193E+05	4.053E+04	N/A
F15	1.578E+06	2.641E+06	1.146E+07	4.729E+04	**3.304E−18**	**3.772E+04**	3.699E+03	4.622E+04	2.891E+04	N/A
F16	6.242E+03	1.710E+02	6.645E+03	5.700E+03	**3.304E−18**	**3.542E+03**	1.686E+02	3.840E+03	3.064E+03	N/A
F17	4.440E+03	1.861E+02	4.810E+03	3.791E+03	**3.304E−18**	**3.258E+03**	1.389E+02	3.483E+03	2.824E+03	N/A

(Continued)

Table 3.7 CEC-17 Benchmark analysis. (*Continued*)

CEC-17 Benchmark analysis										
	ABC					**L-ABC**				
	MEAN	**SD**	**MAX**	**MIN**	**P-value**	**MEAN**	**SD**	**MAX**	**MIN**	**P-value**
F18	4.559E+07	1.593E+07	7.882E+07	1.125E+07	**3.304E−18**	**1.478E+06**	3.219E+05	2.030E+06	8.086E+05	N/A
F19	1.910E+04	5.363E+03	3.062E+04	8.030E+03	**1.437E−07**	2.381E+04	1.315E+03	2.822E+04	2.105E+04	N/A
F20	4.334E+03	1.694E+02	4.645E+03	3.905E+03	**3.304E−18**	**3.185E+03**	1.318E+02	3.427E+03	2.855E+03	N/A
F21	2.787E+03	1.449E+01	2.819E+03	2.741E+03	1.221E−01	**2.727E+03**	1.462E+02	2.870E+03	2.435E+03	N/A
F22	1.694E+04	3.768E+02	1.756E+04	1.581E+04	**3.304E−18**	**1.277E+04**	2.308E+02	1.333E+04	1.226E+04	N/A
F23	3.182E+03	2.555E+01	3.227E+03	3.102E+03	**3.005E−15**	3.287E+03	6.984E+01	3.380E+03	2.971E+03	N/A
F24	3.417E+03	2.248E+01	3.451E+03	3.358E+03	**1.400E−04**	3.439E+03	2.753E+01	3.512E+03	3.392E+03	N/A
F25	3.124E+03	4.552E+01	3.251E+03	3.032E+03	**2.990E−06**	**3.090E+03**	9.318E+00	3.121E+03	3.073E+03	N/A
F26	8.069E+03	1.973E+02	8.522E+03	7.716E+03	**1.900E−17**	**6.440E+03**	7.915E+02	8.017E+03	4.864E+03	N/A
F27	3.200E+03	5.223E−05	3.200E+03	3.200E+03	**3.304E−18**	**3.200E+03**	1.052E−04	3.200E+03	3.200E+03	N/A
F28	3.300E+03	1.449E−02	3.300E+03	3.300E+03	**3.297E−13**	**3.300E+03**	2.000E−01	3.300E+03	3.299E+03	N/A
F29	8.174E+03	5.167E+02	9.371E+03	7.204E+03	**3.304E−18**	**4.299E+03**	1.294E+02	4.592E+03	4.063E+03	N/A
F30	1.970E+08	8.541E+07	4.618E+08	4.224E+07	**3.304E−18**	**1.991E+05**	5.626E+04	3.336E+05	7.067E+04	N/A

3.6.1 Convergence Rate Test

A comparison of the modified version of ABC with other meta-heuristic algorithms in terms of convergence rate is represented by convergence graphs. Convergence graphs of some functions are denoted in Figures 3.4, 3.5, and 3.6. This comparison helps to determine which algorithm provides optimal solutions to the optimization problems in minimum time. The median value of each function for 20 runs is estimated which is used to plot the convergence graph. In the convergence graph, the number of iterations is denoted by x-axis while the best score is represented on y-axis. Figures 3.4, 3.5, and 3.6 represent the convergence graph of unimodal benchmark test functions, multimodal benchmark test functions, and fixed-dimension multimodal benchmark test functions, respectively. From the convergence graph analysis, it is seen that modified version of ABC achieves the optimal solutions as fast as possible in comparison to other meta-heuristics algorithms. From this comparative graphical analysis, it is proved that the convergence rate of modified version of ABC is fast in comparison to others which proves that the modified version is acceptable.

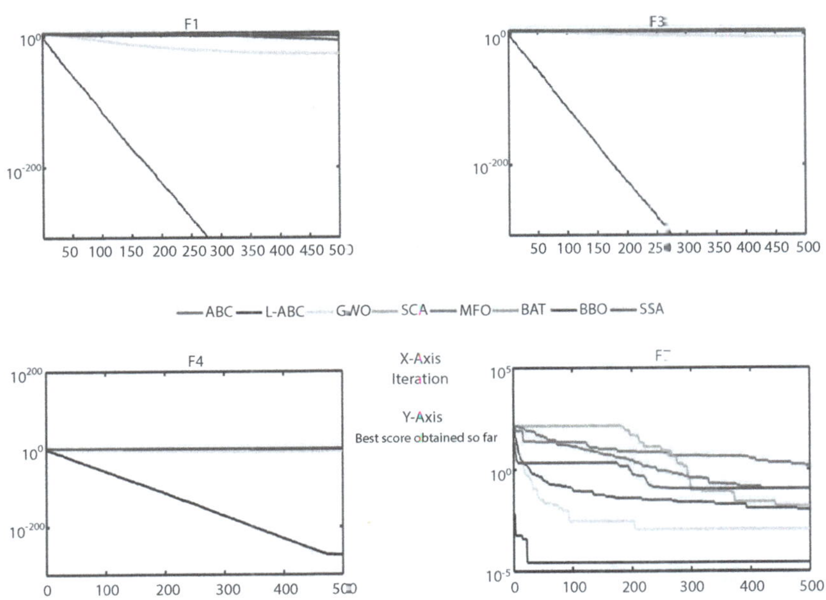

Figure 3.4 Convergence analysis of unimodal functions.

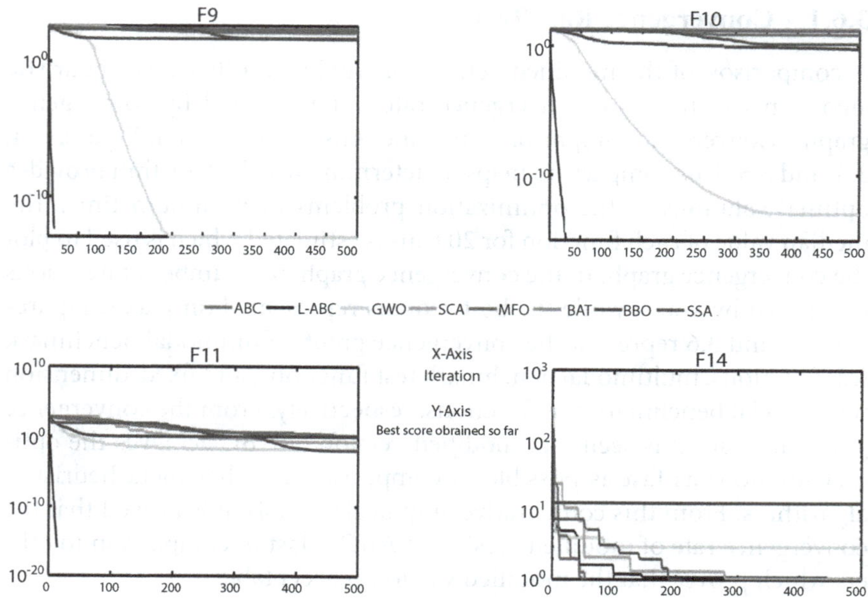

Figure 3.5 Convergence analysis of multimodal functions.

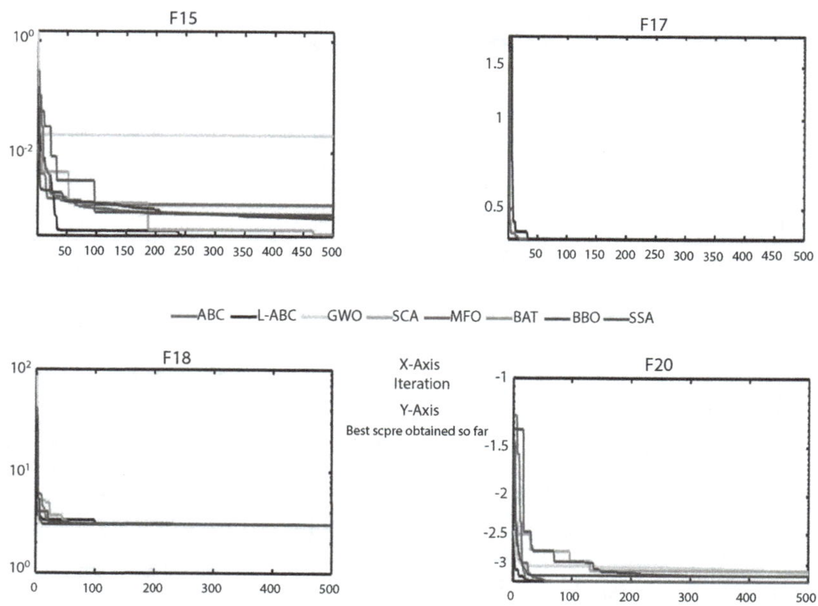

Figure 3.6 Convergence analysis of fixed-dimension functions.

3.6.2 Box Plot Analysis

Box plot analysis is a data distribution tool that splits the data for each and every run. Box plot analysis was used to examine the stability of the solution by a modified version of ABC in comparison to other metaheuristic algorithms. Box plot analysis is a visualization based approach that depends on two parameters, first one is "interquartile range," which is evaluated by the thickness of the box plot and the second one is "range of box plot", this range lies between the minimum and maximum value. To check the solution quality, a modified version of ABC is compared with parent ABC which is illustrated in Figure 3.7. After observing these figures, it is clear that the modified version of ABC provides optimum results in comparison to parent ABC which proves that modified version of ABC provides promising results to solve real-time optimization problems.

3.6.3 Wilcoxon Rank Sum Test

To examine the statistical significance of the modified version of ABC, a nonparametric approach is used, known as the Wilcoxon rank-sum test. The approach of this test is based on the pair of the algorithms in which data samples of these algorithms are compared with each other.

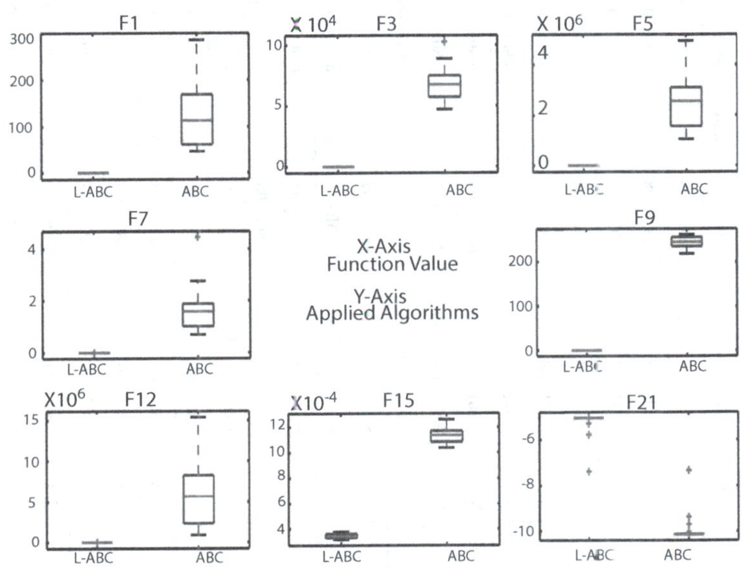

Figure 3.7 Box plot analysis of proposed algorithm.

Table 3.8 Wilcoxson rank-sum test.

Wilcoxson rank sum test															
Function	LABC	ABC	Status	GWO	Status	SCA	Status	MFO	Status	BAT	Status	BBO	Status	SSA	Status
F1	N/A	8.01E–09	⊕	8.01E–09	⊕	8.01E–09	⊕	8.01E–09	⊕	8.01E–09	⊕	8.01E–09	⊕	8.01E–09	⊕
F2	N/A	6.80E–08	⊕	6.80E–08	⊕	6.80E–08	⊕	6.80E–08	⊕	6.80E–08	⊕	6.80E–08	⊕	6.80E–08	⊕
F3	N/A	8.01E–09	⊕	8.01E–09	⊕	8.01E–09	⊕	8.01E–09	⊕	8.01E–09	⊕	8.01E–09	⊕	8.01E–09	⊕
F4	N/A	6.80E–08	⊕	6.80E–08	⊕	6.80E–08	⊕	6.80E–08	⊕	6.80E–08	⊕	6.80E–08	⊕	6.80E–08	⊕
F5	N/A	6.80E–08	⊕	1.80E–06	⊕	6.80E–08	⊕	6.80E–08	⊕	6.80E–08	⊕	6.80E–08	⊕	3.75E–04	⊕
F6	N/A	6.80E–08	⊕	6.80E–08	⊕	6.80E–08	⊕	1.06E–02	⊕	6.80E–08	⊕	9.09E–02	⊖	6.80E–08	⊕
F7	N/A	6.80E–08	⊕	6.80E–08	⊕	6.80E–08	⊕	6.80E–08	⊕	6.80E–08	⊕	6.80E–08	⊕	6.80E–08	⊕
F8	N/A	6.80E–08	⊕	4.60E–04	⊕	6.80E–08	⊕	1.38E–06	⊕	2.85E–01	⊖	2.69E–06	⊕	6.87E–04	⊕
F9	N/A	8.01E––09	⊕	7.51E–09	⊕	8.01E–09	⊕	8.01E–09	⊕	8.01E–09	⊕	8.01E–09	⊕	8.01E–09	⊕
F10	N/A	8.01E–09	⊕	7.37E–09	⊕	8.01E–09	⊕	8.01E–09	⊕	8.01E–09	⊕	8.01E–09	⊕	8.01E–09	⊕
F11	N/A	8.01E–09	⊕	4.53E–03	⊕	8.01E–09	⊕	8.01E–09	⊕	8.01E–09	⊕	8.01E–09	⊕	8.01E–09	⊕
F12	N/A	6.80E–08	⊕	1.79E–04	⊕	6.80E–08	⊕	6.80E–08	⊕	6.80E–08	⊕	6.80E–08	⊕	6.80E–08	⊕
F13	N/A	6.80E–08	⊕	6.80E–08	⊕	2.06E–06	⊕	1.20E–06	⊕	6.80E–08	⊕	6.80E–08	⊕	3.38E–04	⊕
F14	N/A	1.56E–01	⊖	7.11E–03	⊖	1.06E–02	⊖	1.00E+00	⊖	6.80E–08	⊕	1.60E–05	⊕	1.09E–03	⊕
F15	N/A	6.80E–08	⊕	1.25E–05	⊕	1.92E–07	⊕	6.79E–08	⊕	6.80E–08	⊕	6.80E–08	⊕	6.80E–08	⊕
F16	N/A	6.80E–08	⊕	6.80E–08	⊕	6.80E–08	⊕	8.01E–09	⊕	6.80E–08	⊕	6.71E–08	⊕	6.80E–08	⊕
F17	N/A	1.20E–06	⊕	6.80E–08	⊕	6.80E–08	⊕	5.64E–08	⊕	6.80E–08	⊕	6.77E–08	⊕	6.80E–08	⊕
F18	N/A	6.80E–08	⊕	1.20E–06	⊕	6.80E–08	⊕	4.46E–08	⊕	1.43E–07	⊕	1.60E–05	⊕	6.80E–08	⊕
F19	N/A	6.80E–08	⊕	1.00E+00	⊕	6.80E–08	⊕	8.01E–09	⊕	6.80E–08	⊕	5.57E–08	⊕	6.80E–08	⊕
F20	N/A	6.80E–08	⊕	2.85E–01	⊕	6.80E–08	⊕	1.00E+00	⊕	6.80E–08	⊕	1.00E+00	⊖	1.08E–01	⊖
F21	N/A	7.90E–08	⊕	2.30E–05	⊕	6.80E–08	⊕	2.97E–01	⊖	6.80E–08	⊕	3.94E–01	⊖	1.72E–01	⊖
F22	N/A	6.80E–08	⊕	6.80E–08	⊕	9.13E–07	⊕	6.74E–01	⊖	6.80E–08	⊕	9.25E–01	⊖	9.79E–03	⊕

Table 3.9 Scalability test analysis.

Scalability test							
		Dim = 10		**Dim = 20**		**Dim = 50**	
Fun		**ABC**	**NEW-ABC**	**ABC**	**NEW-ABC**	**ABC**	**NEW-ABC**
F1	**Mean**	1.88E−05	**1.77E−162**	2.05E−01	**0.00E+00**	1.45E+04	**0.00E+00**
	SD	1.45E−05	8.01E−162	2.18E−01	0.00E+00	2.58E+03	0.00E+00
F2	**Mean**	7.25E−06	**1.01E−96**	1.01E−02	**1.10E−209**	8.53E+08	**3.54E−290**
	SD	5.59E−06	3.78E−96	4.94E−03	0.00E+00	2.64E+09	0.00E+00
F3	**Mean**	2.28E+02	**1.92E−169**	1.88E+04	**0.00E+00**	2.46E+05	**0.00E+00**
	SD	1.01E+02	0.00E+00	3.26E+03	0.00E+00	5.06E+04	0.00E+00
F4	**Mean**	3.75E+00	**4.15E−82**	3.43E+01	**3.20E−161**	8.94E+01	**8.33E−277**
	SD	8.68E−01	1.13E−81	3.56E+00	1.24E−160	3.00E+00	0.00E+00
F5	**Mean**	1.91E+01	**8.69E+00**	8.95E+03	**1.85E+01**	2.08E+08	**4.80E+01**
	SD	1.69E+01	5.63E−02	5.79E+03	1.20E−01	6.10E+07	1.96E−01
F6	**Mean**	1.78E−05	3.42E−01	2.66E−01	1.12E+00	1.53E+04	**3.32E+00**
	SD	1.27E−05	1.31E−01	4.45E−01	2.69E−01	3.76E+03	6.10E−01

(*Continued*)

Table 3.9 Scalability test analysis. (*Continued*)

Scalability test							
		Dim = 10		**Dim = 20**		**Dim = 50**	
Fun		**ABC**	**NEW-ABC**	**ABC**	**NEW-ABC**	**ABC**	**NEW-ABC**
F7	**Mean**	1.31E−02	**5.83E−05**	8.73E−02	**4.65E−05**	1.33E+02	**3.59E−05**
	SD	4.31E−03	5.30E−05	3.27E−02	3.43E−05	3.23E+01	3.36E−05
F8	**Mean**	−2.81E+60	−3.43E+03	−2.08E+62	−5.40E+03	−1.85E+60	−9.33E+03
	SD	1.00E+61	1.94E+02	9.21E+62	5.58E+02	4.68E+60	7.23E+02
F9	**Mean**	3.05E+01	**0.00E+00**	1.26E+02	**0.00E+00**	5.89E+02	**0.00E+00**
	SD	4.81E+00	0.00E+00	1.07E+01	0.00E+00	2.96E+01	0.00E+00
F10	**Mean**	3.29E−02	**8.88E−16**	2.12E+00	**8.88E−16**	1.89E+01	**8.88E−16**
	SD	2.48E−02	0.00E+00	7.12E−01	0.00E+00	7.09E−01	0.00E+00
F11	**Mean**	4.28E−01	**0.00E+00**	8.22E−01	**0.00E+00**	1.23E+02	**0.00E+00**
	SD	1.04E−01	0.00E+00	1.28E−01	0.00E+00	2.55E+01	0.00E+00
F12	**Mean**	3.67E−04	2.68E−02	2.06E+01	**3.12E−02**	5.38E+08	**2.05E−02**
	SD	8.28E−04	9.25E−03	6.23E+00	7.91E−03	1.96E+08	4.97E−03
F13	**Mean**	4.08E−03	3.45E−01	1.84E+03	**1.36E+00**	9.76E+08	**4.95E+00**
	SD	4.40E−03	7.59E−02	3.63E+03	1.65E−01	3.05E+08	1.95E−02

This comparison helps to determine that algorithm is statistically significant or not. This test works at a 5% significance level. This significance level criterion is predefined, which are described as

i. p-value ≤ 0.01: If LABC variations are highly essential
ii. 0.01 < p-value ≤ 0.05: When LABC difference is at significant level.
iii. p-value > 0.05: when LABC difference is not at significant level.

Results of the Wilcoxon rank-sum test are depicted in Table 3.8. In these results, ⊕ is used when p-value is less than the 0.05 significance level, while ⊖ is used when p-value is greater than 0.05 significance level. We have observed that optimization outcomes of LABC are distinct from parent ABC. After inspecting these results it is evident that the modified version of ABC is statistically significant.

3.6.4 Scalability Test

A scalability test is used to check that applied modification is applicable to solve the multidimension problems or not. In this test, different dimensions 10, 20, and 50 are used to compare the modified version of ABC with the parent ABC algorithm. This scalability test process depends on the numerical values of mean, minimum, standard deviation, and maximum values which are depicted in Table 3.9.

3.7 Design Analysis of Harmonic Estimator

In this segment, we use modified version of ABC to design an efficient and powerful harmonic estimator which gives an accurate estimation of phase and amplitude component of power harmonics. To design this type of harmonic estimator we take into account four problems, which are used to validate the effectiveness of harmonic estimator.

3.7.1 Assessment of Harmonic Estimator Design Problem 1

Harmonic estimator design problem 1 and problem 2 are taken from the literature [44]. Waveform expression of harmonic estimator design problem 1 is written as

Table 3.10 Analysis of harmonics with competitor algorithms on problem 1.

Analysis of harmonics with competitor algorithms on problem 1							
		Harmonics order					
Algorithms	**Parameters**	**Fun.**	**2nd**	**3rd**	**4th**	**5th**	**6th**
Original	**Phase**	0.25	0.27	0.29	0.20	0.30	0.40
	Freq.	5	10	15	20	25	30
	Amp.	6.00	5.00	4.00	3.00	2.00	1.00
ABC	**Phase**	0.2500	0.2700	0.2900	0.2000	0.3000	0.4000
	% Phase Error	1.100E−03	7.000E−03	5.000E−03	1.000E−04	2.400E−03	0.000E+00
	Amp.	6.0050	4.9999	4.0001	3.0001	2.0004	1.0003
	% Amp. Error	8.400E−03	2.700E−03	2.000E−03	3.000E−03	1.820E−02	2.550E−02
BAT	**Phase**	0.2501	0.2700	0.2901	0.1999	0.3000	0.4010
	% Phase Error	6.400E−03	9.000E−03	3.200E−03	5.600E−03	6.240E−02	2.440E−02
	Amp.	6.0015	5.0006	4.0014	3.0010	2.0006	1.0009
	% Amp. Error	9.000E−03	5.400E−03	3.800E−03	1.090E−02	9.600E−03	2.180E−02
Proposed Algorithm	**Phase**	0.2500	0.2700	0.2900	0.2000	0.3000	0.4002
	% Phase Error	8.361E−03	**2.630E−04**	**2.893E−03**	4.229E−03	9.939E−03	6.120E−02
	Amp.	5.9998	5.0000	4.0000	2.9999	1.9999	0.9999
	% Amp. Error	**2.599E−03**	**4.367E−04**	**1.715E−04**	**2.606E−03**	7.097E−03	**5.387E−03**

Table 3.11 Analysis of harmonics with competitor algorithms on problem 1 with noise.

Analysis of harmonics with competitor algorithms on problem 1 with noise							
		Harmonics order					
Algorithms	Parameters	Fun.	2nd	3rd	4th	5th	6th
Original	Phase	0.25	0.27	0.29	0.20	0.30	0.40
	Freq.	5	10	15	20	25	30
	Amp.	6.00	5.00	4.00	3.00	2.00	1.00
P2 With 10dB Noise	Phase	2.50E−01	2.70E−01	2.90E−01	2.01E−01	3.00E−01	4.00E−01
	% Phase Error	3.16E−03	1.06E−03	1.16E−02	5.30E−01	5.21E−02	7.84E−02
	Amp.	6.00E+00	4.99E+00	4.00E+00	3.00E+00	2.00E+00	1.00E+00
	% Amp. Error	5.87E−02	1.09E−01	9.12E−02	1.38E−01	2.38E−01	1.89E−01
P2 With 20dB Noise	Phase	2.50E−01	2.70E−01	2.90E−01	2.01E−01	3.00E−01	4.00E−01
	% Phase Error	1.61E−03	4.09E−04	7.99E−03	5.05E−01	4.85E−03	1.20E−01
	Amp.	6.00E+00	5.00E+00	4.00E+00	2.99E+00	2.00E+00	1.00E+00
	% Amp. Error	6.01E−02	6.83E−02	1.01E−01	1.76E−01	1.66E−01	1.71E−02

(Continued)

Table 3.11 Analysis of harmonics with competitor algorithms on problem 1 with noise. (*Continued*)

Analysis of harmonics with competitor algorithms on problem 1 with noise							
		Harmonics order					
Algorithms	**Parameters**	**Fun.**	**2nd**	**3rd**	**4th**	**5th**	**6th**
P2 With 30dB Noise	**Phase**	2.50E−01	2.70E−01	2.90E−01	2.00E−01	3.00E−01	4.01E−01
	% Phase Error	1.44E−03	7.47E−04	4.35E−03	4.40E−03	4.47E−03	1.33E−01
	Amp.	6.00E+00	5.00E+00	4.00E+00	3.00E+00	2.00E+00	9.96E−01
	% Amp. Error	5.67E−02	8.60E−02	9.21E−02	1.50E−01	1.97E−01	3.88E−01
P2 With 40dB Noise	**Phase**	2.50E−01	2.70E−01	2.90E−01	2.00E−01	3.00E−01	4.01E−01
	% Phase Error	8.29E−04	1.30E−02	3.54E−03	2.37E−01	6.07E−02	1.31E−01
	Amp.	6.00E+00	5.00E+00	4.00E+00	3.00E+00	2.00E+00	9.99E−01
	% Amp. Error	5.41E−02	7.23E−02	9.30E−02	1.48E−01	1.83E−01	5.42E−02

$$C_3(t) = \frac{12}{2}\sin(10\neq t + 0.25) + \frac{10}{2}\sin(20\neq t + 0.27) + \frac{8}{2}\sin(30\neq t + 0.29)$$
$$+ \frac{6}{2}\sin(40\neq t + 0.2) + \frac{4}{2}\sin(50\neq t + 0.3) + \sin(60\neq t + 0.4)$$

Problem 1 is a combination of fundamental (5Hz), 2nd, 3rd, 4th, 5th, and 6th harmonic components of harmonics. Problem 1, is mainly focused on the exact estimation of phase and amplitude components of power harmonics.

Numerical results of harmonic problem 1 are recorded in Table 3.10 and Table 3.11. Where Table 3.10 describes a comparison of ABC, BAT algorithm with a modified version of ABC in terms of phase, amplitude, % phase error, and % amplitude error. From this analysis, it is observed that the modifiedversion of ABC provides optimal results of power harmonics. In Table 3.10, results are reported in the absence of noise component. Table 3.11, accumulated the numerical results in the presence of noise signals, these noise signals are 10 dB, 20 dB, 30 dB, and 40 dB. These numerical results help to conclude that noise signals disturb the phase, amplitude

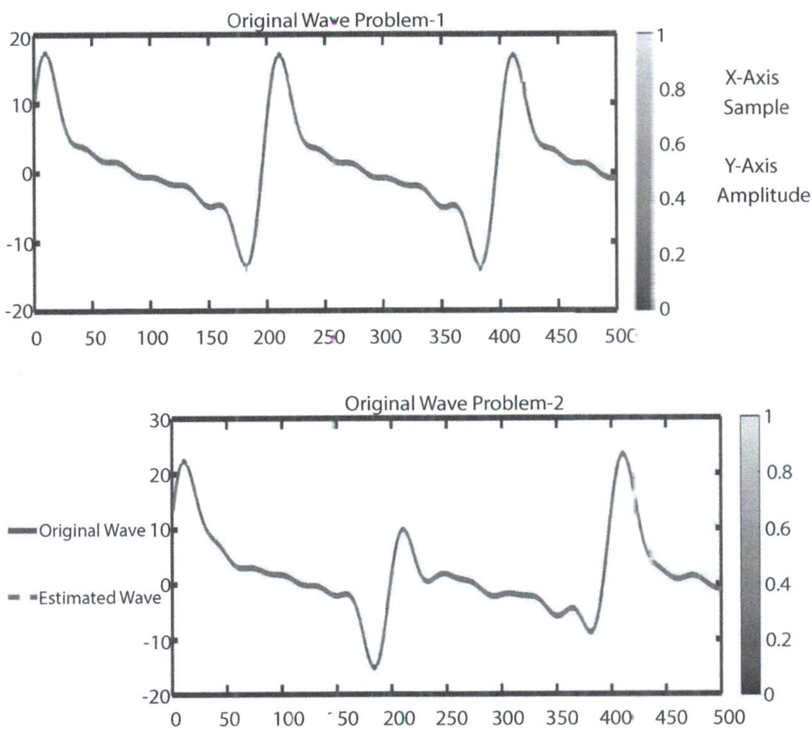

Figure 3.8 Estimated waves of problem 1 and problem 2 under no noise.

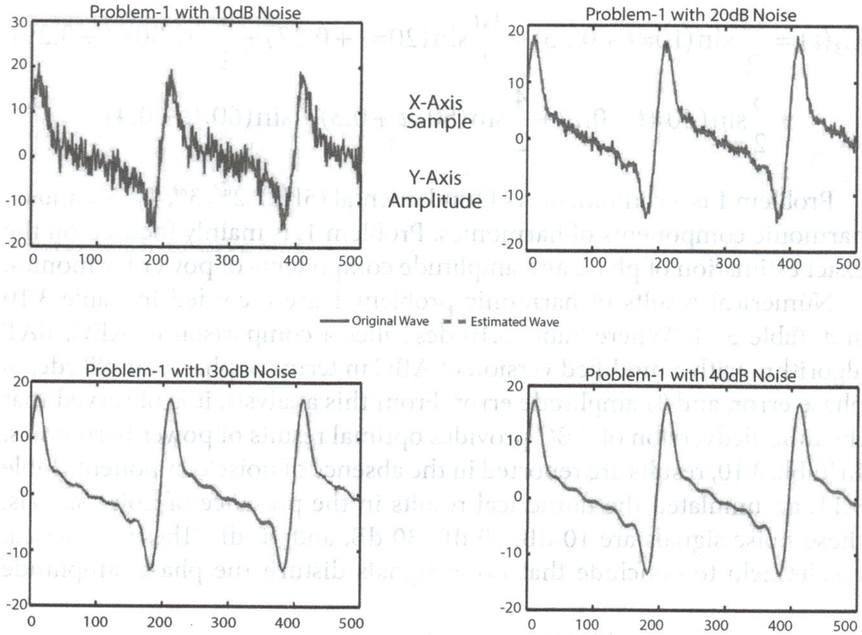

Figure 3.9 Estimated waves of problem 1 under noise.

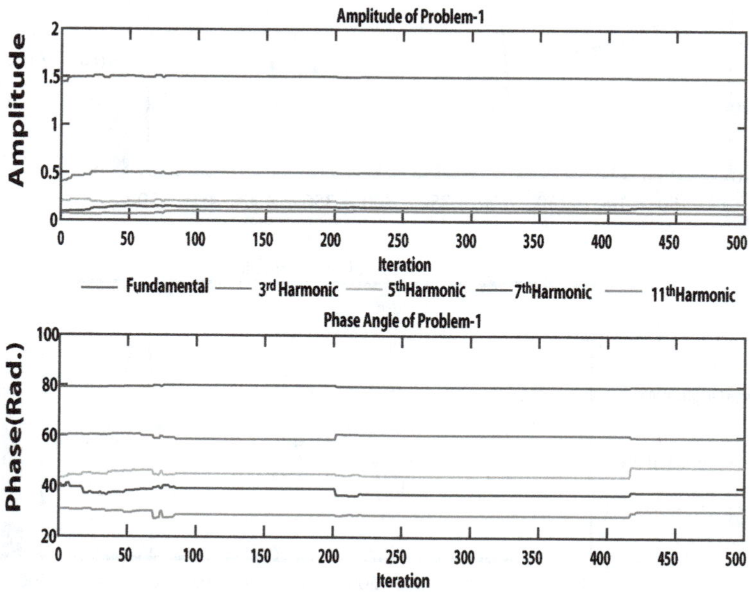

Figure 3.10 Trajectory analysis of problem 1 phase and amplitude.

values, and opted wave of problem 1 with the help of the optimization process, which is illustrated in Figure 3.8.

Figure 3.9 is used to represent the effect of 10 dB, 20 dB, 30 dB, 40 dB noise signals, on the original signal of problem 1. From this fig, it is observed that 10 dB, 20 dB noise signals are more effective on the original wave while 30 dB and 40 dB noise signals are less effective on the original signal. Trajectory analysis of problem 1, is reported in Figure 3.10, for phase and amplitude values respectively. After observing the results of problem 1, it is concluded that the modified version of ABC is useful to design an efficient and powerful harmonic estimator with noise and without noise signals. This newly developed harmonic estimator is powerful tool that is helpful in practical use in power plants.

3.7.2 Assessment of Harmonic Estimator Design Problem 2

Waveform expression of harmonic estimator design problem 2 given in [44] is written as

$$C_4(t)=\frac{7}{2}\sin(6\pi t+0.23)+\frac{12}{2}\sin(10\pi t+0.25)+\frac{5}{2}\sin(16\pi t+0.35)$$
$$+\ldots\frac{10}{2}\sin(20\pi t+0.27)+\frac{3}{2}\sin(26\pi t+0.21)+\frac{8}{2}\sin(30\pi t+0.29)$$
$$+\ldots\frac{6}{2}\sin(40\pi t+0.2)+\frac{4}{2}\sin(50\pi t+0.3)+\sin(60\pi t+0.4)$$

Problem 2 is a combination of sub, fundamental (5Hz), inter 1st, 2nd, inter-2, 3rd, 4th, 5th, and 6th harmonic components. Problem 2, is mainly focused on the exact estimation of phase and amplitude components of power harmonics.

Numerical results of problem 2 are accumulated in Tables 3.12 and 3.13. A comparison of ABC, BAT algorithm with a modified version of ABC are reported in Table 3.12, these results are recorded in terms of phase, amplitude value, % error analysis respectively for power harmonics of problem 2. From these results, it is evident that the modified version of ABC is capable to estimate the exact estimation of phase and amplitude values, while modified version of ABC also has the capability to reach an optimum value of % error values of phase and amplitude respectively. In Table 3.12 results are reported in absence of noise signals. Table 3.13 is used to accumulate the numerical results of the problem 2 including noise signals. These noise signals used to test that newly developed harmonic estimator is efficient to handle the harmonic problem when the noise signals are taken into account.

Table 3.12 Analysis of harmonics with competitor algorithms on problem 2.

Analysis of harmonics with competitor algorithms on problem 2										
		Harmonics order								
Algorithms	**Parameters**	**Sub.**	**Fun.**	**Int. 1**	**2nd**	**Int. 2**	**3rd**	**4th**	**5th**	**6th**
Original	**Phase**	0.23	0.25	0.35	0.27	0.21	0.29	0.20	0.30	0.40
	Freq.	3	5	8	10	13	15	20	25	30
	Amp.	3.5	6.00	2.50	5.00	1.50	4.00	3.00	2.00	1.00
ABC	**Phase**	2.30E–01	2.50E–01	3.50E–01	2.70E–01	2.10E–01	2.90E–01	2.00E–01	3.00E–01	4.00E–01
	% Phase Error	2.29E–02	9.00E–03	1.96E–02	3.48E–02	1.00E–01	2.33E–02	1.70E–03	5.04E–02	1.03E–02
	Amp.	3.50E+00	6.00E+00	2.50E+00	5.00E+00	1.50E+00	4.00E+00	3.00E+00	2.00E+00	1.00E+00
	% Amp. Error	1.27E–02	1.47E–02	2.78E–02	1.28E–02	7.90E–02	9.50E–03	7.79E–02	6.18E–02	1.10E–01
BAT	**Phase**	2.29E–01	2.50E–01	3.51E–01	2.70E–01	2.13E–01	2.91E–01	2.00E–01	2.98E–01	3.99E–01
	% Phase Error	6.46E–01	1.74E–02	3.12E–01	2.04E–01	1.54E+00	1.75E–01	7.21E–02	6.02E–01	2.89E–01
	Amp.	3.52E+00	6.00E+00	2.51E+00	5.00E+00	1.51E+00	4.01E+00	3.00E+00	2.01E+00	1.01E+00
	% Amp. Error	6.32E–01	5.30E–03	3.28E–01	3.53E–02	6.42E–01	2.20E–01	2.68E–02	4.68E–01	5.42E–01
Proposed Algorithm	**Phase**	2.30E–01	2.50E–01	3.50E–01	2.70E–01	2.10E–01	2.90E–01	2.00E–01	3.00E–01	4.00E–01
	% Phase Error	**0.00E+00**	**0.00E+00**	5.00E–02	**0.00E+00**	**6.07E–02**	**1.04E–02**	**0.00E+00**	**1.19E–02**	4.68E–02
	Amp.	3.50E+00	6.00E+00	2.50E+00	5.00E+00	1.50E+00	4.00E+00	3.00E+00	2.00E+00	1.00E+00
	% Amp. Error	**5.33E–03**	**1.95E–03**	**8.28E–03**	**9.83E–03**	**1.99E–02**	**3.53E–03**	**4.37E–03**	**1.09E–02**	**1.62E–02**

Table 3.13 Analysis of harmonics with competitor algorithms on problem-2 with noise.

Analysis of harmonics with competitor algorithms on problem-2 with noise signals										
		Harmonics order								
Algorithms	Parameters	Sub.	Fun.	Int. 1	2nd	Int. 2	3rd	4th	5th	6th
Original	Phase	0.23	0.25	0.35	0.27	0.21	0.29	0.20	0.30	0.40
	Freq.	3	5	8	10	13	15	20	25	30
	Amp.	3.5	6.00	2.50	5.00	1.50	4.00	3.00	2.00	1.00
P1 With 10dB Noise	Phase	2.30E−01	2.50E−01	3.52E−01	2.70E−01	2.10E−01	2.89E−01	2.00E−01	3.00E−01	4.00E−01
	% Phase Error	0.00E+00	0.00E+00	4.92E−01	0.00E+00	4.34E−02	1.80E−01	0.00E+00	1.60E−02	9.26E−02
	Amp.	3.50E+00	6.00E+00	2.51E+00	5.01E+00	1.50E+00	3.99E+00	3.00E+00	2.00E+00	1.00E+00
	% Amp. Error	6.23E−02	8.04E−02	2.10E−01	2.05E−01	2.49E−02	1.45E−01	1.30E−02	8.30E−02	2.43E−01
P1 With 20dB Noise	Phase	2.30E−01	2.50E−01	3.54E−01	2.70E−01	2.10E−01	2.89E−01	2.00E−01	3.00E−01	3.99E−01
	% Phase Error	0.00E+00	0.00E+00	1.03E+00	0.00E+00	4.40E−02	2.23E−01	0.00E+00	1.56E−02	1.36E−01
	Amp.	3.50E+00	6.00E+00	2.51E+00	5.01E+00	1.50E+00	3.99E+00	3.00E+00	2.00E+00	1.00E+00
	% Amp. Error	5.10E−02	8.32E−02	2.22E−01	2.34E−01	3.63E−02	1.63E−01	1.61E−02	6.44E−02	2.79E−01

(Continued)

Table 3.13 Analysis of harmonics with competitor algorithms on problem-2 with noise. *(Continued)*

Analysis of harmonics with competitor algorithms on problem-2 with noise signals										
Algorithms	**Parameters**	**Harmonics order**								
		Sub.	**Fun.**	**Int. 1**	**2nd**	**Int. 2**	**3rd**	**4th**	**5th**	**6th**
P1 With 30dB Noise	**Phase**	2.30E−01	2.50E−01	3.53E−01	2.70E−01	2.10E−01	2.89E−01	2.00E−01	3.00E−01	3.99E−01
	% Phase Error	0.00E+00	0.00E+00	7.47E−01	0.00E+00	4.13E−02	1.80E−01	0.00E+00	1.86E−02	1.33E−01
	Amp.	3.50E+00	5.99E+00	2.51E+00	5.01E+00	1.50E+00	3.99E+00	3.00E+00	2.00E+00	1.00E+00
	% Amp. Error	5.06E−02	1.21E−01	2.33E−01	1.15E−01	3.13E−02	1.84E−01	1.70E−02	7.95E−02	1.95E−01
P1 With 40dB Noise	**Phase**	2.30E−01	2.50E−01	3.53E−01	2.70E−01	2.10E−01	2.90E−01	2.00E−01	3.00E−01	3.99E−01
	% Phase Error	0.00E+00	0.00E+00	8.99E−01	0.00E+00	4.24E−02	1.65E−01	0.00E+00	1.54E−02	1.34E−01
	Amp.	3.50E+00	5.99E+00	2.51E+00	5.00E+00	1.50E+00	3.99E+00	3.00E+00	2.00E+00	1.00E+00
	% Amp. Error	3.81E−02	1.11E−01	3.18E−01	4.71E−02	2.91E−02	2.19E−01	1.44E−02	7.88E−02	2.65E−01

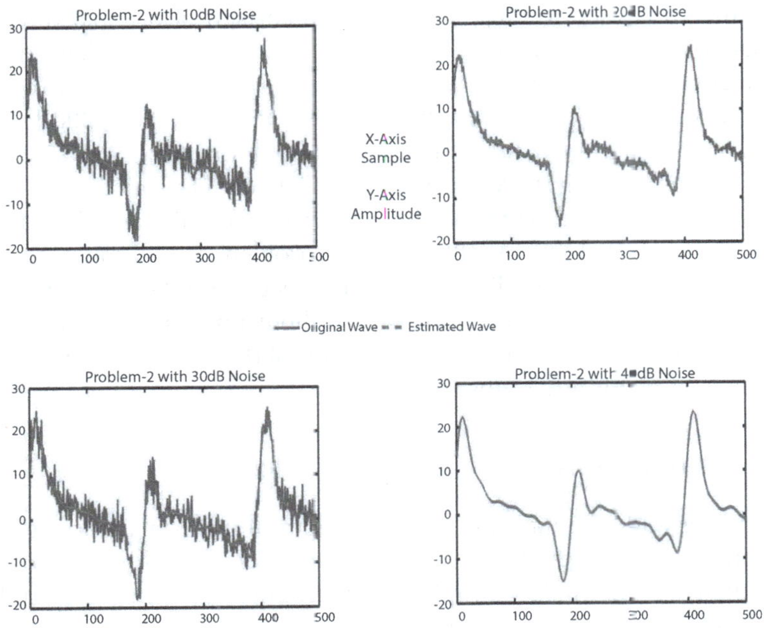

Figure 3.11 Estimated waves of problem-2 under noise.

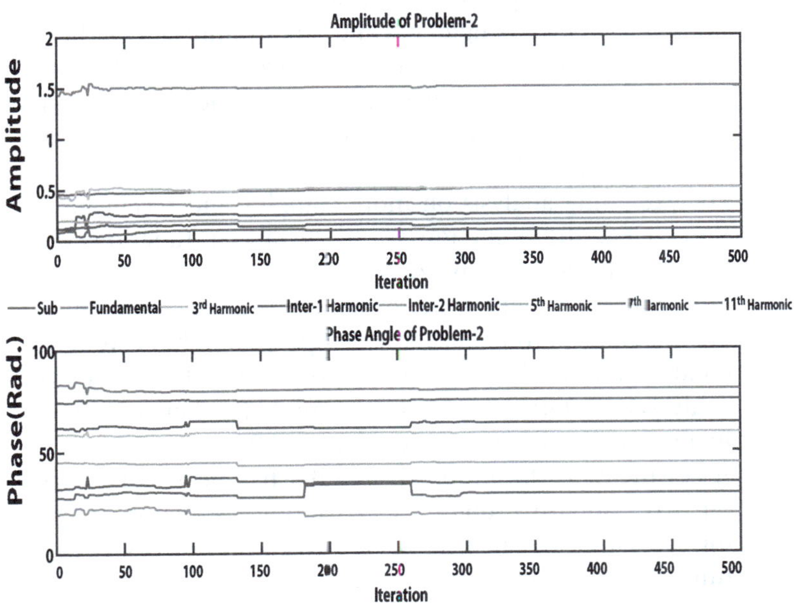

Figure 3.12 Trajectory analysis of problem-2 phase and amplitude.

Figure 3.8 is used to illustrate a fair comparison of original wave and estimated wave of problem 2. From this assessment, it is evident that the estimated wave and original wave are overlapping with each other. To check the effect of noise signals, we test these noise signals (10dB, 20dB, 30dB, 40dB) with the original wave of problem 2. Estimated waves of original signal and noise signals are illustrated in Figure 3.11. Trajectory analysis of phase and amplitude components of power harmonics are illustrated in Figure 3.12. After observing the results of problem 1, it is concluded that the modified version of ABC is useful to design an efficient and powerful harmonic estimator with noise and without noise signals. This newly developed harmonic estimator is powerful tool that is helpful for practical use in power plants and industries.

3.8 Conclusion

The chapter addressed an important problem of power networks namely harmonic components identification. A Laplacian factor-driven Artificial Bee Colony (L-ABC) algorithm has been proposed in the chapter and the same has been applied on power network problems. Following conclusions are noteworthy from this work

- A detailed investigation on the standard benchmark functions that possess single and multiple minimas have been considered for judging the performance of the L-ABC. We have observed that optimization performance on the standard functions are satisfactory as compared with the contemporary algorithms.
- For extending analysis on more complex and recent functions, CEC-2017 function set of 29 diverse functions has been employed. We have observed that for these functions also, the optimization performance is quite competitive. For establishing the efficacy of proposed method, we have conducted tests like convergence property analysis, box plot analysis and statistical significance tests. Further, a scalability test has also been considered for L-ABC.
- Assessment of the proposed L-ABC has been conducted on two standard problems of harmonic estimator design, for evaluating the algorithm, we constructed a least square function and minimize the error between predicted and actual value. We observed that performance of L-ABC

is competitive as compared with some state of the art algorithms.

References

1. Mirjalili, S., The ant lion optimizer. *Adv. Eng. Softw.*, 83, 80–98, 2015.
2. Kennedy, J. and Eberhart, R., Particle swarm optimization, in: *Proceedings of ICNN'95-International Conference on Neural Networks*, vol volume 4, IEEE, pp. 1942–1948, 1995.
3. Mirjalili, S., Mirjalili, S.M., Lewis, A., Grey wolf optimizer. *Adv. Eng. Softw.*, 69, 46–61, 2014.
4. Priyadarshi, N., Ramachandaramurthy, V.K., Padmanaban, S., Azam, F., An ant colony optimized mppt for standalone hybrid pv-wind power system with single cuk converter. *Energies*, 12, 1, 167, 2019.
5. Rao, R.V., Savsani, V.J., Vakharia, D.P., Teaching–learning-based optimization: a novel method for constrained mechanical design optimization problems. *Comput. Aided Des.*, 43, 3, 303–315, 2011.
6. Geem, Z.W., Kim, J.H., Loganathan, G.V., A new heuristic optimization algorithm: Harmony search. *Simulation*, 76, 2, 60–68, 2001.
7. Ghorbani, N. and Babaei, E., Exchange market algorithm. *Appl. Soft Comput.*, 19, 177–187, 2014.
8. Al-Betar, M.A., Alyasseri, Z.A.A., Awadallah, M.A., Doush, I.A., Coronavirus herd immunity optimizer (CHIO). *Neural Comput. Appl.*, 33, 10, 5011–5042, 2021.
9. Erol, O.K. and Eksin, I., A new optimization method: big bang–big crunch. *Adv. Eng. Softw.*, 37, 2, 106–111, 2006.
10. Hsiao, Y.-T., Chuang, C.-L., Jiang, J.-A., Chien, C.-C., A novel optimization algorithm: space gravitational optimization, in: *IEEE International Conference on Systems, Man and Cybernetics*, vol. 3, IEEE, pp. 2323–2328, 2005.
11. Formato, R.A., Central force optimization. *Prog. Electromagn. Res.*, 77, 425–491, 2007.
12. Tayarani-N., M.H. and Akbarzadeh-T, M.R., Magnetic optimization algorithms a new synthesis, in: *IEEE Congress on Evolutionary Computation (IEEE World Congress on Computational Intelligence)*, IEEE, pp. 2659–2664, 2008.
13. Singh, N. and Singh, S.B., A novel hybrid gwo-sca approach for optimization problems. *Eng. Sci. Technol. an Int. J.*, 20, 6, 1586–1601, 2017.
14. Shelokar, P.S., Siarry, P., Jayaraman, V.K., Kulkarni, B.D., Particle swarm and ant colony algorithms hybridized for improved continuous optimization. *Appl. Math. Comput.*, 188, 1, 129–142, 2007.

15. Padmanaban, S., Priyadarshi, N., Bhaskar, M.S., Holm-Nielsen, J.B., Ramachandaramurthy, V.K., Hossain, E., A hybrid anfis-abc based mppt controller for pv system with anti-islanding grid protection: Experimental realization. *IEEE Access*, 7, 103377–103389, 2019.
16. Chang, G.W., Chen, C.I., Liu, Y.J., Wu, M.C., Measuring power system harmonics and interharmonics by an improved fast Fourier transform-based algorithm. *IET Gener. Transm. Dis.*, 2, 2, 192–201, 2008.
17. Wang, Z. and Hunt, B.R., The discretew transform. *Appl. Math. Comput.*, 16, 1, 19–48, 1985.
18. Yazdani, D., Bakhshai, A., Joos, G., Mojiri, M., A real-time three-phase selective-harmonic-extraction approach for grid-connected converters. *IEEE Trans. Ind. Electron.*, 56, 10, 4097–4106, 2009.
19. Abdelsalam, A.A., Eldesouky, A.A., Sallam, A.A., Classification of power system disturbances using linear kalman filter and fuzzy-expert system. *Int. J. Electr. Power Energy Syst.*, 43, 1, 688–695, 2012.
20. Zouidi, A., Fnaiech, F., AL-Haddad, K., Rahmani, S., Artificial neural networks as harmonic detectors, in: *IECON 2006-32nd Annual Conference on IEEE Industrial Electronics*, IEEE, pp. 2889–2892, 2006.
21. Brown, M. and Rogers, S.J., User identification via keystroke characteristics of typed names using neural networks. *Int. J. Man Mach. Stud.*, 39, 6, 999–1014, 1993.
22. Nguyen, T.T., Parametric harmonic analysis. *IEE Proc. Gener. Transm. Distrib.*, 144, 1, 21–25, 1997.
23. Wang, M. and Sun, Y., A practical method to improve phasor and power measurement accuracy of dft algorithm. *IEEE Trans. Power Deliv.*, 21, 3, 1054–1062, 2006.
24. Karaboga, D. and Basturk, B., A powerful and efficient algorithm for numerical function optimization: Artificial bee colony (abc) algorithm. *J. Global Optim.*, 39, 3, 459–471, 2007.
25. Zhu, G. and Kwong, S., Gbest-guided artificial bee colony algorithm for numerical function optimization. *Appl. Math. Comput.*, 217, 7, 3166–3173, 2010.
26. Tsai, P.-W., Pan, J.-S., Liao, B.-Y., Chu, S.-C., Enhanced artificial bee colony optimization. *Int. J. Innov. Comput. Inf. Control*, 5, 12, 5081–5092, 2009.
27. Sharma, N., Sharma, H., Sharma, A., Bansal, J.C., Modified artificial bee colony algorithm based on disruption operator, in: *Proceedings of Fifth International Conference on Soft Computing for Problem Solving*, Springer, pp. 889–900, 2016.
28. Awadallah, M.A., Al-Betar, M.A., Bolaji, A.L.A., Doush, I.A., Hammouri, A.I., Mafarja, M., Island artificial bee colony for global optimization. *Soft Comput.*, 24, 17, 13461–13487, 2020.
29. Awadallah, M.A., Al-Betar, M.A., Bolaji, A.L., Alsukhni, E.M., Al-Zoubi, H., Natural selection methods for artificial bee colony with new versions of onlooker bee. *Soft Comput.*, 23, 15, 6455–6494, 2019.

30. Awadallah, M.A., Bolaji, A.L., Al-Betar, M.A., A hybrid artificial bee colony for a nurse rostering problem. *Appl. Soft Comput.*, 35, 726–739, 2015.
31. Bergmans, P., A simple converse for broadcast channels with additive white gaussian noise (corresp.). *IEEE Trans. Inf. Theory*, 20, 2, 279–280, 1974.
32. Kabalci, Y., Kockanat, S., Kabalci, E., A modified abc algorithm approach for power system harmonic estimation problems. *Electr. Power Syst. Res.*, 154, 160–173, 2018.
33. Kumar, R., Directed bee colony optimization algorithm. *Swarm Evol. Comput.*, 17, 60–73, 2014.
34. Saxena, A., Soni, B.P., Kumar, R., Gupta, V., Intelligent grey wolf optimizer–development and application for strategic bidding in uniform price spot energy market. *Appl. Soft Comput.*, 69, 1–13, 2018.
35. Shekhawat, S. and Saxena, A., Development and applications of an intelligent crow search algorithm based on opposition based learning. *ISA Trans.*, 2019.
36. Dinkar, S.K. and Deep, K., Opposition based laplacian ant lion optimizer. *J. Comput. Sci.*, 23, 71–90, 2017.
37. Zhang, X., Wang, D., Chen, H., Mao, W., Liu, S., Liu, G., Dou, Z., Improved laplacian biogeography-based optimization algorithm and its application to qap. *Complexity*, 2020, 1–19, 2020.
38. Dinkar, S.K., Deep, K., Mirjalili, S., Thapliyal, S., Opposition-based laplacian equilibrium optimizer with application in image segmentation using multi-level thresholding. *Expert Syst. Appl.*, 174, 114766, 2021.
39. Mirjalili, S., Sca: A sine cosine algorithm for solving optimization problems. *Knowl. Based Syst.*, 96, 120–133, 2016.
40. Mirjalili, S., Moth-flame optimization algorithm: A novel nature-inspired heuristic paradigm. *Knowl. Based Syst.*, 89, 228–249, 2015.
41. Yang, X.-S., Bat algorithm for multi-objective optimisation. *Int. J. Bio-Inspired Computation*, 3, 5, 267–274, 2012, arXiv preprint arXiv:1203.6571.
42. Simon, D., Biogeography-based optimization. *IEEE Trans. Evol. Comput.*, 12, 6, 702–713, 2008.
43. Mirjalili, S., Gandomi, A.H., Mirjalili, S.Z., Saremi, S., Faris, H., Mirjalili, S.M., Salp swarm algorithm: A bio-inspired optimizer for engineering design problems. *Adv. Eng. Softw.*, 114, 163–191, 2017.
44. Kockanat, S., Acceleration harmonics estimation and elimination with mabc-rls algorithm: Simulation and experimental analyses on shaking table. *Appl. Soft Comput.*, 92, 106377, 2020.

4

Applications of Cuckoo Search Algorithm in Reliability Optimization

V. Kaviyarasu[1]* and V. Suganthi[2]

[1]Department of Statistics, Bharathiar University, Coimbatore, India
[2]Department of Computer Science, Sri Ramakrishna College of Arts & Science, Coimbatore, India

Abstract

In today's world, areas of science, engineering, and others have their own applications in the field of optimization. It involves the usage of mathematical modeling. In each search space, finding an acceptable solution to an objective function is the main aim. Optimization algorithms have been classified into two types, such as deterministic and stochastic. Deterministic offers only a theoretical solution in solving optimization problems but stochastic provides a much faster solution in finding a global optimum. Reliability optimization says how the software product can be trusted. The comparison of different optimization algorithms can be made for the best one to support in terms of software reliability. Cuckoo Search (CS) algorithm is a novel nature-inspired algorithm. It is used to solve optimization problems that are complex in nature. The algorithm depends on the brood-parasitic strategy of Cuckoo species. The usage of Levy flights is used to produce new candidate resolutions. It can improve the relationship between exploration and exploitation toward the potential of searching. It is used in solving engineering problems, such as embedded systems, distribution of networks, and scheduling problems. In this article, a study of the reliability of the software at static and runtime is performed, and the results are discussed. As per the result, a conclusion is drawn.

***Keywords*:** Optimization, search space, Cuckoo search algorithm, objective, exploration, reliability, evolutionary algorithm, levy flights

**Corresponding author*: kaviyarasu@buc.edu.in

S. C. Malik, Deepak Sinwar, Ashish Kumar, S. R. Gadde, Prasenjit Chatterjee and Bui Thanh Hung (eds.) Computational Intelligence in Sustainable Reliability Engineering, (97–108) © 2023 Scrivener Publishing LLC

4.1 Introduction

Each and every evolutionary algorithm (EA) needs to address the concern about exploration and exploitation of the search space. When the search space is visited as new points, then it is termed as exploration. Exploitation means the process that refines the points within the locality of already visited locations, so that the quality of the solution can be improved [3]. An optimization problem finds the best solution out of all feasible solutions. The output received from the optimization problem can be either maximum or minimum based on the input values received from a set of values. In a given search space, finding an acceptable solution of an objective function is the main aim.

$$\text{Minimize } f(x) \tag{4.1}$$

$$\text{Subject to } = g_i\,(x) \leq 0 \text{ i=1, 2 ,....m} \tag{4.2}$$

$$h_j\,(x) = 0 \text{ j=1, 2 ,... p} \tag{4.3}$$

Optimization algorithms have been classified into two types as deterministic and stochastic. Deterministic offers only a theoretical solution in solving optimization problems but stochastic provides mush faster solution in finding a global optimum.

Cuckoo Search algorithm is one of the meta-heuristic optimization algorithms. It is used to solve optimization problems [1]. Cuckoo Search (CS) algorithm is a novel nature-inspired algorithm. It is used to solve optimization problems, which are complex in nature. Algorithm depends on brood-parasitic strategy of Cuckoo species. The usage of Levy flights is used to produce new candidate resolutions. It is used in solving engineering problems, such as embedded systems, distribution of networks, and scheduling problems [9].

4.2 Cuckoo Search Algorithm

4.2.1 Performance of Cuckoo Search Algorithm

The performance of the Cuckoo Search (CS) algorithm is better than the other intelligent optimization algorithms because of the factors, such as less parameter settings, simple operation. The process of CS is clear. The unique characteristic of CS has attracted the responsiveness of people. Many researchers have made a research based on three factors, such as improving the position update mechanism or adjustment in parameter settings or application research. Hence, the CS algorithm is selected in the implementation of reliability of the software.

4.2.2 Levy Flights

In Cuckoo Search algorithm [2], Levy flight is the one which is an important aspect. Regular characteristics of Levy flight has been adopted by many of the animals, as well as insects. This has been proven by many of the studies. The concept of Levy flight was introduced by Viswanathan and his contemporaries [10].

Cuckoo Search algorithm is dependent on three rules as follows

1. Only one egg is laid at a time by a cuckoo and dumps it in a nest, which is randomly chosen.
2. The nests, which have a high quality, will be forwarded to the next generations.
3. The host nest is fixed with certain numbers, it can determine a strange egg.

4.2.3 Software Reliability

Reliability is an important concept which is used to measure the quality of the software. The concept can be applied in a situation where software is expected to behave in a certain way. It is one of the metrics that measures the quality of the software [4]. It relies upon user-oriented quality factors. When the users of the software are experiencing the failure not often, then the system is more reliable. It is considered to be better than the one which fails frequently. A system is said to be highly reliable when it is without faults. Software reliability is defined in terms of three concepts

1. Fault
2. Failure
3. Time.

The reliability of the software can be calculated using the expression specified below,

$$reliability = \frac{failurerate(f_r)}{execution\ time} \tag{4.4}$$

4.3 Modified Cuckoo Search Algorithm (MCS)

The discussion starts with Cuckoo Search algorithm along with its advantages and disadvantages. The disadvantages can be overcome through the modification in the existing Cuckoo Search algorithm.

Advantages of Cuckoo Search Algorithm

1. Excellent global search ability,
2. Avoids falling into local optimum,
3. Good applicability dealing with multi objective problems,
4. Efficient for optimal algorithms.

Limitations

1. Slower convergence,
2. Convergence rate is affected by Levi flight and may be slight slower.

The limitations can be overcome through the modifications performed in Cuckoo Search algorithm [8]. There are lots of improvements can be performed, but the modifications are based on the parameters that are needed for the improvements that support the reliability of the software. The modifications that are made in the algorithm make an improvement in the way they produces the better result in terms of software reliability.

Improvements in Levy flights
Levy flights procedure can be improved through various methods, such as

1. Introducing Gauss distribution
2. TLCS algorithm
3. Assigning a dynamic weight factor
4. Out of bound project strategy

These are some of the methods in which improvements have been performed in Cuckoo Search algorithm. In the proposed approach, the implementation of gauss distribution along with assigning a dynamic weight factor is performed. When both are combined, there is much improvement in Levy flights to get implemented in finding the best solutions. The procedure is given below:

Step 1: Initialization Phase
The population (mi, where i=1,2,...n) of host nest is originated randomly.
Step 2: Generating New Cuckoo Phase
A cuckoo is chosen at random through the means of Levy flights, which produces the new solutions. The cuckoo that is produced is assessed through the objective function which finds the quality of the solution.
Step 3: Fitness Evaluation Phase
Assess the fitness function based on the equation and after that choose the best one.

$$P_{\max} = \frac{P_S}{p_T} \tag{4.5}$$

$$fitness = maximum\ popularity = P_{\max} \tag{4.6}$$

where
P_S, selected population
P_T, total population

Step 4: updation phase
The Levy flights used for ordinary Cuckoo Search algorithm is,

$$m_i^* = m_i^{(t+1)} = m_i^{(t)} + \alpha \oplus Levy(n) \tag{4.7}$$

The ordinary search equation is modified by the application of gauss distribution. The Levy flights have more effect when gauss distribution is applied. The speed and efficiency in finding the search space gets improved when there is an improvement in the existing algorithm by applying a dynamic weight factor to the Levy flights.

By altering the above equation, Levy flight equation by employing the Gauss distribution is shown below,

$$m_i^* = m_i^{(t+1)} = m_i^{(t)} + \alpha \oplus \sigma_s \tag{4.8}$$

where

$$\sigma_s = \sigma_0 \exp(-\mu\kappa) \quad (4.9)$$

σ_0, μ, constants
K, current generation

Step 5: Reject Worst Nest Phase
The novel ones are selected out of the solutions [7]. The ones which are worst are thrown away. The best solutions are selected based on their fitness value. When the solutions are reached, it is treated as optimal solutions.
Step 6: Stopping Criterion Phase
Until the maximum iteration is accomplished, this process is replicated.

4.4 Optimization in Module Design

Modularity is a concept for managing large systems by breaking them down into a series of interdependent and independent modules. The advantages of modularity are frequently realized through module independence, which allows for autonomous development and hence reduces overall lead time due to the sharing of similar modules across products in a product, family, time and economies of scale can be achieved.

An open source software balloon tool tip is used for accessing the quality metrics during the design time. These metrics are categorized into dependable and nondependable packages based on the values that are received from the metrics.

The software can be considered as the maintainable one when there are more number of dependable packages. From Table 4.1, it is clear that the number of dependable packages are more than that of nondependable

Table 4.1 Percentage of dependable and nondependable packages.

Open source software	Nondependent package (A=0) and (I=1) (Np)	Dependent package (Ce=0) and (I=0) (Dp)	Percentage of non-dependable package (Np/Tp)	Percentage of dependable package (Dp/Tp)	Total package (Tp)
Balloon tool tip	1	9	10	90	10

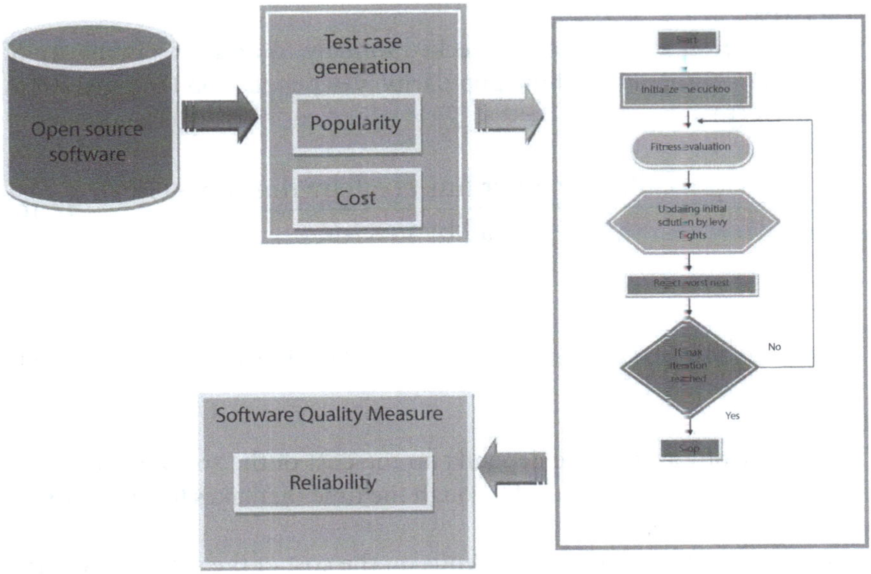

Figure 4.1 Optimization process.

packages. Hence, the software can be justified as maintainable. At the design time, the number of dependable packages can assure the quality of the software, and it can be executed through the following Figure 4.1.

4.5 Optimization at Dynamic Implementation

Generation of test case:
Test case is the one, which is fed into the system for testing with different values. When there are different input values, the classes will be checked against each and every value, so that it ensures no fault to occur [5]. The factors considered for the test cases are cost and execution time.

The generation of test case is based on two parameters:

1. Popularity
2. Cost.

The result obtained from the test case is fed into the adopted Cuckoo Search algorithm to access the quality of the software.

Popularity

The function in a class can be checked initially to find the popularity of the software. Evaluation of the popularity is dependent on the function in each class (F)

$$F = \frac{\text{called by other fun} + (\text{call by this fun} - 1)}{\text{call by this fun}} \tag{4.10}$$

$$\text{Popularity} = \text{sum of function } (F) \text{ in each class} \tag{4.11}$$

Cost

The quality of the software depends on the cost of the software. The estimation of the cost is calculated through the time that takes for the software to be implemented.

The quality of the software depends on the cost of the software [6]. The cost of the software is estimated based on the execution time of the software and value of the error that is generated.

$$Cost = C_{nt} + C_a(1+r)f_r(t) + C_b\left[f_r(t_j) - (1+r)f_r(t)\right] + C_c\left(\int_0^t x(t)dt\right) \tag{4.12}$$

where cost of adopting a new automated testing tools;

r is directly proportional to cost;

Ca, cost of correcting the error during the testing;

Cb, cost of correcting an error during process;

Cc, cost of testing per unit testing expenditure;

Fr (t), failure rate.

4.6 Comparative Study of Support of Modified Cuckoo Search Algorithm

Among the search algorithm the modified Cuckoo Search Algorithm gives a better solution toward the optimization problem. In this comparative

Table 4.2 Efficiency analysis.

S. no.	1	2	3	4	5	6	7	8	9	10
R-E*	12.93	9.61	8.37	5.19	3.66	3.42	3.22	2.23	1.72	1.66
C-E**	19.76	9.39	6.63	3.92	2.72	2.40	1.80	1.60	1.55	1.65
F-E***	2	1.5	1.5	3	1.2	1	1.3	1.4	1.5	1.3
S. no.	11	12	13	14	15	16	17	18	19	20
R-E*	1.57	1.59	1.68	0.41	0.44	0.48	0.49	0.47	0.28	0.21
C-E**	1.17	1.18	1.00	1.00	0.72	0.70	0.78	0.76	0.78	0.76
F-E***	1.5	1.3	1.2	1.5	1.3	1.2	1.3	1.3	1.2	1.0

Reliability-Efficiency*, Cost-Efficiency**, Function-Efficiency***

study, few parameters like time, cost, reliability rate, and its failures are compared for the data set in Table 4.2.

The efficiency analysis is carried out among the modified Cuckoo Search algorithm and PSO algorithm for a set of 20 data sets.

Here efficiency analysis is calculated for the reliability rate and its failures of the open source software. If the value obtained by E_1/E_2 is greater than 1, E_1 is efficient than E_2 based on this given interpretation.

E1 is the result obtained by the value generated by MCS algorithm and E_2 is obtained by PSO algorithm. The results obtained by the efficiency algorithm yields the better result for MCS algorithm than the PSO algorithm.

4.7 Results and Discussions

Balloon tooltip Open source software version 2.1 is used for the proposed work. The fitness value is calculated using the optimization technique for generation of the test case. The computational cost of the software is measured. Reliability measure is calculated using the formula. The first step is to generate test case values and for the test cases, optimization is carried out. When the optimization is carried over, then the output generated will be optimized one. The process of optimization helps in selecting the test cases, which are required to support the system. The test cases that are not optimized is not considered for the generation of the result. For the results and discussions, a comparative study of particle swarm optimization (PSO) and MCS is taken. PSO was designed on the aspect of social activities of birds in a group. Each particle flies with a velocity which is adjusted by flying memory and the distance between the neighbors.

Each particle has its objective function value, which is decided by a fitness function in Table 4.3.

The fitness values of existing PSO algorithm is compared with that of proposed Cuckoo Search Algorithm. The fitness value for the PSO algorithm is calculated using the formula

$$\text{Finess } f = \frac{g_s}{P_s}$$

The fitness value for the MCS algorithm is calculated using the expression

$$P_{\max} = \frac{P_S}{p_T}$$

$$\textit{fitness} = \textit{maximum popularity} = P_{\max}$$

where P_S is selected population, P_T, total population

Table 4.4 produces the reliability and cost value attained from the proposed method. The result has been achieved for various iterations and given in Figure 4.2.

Table 4.3 Fitness value for different iterations.

No. of iterations	Fitness value using PSO	Fitness value using MCS
5	0.87	2.31
10	0.95	2.37
15	1.22	2.62
20	1.44	2.64
25	1.69	3.241

Table 4.4 Reliability and computational cost (CC) estimate in relation to execution time.

No. of iterations	Reliability value	CC
5	0.07634	20275.39
10	0.08078	21157.64
15	0.09553	24670.33
20	0.07786	21170.11
25	0.07079	19753.80

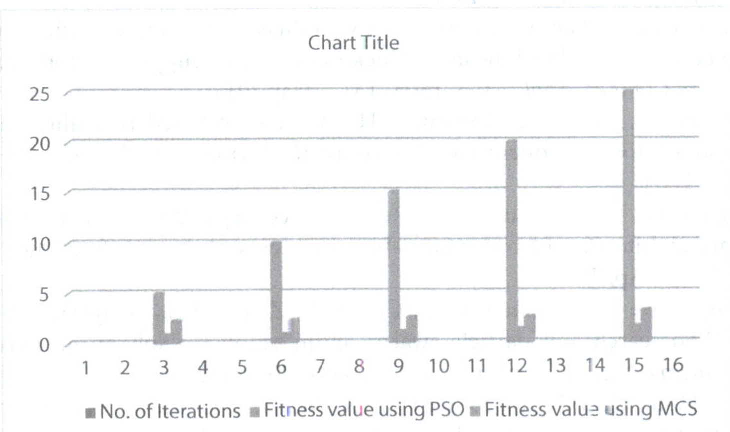

Figure 4.2 Graphical representation of fitness values for MCS and PSO.

4.8 Conclusion

The study has revealed the comparison of two algorithms, one as PSO and the other as modified Cuckoo Search algorithm. In MCS, the results are comparatively better. The results are proven with the help of the software. The testing is also performed in terms of hypothesis to prove that the proposed algorithm (MCS) is performing better than PSO algorithm. In terms of software reliability, the efficiency analysis shows that cost and time may be saved when executing the MCS, which can be attained from the proposed method and yields the better result.

References

1. Yang, X.-S. and Deb, S., Cuckoo search via Levy flights, in: *Proc. World Congr. Nature Biologically Inspired Comput. (NaBIC)*, Coimbatore, India, pp. 210–214, Dec. 2009.
2. Ouyang, X., Zhou, Y., Luo, Q., Chen, H., A novel discrete cuckoo search algorithm for spherical traveling salesman problem. *Appl. Math. Inf. Sci.*, 7, 2, 777–784, Mar. 2013.
3. Cheng, Z., Wang, J., Zhang, M., Song, H., Chang, T., Bi, Y., Sun, K., Improvement and application of adaptive hybrid cuckoo search algorithm. *IEEE Access*, 7, 145489–145515, 2019.
4. Agasthian, A., Pamula, R., Kumaraswamidhas, L.A., Fault classification and detection in wind turbine using Cuckoo-optimized support vector machine. *Neural Comput. Appl.*, 31, 5, 1503–1511, May 2019.
5. Yacoub, S., Cukic, B., Ammar, H.H., A scenario-based reliability analysis approach for component-based software. *IEEE Trans. Reliab.*, 53, 4, 465–480, Dec. 2004.
6. Huang, G., Mei, H., Yang, F.-Q., Runtime recovery and manipulation of software architecture of component-based systems. *Autom. Softw. Eng.*, 13, 2, 257–281, April 2006.
7. Chi, R., Su, Y.X., Zhang, D.H., Chi, X.X., Zhang, H.J., A hybridization of cuckoo search and particle swarm optimization for solving optimization problems. *Neural Comput. Appl.*, 31, 653–670, Jan. 2019.
8. Ma, H.-S., Li, S.-X., Li, S.-F., Lv, Z.-N., Wang, J.-S., An improved dynamic self-adaption Cuckoo search algorithm based on collaboration between subpopulations. *Neural Comput. Appl.*, 31, 5, 1375–1389, May 2019.
9. Ong, P. and Zainuddin, Z., Optimizing wavelet neural networks using modified Cuckoo search for multi-step ahead chaotic time series prediction. *Appl. Soft Comput.*, 80, 374–386, Jul. 2019.
10. Viswanathan, G.M., Buldyrev, S.V., Havlin, S. *et al.*, Optimizing the success of random searches. *Nature*, 401, 6756, 911–914, 1999.

5

Series-Parallel Computer System Performance Evaluation with Human Operator Using Gumbel-Hougaard Family Copula

Muhammad Salihu Isa[1,2*], Ibrahim Yusuf[3], Uba Ahmad Ali[3] and Wu Jinbiao[1]

[1]School of Mathematics and Statistics, Central South University, Changsha, China
[2]Department of Mathematics, Federal University Dutse, Dutse, Nigeria
[3]Department of Mathematical Sciences, Bayero University Kano, Kano, Nigeria

Abstract
This chapter examined the performance of a computer network made up of four workstations, three hubs, and two routers. Subsystem 1 consists of workstations A_1, A_2, A_3, and A_4 that are parallel and linked to subsystem 2 (hubs B_1, B_2, and B_3), while subsystem 3 router consists of C_1 and C_2 that are also parallel to each other. All subsystems are managed by the human operator H. To analyze the reliability of the system, the partial differential equations are derived from the system's schematic diagram in which reliability measure of system strength, such as reliability, availability, MTTF, and cost function are computed. Time to failure of devices, such as workstation, hub, and router, obeys exponential distribution, whereas the corresponding repair time follows two different distributions, namely general and copula distribution. Tables and graphs show some of the most important findings.

***Keywords*:** Performance, workstation, Gumbel-Hougaard family copula, hub and routers

**Corresponding author*: salihu.muhd.isa@gmail.com

S. C. Malik, Deepak Sinwar, Ashish Kumar, S. R. Gadde, Prasenjit Chatterjee and Bui Thanh Hung (eds.) Computational Intelligence in Sustainable Reliability Engineering, (109–128) © 2023 Scrivener Publishing LLC

5.1 Introduction

Failures in computer systems can be classified as software, hardware, or both. Various techniques for improving computer system performance have been proposed. Backup is one of the techniques used to improve and predict system strength and effectiveness, resulting in increased system safety, quality, output, and income mobilization.

The technique of unit-wise redundancy in cold standby mode is also used in computer systems. Each computer system contains programs that run on many different computers that are linked by a network, which has become extremely complicated and difficult to rely on. Reliability refers to a system's ability to perform its intended function under specified conditions for a specified period of time.

Many researchers have proposed various types of studies/mathematical models to improve computer system reliability, and their operations have been proclaimed to perform better. Wu [11] for example, discussed the practical performance modeling of distributed file systems. Malik [7] investigated various models of computer network with unit-by-unit standby and different repair policies. However, component-wise redundancy has been shown to be more reliable than unit-wise redundancy.

Kumar *et al.* [5] investigated the computer system performance including hardware fault detection. Gahlot *et al.* [8] used a copula linguistic approach to compare the performance of (k-out-of-n: G/F). There are two types of repairable systems, each with a different type of failure and a different type of repair, and they concluded that copula repair policy outperforms general repair policy. Garg [1] studied the prediction of abnormal behavior in critical engineering systems in the presence of haze. Garg [2] discussed how to analyze the reliability of industrial systems using fuzzy Kolmogorov differential equations. Garg [3] presented a series-parallel system reliability analysis using credibility theory and various types of intuitionistic fuzzy numbers.

Niwas and Garg [10] examined the profitability and dependability of a cost-free warranty policy-based industrial system. Yusuf *et al.* [13] investigated the financial gain analysis of a series-parallel system. Yusuf [12] presents modeling the reliability of a system with parallel components that exhibits two types of preventive maintenance. Isa *et al.* [4] concentrate on the cost-benefit analysis of three different series parallel with dynamo configurations. In Lado *et al.* [6], Copula linguistics was used to evaluate the performance of a repairable system in series configuration under various types of failure and repair policies. Monika *et al.* [9] discussed on stochastic analysis of a two units' complex repairable system with switch and human failure using Copula approach.

In this research work of computer system, we observed the three subsystems connected in series arrangement, which are workstation, hub, and routers. The model has considered workstation as subsystem 1, hub as subsystem 2 and router as subsystem 3. The main contribution of this paper is that we further classified all the system failures due to human error, and workstation to hub interaction failures, into three groups that is workstation-induced hub failures and hub-induced router failures.

There has been little or no other research that has included the aforementioned failure categories of workstation failures, hub failures, router failures, and workstation-to-router interaction failures. Existing literature has investigated the interaction of a workstation/hub or a hub/router in a network of computers, where workstation failure either causes the failure of a hub and vice versa, or causes the failure of a router and vice versa. The study captured the impact of human operators and replication on computer reliability, which may limit the computer system's work as well as its reliability/availability. A system failure can occur as a result of the failure of a workstation, two hubs, or two routers (as in Figures 5.1 and 5.2).

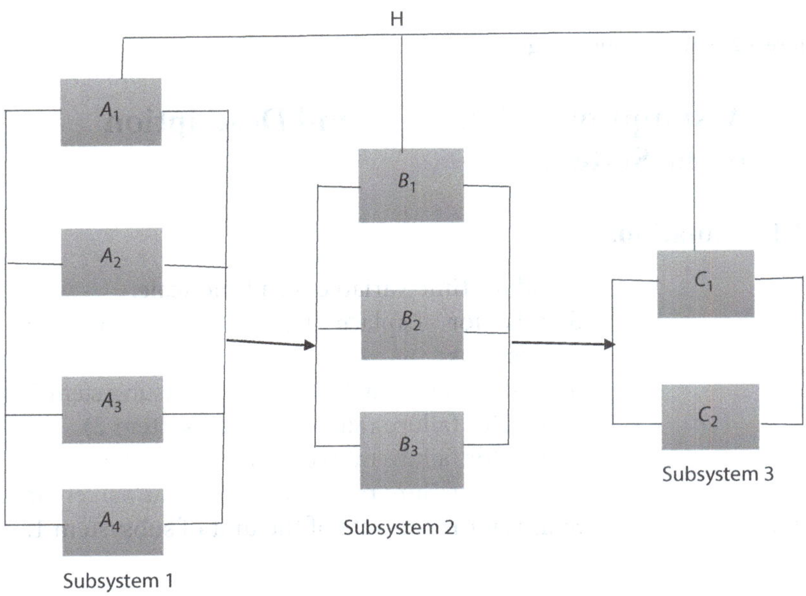

Figure 5.1 Reliability block diagram of system.

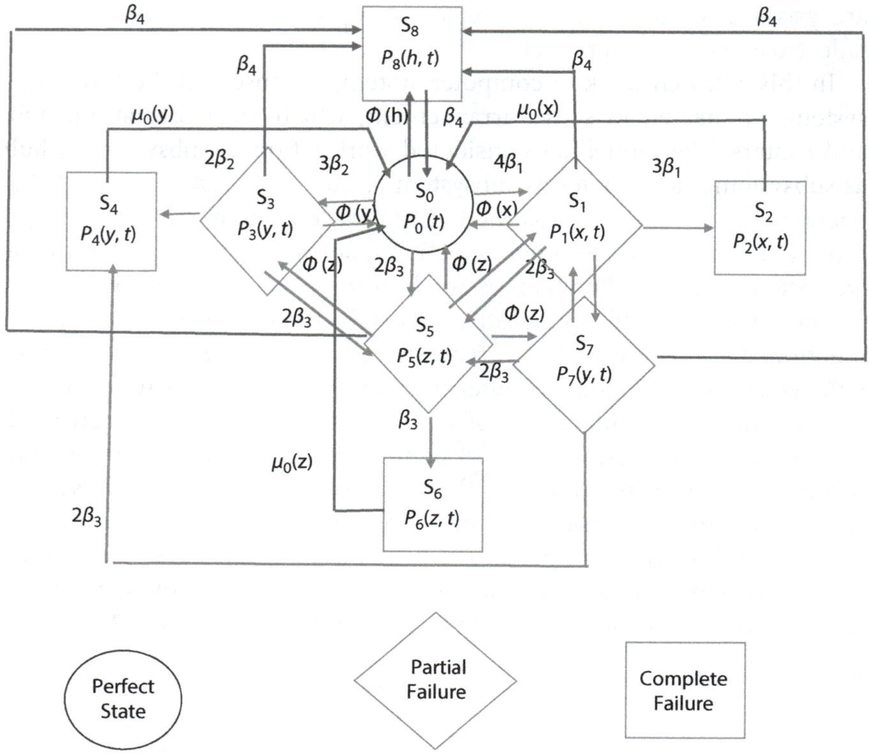

Figure 5.2 State transition diagram.

5.2 Assumptions, Notations, and Description of the System

5.2.1 Notations

t:	Stand for time variable on a time scale.
s:	Stands for Laplace transform variable for all expressions
β_1:	Stands for failure rate of workstation (subsystem 1).
β_2:	Stands for failure rate of hub (subsystem 2).
β_3:	Stands for failure rate of router (subsystem 3).
β_4:	Stands for failure rate as a result of human error.
$\Phi(x)$:	Stands for repair rate of the unit of subsystem 1.

$\Phi(y)$:	Stands for repair rate of the unit of subsystem 2.
$\Phi(z)$:	Stands for repair rate of the unit of subsystem 3.
$\Phi(h)$:	Stands for repair rate of human error.
H:	Stands for human operator of the system.
$\mu0\ (x)/\mu_0\ (y)/\mu_0\ (z)$:	Stands for repair rates for complete failed states
WS/ HB / RT:	stands for workstation, hub and router
$WS_G/HB_G/RT_G$:	stands for workstation, hub and router are good
$WS_F/HB_F/RT_F$:	stands for workstation, hub and router are in failure mode
$WS_{SB}/HB_{SB}/RT_{SB}$:	stands for workstation, hub and router are on standby
$WS_{ID}/HB_{ID}/RT_{ID}$:	stands for workstation, hub and router are Idle
SF_{HO}:	stands for system failure due to human operator
$p_i\ (t)$:	stands for probability that the system is in S_i state at instants for $i = 0$ to 8
$\overline{P}(s)$:	stands for Laplace transformation of state transition probability $p\ (t)$
$P_i\ (x, t)$:	stands for probability that a system is in state S_i for $i = 1$.......8, the system under repair and elapse repair time is (x, t) with repair variable x and time variable t
$P_i\ (y, t)$:	stands for probability that a system is in state S_i for $i = 1$.......8, the system under repair and elapse repair time is (y, t) with repair variable x and time variable t
$P_i\ (z, t)$:	Stands for probability that a system is in state S_i for $i = 1$.......8, the system under repair and elapse repair time is (z, t) with repair variable z and time variable t
$E_p\ (t)$:	Stands for expected profit during the time interval [0, t)
K_1, K_2:	Stands for revenue and service cost per unit time, respectively.
$\mu_0\ (x)$:	Stands for expression of joint probability (failed state S_i to good state S_0) according to Gumbel-Hougaard family copula definition

$$\mu_0(x) = c_\theta(u_1, u_2(x)) = \exp\left(x^\theta + \{\log\phi(x)^\theta\}^{\frac{1}{\theta}}\right) 1 \le \theta \le \infty.$$

Where $\mu_1 = \phi(x)$, and $\mu_2 = e^x$

5.2.2 Assumptions

1. Each failure is repairable.
2. Workstation are identical to each other.
3. Hub and routers are identical to each other.
4. Each computer system failed independent of the other.
5. Computer system works simultaneously and independently.

5.2.3 Description of the System

- ✓ Subsystem 1: Workstation A_1, Workstation A_2, Workstation A_3 and Workstation A_4 connected in series with hub
- ✓ Subsystem 2: Hub B_1 , Hub B_2 and Hub B_3 connected to router
- ✓ Subsystem 3: Consist of Router C_1, Router C_2 and Router C_3.

Initially, the system happens to be in perfect working order, with all subsystems functioning properly. When a unit from subsystems 1, 2, or 3 develops a failure state, the system is said to be in minor partial failure and remains operational, while the failed unit is immediately sent for repairs. System failure occurs when two units of subsystem 1 fail, or when two units of subsystem 2 and two units of subsystem 3 fail, or when human failure occurs, which is likely at all states. General distribution is used to repair minor/completely failed states, while Gumbel-Hougaard family copula distribution is used to repair completely failed states. Table 5.1 describes the system's different states. The system described above is a classic example of a workstation-router computer system with a geographically distanced separated hub typically in a cloud computing setting that provides similar services to workstation located on different continents.

System failures are inevitable and can be caused as a result of natural disaster, human error, hardware/software failure or cyber-attack. System failures are damaging and sometimes the cost of repair is more than the cost of rebuilding the system. Failure can be damaging as it may cause data loss or corruption which cannot sometimes be recreated or corrected. For instance, in times of world pandemics, such as the recent corona virus, World Health Organization (WHO) communicated and exchanges data with all countries around the world. If WHO uses the cloud data architecture with dedicated hubs for each region on the African continent, imaging how catastrophic and damaging a data loss could be. If it were a business venture, the same applies. For these reasons, critical systems must be reliable, and their analyses are investigated.

Table 5.1 Transitional states of the system.

State	Subsystem 1				Subsystem 2			Subsystem 3		Systems' status
	WS	WS	WS	WS	HB	HB	HB	RT	RT	
S_0	WS_G	WS_G	WS_G	WS_{SB}	HB_G	HB_G	HB_G	RT_G	RT_{SB}	Operational
S_1	WS_F	WS_G	WS_G	WS_G	HB_G	HB_G	HB_{SB}	RT_G	RT_{SB}	Operational
S_2	WS_F	WS_F	WS_{ID}	WS_{ID}	HB_{ID}	HB_{ID}	HB_{SB}	RT_{ID}	RT_{SB}	Down
S_3	WS_G	WS_G	WS_G	WS_{SB}	HB_F	HB_G	HB_G	RT_G	RT_{SB}	Operational
S_4	WS_{ID}	WS_{ID}	WS_{ID}	WS_{ID}	HB_F	HB_F	HB_{ID}	RT_{ID}	RT_{SB}	Down
S_5	WS_G	WS_G	WS_G	WS_{SB}	HB_G	HB_G	HB_{SB}	RT_F	RT_G	Operational
S_6	WS_{ID}	WS_{ID}	WS_{ID}	WS_{SB}	HB_{ID}	HB_{ID}	HB_{SB}	RT_F	RT_F	Down
S_7	WS_F	WS_G	WS_G	WS_G	HB_G	HB_G	HB_{SB}	RT_F	RT_G	Operational
S_8	SF_{HO}									Down

5.3 Reliability Formulation of Models

Using the method adopted in Gahlot *et al.* (2018), Lado *et al.* (2018) and Abdulkareem and Singh (2019) the following partial differential equations:

$$\left(\frac{\delta}{\delta t}+4\beta_1+3\beta_2+2\beta_3+\beta_4\right)P_0(t)=\int_0^\infty \phi(x)p_1(x,t)dx+\int_0^\infty \phi(y)p_3(y,t)dy+\int_0^\infty \phi(h)p_8(h,t)dh+\int_0^\infty \phi(z)p_5(z,t)dz$$
$$+\int_0^\infty \mu_0(x)p_2(x,t)dx+\int_0^\infty \mu_0(y)p_4(y,t)dy+\int_0^\infty \mu_0(z)p_6(z,t)dz, \tag{5.1}$$

$$\left(\frac{\delta}{\delta t}+\frac{\delta}{\delta x}+3\beta_1+3\beta_2+2\beta_3+\beta_4+\phi(x)\right)P_1(x,t)=0, \tag{5.2}$$

$$\left(\frac{\delta}{\delta t}+\frac{\delta}{\delta x}+\mu_0(x)\right)P_2(x,t)=0, \tag{5.3}$$

$$\left(\frac{\delta}{\delta t}+\frac{\delta}{\delta y}+2\beta_2+2\beta_3+\beta_4+\phi(y)\right)P_3(y,t)=0, \tag{5.4}$$

$$\left(\frac{\delta}{\delta t}+\frac{\delta}{\delta y}+\mu_0(y)\right)P_4(y,t)=0, \tag{5.5}$$

$$\left(\frac{\delta}{\delta t}+\frac{\delta}{\delta z}+\beta_3+\beta_4+4\phi(z)\right)P_5(z,t)=0, \tag{5.6}$$

$$\left(\frac{\delta}{\delta t}+\frac{\delta}{\delta z}+\mu_0(z)\right)P_6(z,t)=0, \tag{5.7}$$

$$\left(\frac{\delta}{\delta t}+\frac{\delta}{\delta y}+2\beta_2+2\beta_3+\beta_4+\phi(y)\right)P_7(y,t)=0, \tag{5.8}$$

$$\left(\frac{\delta}{\delta t}+\frac{\delta}{\delta h}+\phi(h)\right)P_8(h,t)=0. \tag{5.9}$$

Boundary Conditions

$$P_1(0,t)=4\beta_1 P_0(t) \tag{5.10}$$

$$P_2(0,t)=12\beta_1^2P_0(t), \tag{5.11}$$

$$P_3(0,t)=3\beta_2P_0(t) \tag{5.12}$$

$$P_4(0,t)=6\beta_2^2(1+4\beta_1)P_0(t), \tag{5.13}$$

$$P_5(0,t)=2\beta_3(1+4\beta_1+12\beta_1\beta_2+3\beta_2)P_0(t), \tag{5.14}$$

$$P_6(0,t)=2\beta_3^2(1+4\beta_1+12\beta_1\beta_2+3\beta_2)P_0(t), \tag{5.15}$$

$$P_7(0,t)=12\beta_1\beta_2P_0(t), \tag{5.16}$$

$$P_8(0,t)=[\beta_4 4\beta_1\beta_4+3\beta_2\beta_4+2\beta_3\beta_4(1+4\beta_1+12\beta_1\beta_2+3\beta_2)+12\beta_1\beta_2\beta_4P_0(t). \tag{5.17}$$

5.3.1 Solution of the Model

Using Laplace transformation of equations (5.1) to (5.17) together with initial condition, one can obtain the following results

$$(s+4\beta_1+3\beta_2+2\beta_3+\beta_4)\bar{P}_0(s)=1+\int_0^\infty\phi(x)p_1(x,s)dx+\int_0^\infty\phi(y)p_3(y,s)dy+\int_0^\infty\phi(h)p_8(h,s)dh+\int_0^\infty\phi(z)p_5(z,s)dz+\int_0^\infty\mu_0(x)p_2(x,s)dx+\int_0^\infty\mu_0(y)p_4(y,s)dy+\int_0^\infty\mu_0(z)p_6(z,s)dz, \tag{5.18}$$

$$\left(s+\frac{\delta}{\delta x}+3\beta_1+3\beta_2+2\beta_3+\beta_4+\phi(x)\right)\bar{P}_1(x,s)=0, \tag{5.19}$$

$$\left(s+\frac{\delta}{\delta x}+\mu_0(x)\right)\bar{P}_2(x,s)=0, \tag{5.20}$$

$$\left(s+\frac{\delta}{\delta y}+2\beta_2+2\beta_3+\beta_4+\phi(y)\right)\bar{P}_3(y,s)=0, \tag{5.21}$$

$$\left(s+\frac{\delta}{\delta y}+\mu_0(y)\right)\overline{P}_4(y,s)=0, \tag{5.22}$$

$$\left(s+\frac{\delta}{\delta z}+\beta_3+\beta_4+4\phi(z)\right)\overline{P}_5(z,s)=0, \tag{5.23}$$

$$\left(s+\frac{\delta}{\delta z}+\mu_0(z)\right)\overline{P}_6(z,s)=0, \tag{5.24}$$

$$\left(s+\frac{\delta}{\delta y}+2\beta_2+2\beta_3+\beta_4+\phi(y)\right)\overline{P}_7(y,s)=0, \tag{5.25}$$

$$\left(s+\frac{\delta}{\delta h}+\phi(h)\right)\overline{P}_8(h,s)=0, \tag{5.26}$$

$$\overline{P}_1(0,s)=4\beta_1\overline{P}_0(s), \tag{5.27}$$

$$\overline{P}_2(0,s)=12\beta_1^2\overline{P}_0(s), \tag{5.28}$$

$$\overline{P}_3(0,s)=3\beta_2\overline{P}_0(s), \tag{5.29}$$

$$\overline{P}_4(0,s)=6\beta_2^2(1+4\beta_1)\overline{P}_0(s), \tag{5.30}$$

$$\overline{P}_5(0,s)=2\beta_3(1+4\beta_1+12\beta_1\beta_2+3\beta_2)\overline{P}_0(s), \tag{5.31}$$

$$\overline{P}_6(0,s)=2\beta_3^2(1+4\beta_1+12\beta_1\beta_2+3\beta_2)\overline{P}_0(s), \tag{5.32}$$

$$\overline{P}_7(0,s)=12\beta_1\beta_2\overline{P}_0(s), \tag{5.33}$$

$$\overline{P}_8(0,s)=[\beta_4+4\beta_1\beta_4+3\beta_2\beta_4+2\beta_3\beta_4(1+4\beta_1+12\beta_1\beta_2+3\beta_3)+12\beta_1\beta_2\beta_4]\overline{P}_0(s), \tag{5.34}$$

$$\bar{P}_0(s) = \frac{1}{D(s)}$$

$$\bar{P}_1(s) = \frac{1}{D(s)}\left\{\frac{1-\bar{S}_C(s+3\beta_1+3\beta_2+2\beta_3+\beta_4)}{s+3\beta_1+3\beta_2+2\beta_3+\beta_4}\right\}4\beta_1, \tag{5.35}$$

$$\bar{P}_2(s) = \frac{1}{D(s)}\left\{\frac{1-\bar{S}_{\mu_0}(s)}{s}\right\}12\beta_1^2, \tag{5.36}$$

$$\bar{P}_3(s) = \frac{1}{D(s)}\left\{\frac{1-\bar{S}_Q(s+2\beta_2+2\beta_3+\beta_4)}{s+2\beta_2+2\beta_3+\beta_4}\right\}3\beta_2, \tag{5.37}$$

$$\bar{P}_4(s) = \frac{1}{D(s)}\left\{\frac{1-\bar{S}_{\mu_0}(s)}{s}\right\}6\beta_2^2(1+4\beta_1), \tag{5.38}$$

$$\bar{P}_5(s) = \frac{1}{D(s)}\left\{\frac{1-\bar{S}_{4Q}(s+\beta_3+\beta_4)}{s+\beta_3+\beta_4}\right\}2\beta_3(1+4\beta_1+12\beta_1\beta_2+3\beta_2), \tag{5.39}$$

$$\bar{P}_6(s) = \frac{1}{D(s)}\left\{\frac{1-\bar{S}_{\mu_0}(s)}{s}\right\}2\beta_3^2(1+4\beta_1+12\beta_1\beta_2+3\beta_2), \tag{5.40}$$

$$\bar{P}_7(s) = \frac{1}{D(s)}\left\{\frac{1-\bar{S}_{\mu_0}(s)}{s}\right\}12\beta_1\beta_2, \tag{5.41}$$

$$\bar{P}_8(s) = \frac{1}{D(s)}[\beta_4+4\beta_1\beta_4+3\beta_2\beta_4+2\beta_3\beta_4(1+4\beta_1+12\beta_1\beta_2+3\beta_3)+12\beta_1\beta_2\beta_4]\left\{\frac{1-\bar{S}_Q(s)}{s}\right\}, \tag{5.42}$$

where:

$$D(s) = 4\beta_1 + 3\beta_2 + 2\beta_3 + \beta_4 - \left(\begin{array}{l} 4\beta_1 \left\{ \overline{S}_Q(s + 3\beta_1 + 3\beta_2 + 2\beta_3 + \beta_4) \right\} 12\beta_1^2 \left\{ \overline{S}_{\mu_0}(s) \right\} + 3\beta_2 \left\{ \overline{S}_Q(s + 2\beta_2 + 2\beta_3 + \beta_4) \right\} \\ +2\beta_3(1 + 4\beta_1 + 12\beta_1\beta_2 + 3\beta_2) \left\{ \overline{S}_{4Q}(s + \beta_3 + \beta_4) \right\} + 2\beta_3^2(1 + 4\beta_1 + 12\beta_1\beta_2 + 3\beta_2) \left\{ \overline{S}_{\mu_0}(s) \right\} \\ +6\beta_2^2(1 + 4\beta_1) \left\{ \overline{S}_{\mu_0}(s) \right\} + [\beta_4 + 4\beta_1\beta_4 + 3\beta_2\beta_4 + 2\beta_3\beta_4(1 + 4\beta_1 + 12\beta_1\beta_2 + 3\beta_3) + \\ 12\beta_1\beta_2\beta_4] \left\{ \overline{S}_{\mu_0}(s) \right\} \end{array} \right), \tag{5.43}$$

$$\overline{P}_{UP}(s) = \overline{P}_0(s) + \overline{P}_1(s) + \overline{P}_3(s) + \overline{P}_5(s) + \overline{P}_7(s). \tag{5.44}$$

5.4 Some Particular Cases Based on Analytical Analysis of the Model

5.4.1 Availability Analysis

Following Gumbel-Hougaard family copula distribution for the repair and by setting and taking the values of different parameter as: $\beta_1 = 0.01$, $\beta_2 = 0.02$, $\beta_3 = 0.03$ and $\beta_4 = 0.04$, $\theta(y) = 1$ and $\mu = 1$ and $\varnothing(x) = 1$ in equation (5.43), then taking the inverse Laplace transform, one can obtain the expression for availability as:

Table 5.2 Variation of availability with respect of time.

Time (t)	Availability
0	0.9999
0.01	0.9996
0.02	0.9993
0.03	0.9990
0.04	0.9987
0.05	0.9984
0.06	0.9981
0.07	0.9978
0.08	0.9976
0.09	0.99732

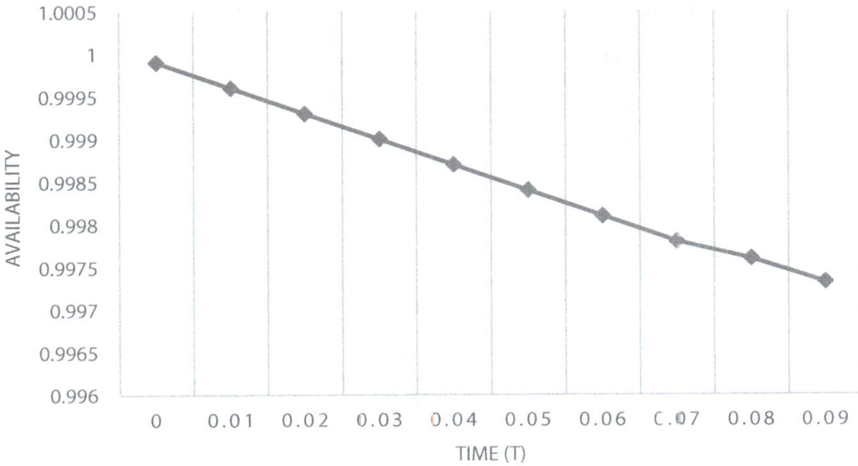

Figure 5.3 Availability against time.

$$P_{up}(t) = -0.01315494254e^{-(4.087902745t)} + 0.02861182898e^{-(2.794659997t)} + 0.0009976085690e^{-(1.263720485t)} + 0.001256451650e^{-(1.164978807t)} + 0.02550289340e^{-(1.029125297t)} + 0.9587813770e^{0.02208733147t}. \quad (5.45)$$

For different values of time t = 0, 0.01, 0.02, 0.03, 0.04, 0.05, 0.06, 0.07, 0.08, 0.09 units of time.

One may obtain different values of $P_{up}(t)$ using equation (5.43) as shown in Table 5.2 and Figure 5.3.

5.4.2 Reliability Analysis

Setting all the repairs rates $\Phi(x)$, $\Phi(y)$, $\mu_0(x)$ and $\mu_0(y)$ in equation (43) to zero with the same values of failure rates as: $\beta_1 = 0.01$, $\beta_2 = 0.02$, $\beta_3 = 0.03$ and $\beta_4 = 0.04$ and then taking inverse Laplace transform, one may get the expression for reliability for system as:

$$R(t) = -0.5088000000e^{-(0.07000000000\,t)} + e^{-(0.1400000000\,t)} + 4.0e^{-(0.1900000000t)} - 4.520800000e^{-(0.2000000000t)} + 0.12000000000 \quad (5.46)$$

For various values of time t = 0, 1, 2, 3, 4, 5, 6, 7, 8, 9 units of time, one may obtain different values of $R(t)$ with the help of equation (5.46) as shown in Table 5.3 and graphical representation in Figure 5.4.

Table 5.3 Reliability variation for copula repair.

Time (t)	Reliability
0	1.0000
1	0.9622
2	0.9151
3	0.8625
4	0.8070
5	0.7509
6	0.6956
7	0.6421
8	0.5910
9	0.5428

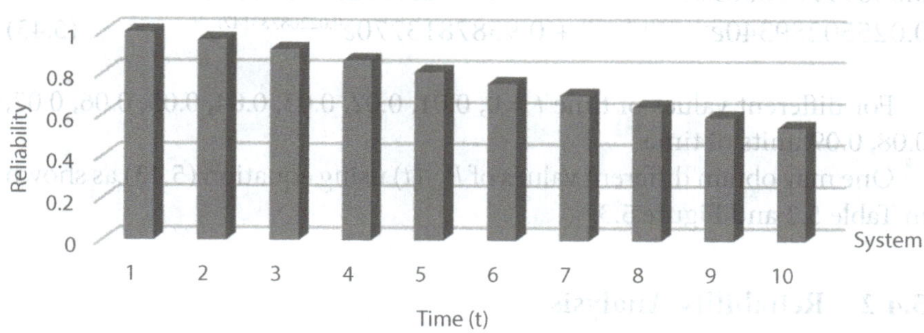

Figure 5.4 Reliability against time.

5.4.3 Mean Time to Failure (MTTF)

Setting all the repairs to zero in equation (5.43) and as $s \to 0$, we obtain MTTF expression as:

$$MTTF = \lim_{s \to 0} \frac{1}{4\beta_1 + 3\beta_2 + 2\beta_3 + \beta_4}\left(1 + \frac{4\beta_1}{3\beta_1 + 3\beta_2 + 2\beta_3 + \beta_4} + \frac{3\beta_2}{2\beta_2 + 2\beta_3 + \beta_4} + \frac{2\beta_3(1 + 4\beta_1 + 12\beta_1\beta_2 + 3\beta_2)}{\beta_3 + \beta_4}\right) \tag{5.47}$$

Setting, $\beta_1 = 0.01$, $\beta_2 = 0.02$, $\beta_3 = 0.03$ and $\beta_4 = 0.04$ and varying β_1, β_2, β_3, and β_4, respectively as 0.01, 0.02, 0.03, 0.04, 0.05, 0.06, 0.07, 0.08, 0.09 in the above equation, one may obtain the variation of MTTF with respect to failure rates as shown in Table 5.4 and Figure 5.5.

Table 5.4 Computation of MTTF corresponding to the various values of failure rates.

Failure $\beta_1/\beta_2/\beta_3/\beta_4$	**MTTF β_1 $\beta_2 = 0.02$, $\beta_3 = 0.03$, $\beta_4 = 0.04$**	**MTTF β_2 $\beta_1 = 0.01$, $\beta_3 = 0.03$, $\beta_4 = 0.04$**	**MTTF β_3 $\beta_1 = 0.01$, $\beta_2 = 0.03$, $\beta_4 = 0.04$**	**MTTF β_4 $\beta_1 = 0.01$, $\beta_2 = 0.02$, $\beta_3 = 0.03$**
0.01	12.92006	14.22453	14.42266	20.28855
0.02	11.55610	12.92006	13.72348	16.98985
0.03	10.45061	11.80858	12.92006	14.66400
0.04	9.54357	10.86564	12.12670	12.92006
0.05	8.78894	10.06206	11.38389	11.55619
0.06	8.15270	9.37218	10.70338	10.45640
0.07	7.60977	8.77507	10.08546	9.54869
0.08	7.14142	8.25409	9.52599	8.78564
0.09	6.73353	7.79609	9.01937	8.13462

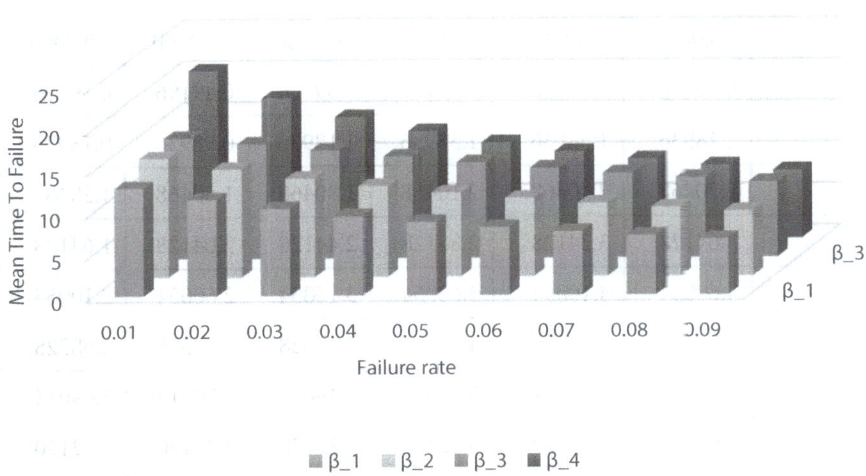

Figure 5.5 Mean time to failure against failure rate.

5.4.4 Cost-Benefit Analysis

Let the service facility be always available, then the expected profit during the interval [0, t).

$$E_p(t) = K_1 \int_0^t P_{up}(t)\,dt - K_2 t$$

where K_1 and K_2 are the revenue generated and service cost per unit time in the interval [0, t). For the same set of parameters as in (5.43), one can obtain (5.48).

$$E_p(t) = K_1\,[0.003218017492e^{-(4.087902745t)} - 0.01023803576e^{-(2.794659997t)} + 0.00007894218546e^{-(1.263720485t)} - 0.001078518890e^{-(1.164978807t)} - 0.02478113547e^{-(1.029125297t)} + 43.40865615e^{0.022087331 47t}] - K_2.t \quad (5.48)$$

Setting $K_1 = 1$ and $K_2 = 0.6, 0.5, 0.4, 0.3, 0.2$, and 0.1, respectively, and varying $t = 0, 1, 2, 3, 4, 5, 6, 7, 8, 9$ units of time, one may obtain the results for expected profit as shown in Table 5.5 and Figure 5.6.

Table 5.5 Cost computation for different value of time.

Time (t)	$E_p(t)$ $K_2 = 0.1$	$E_p(t)$ $K_2 = 0.2$	$E_p(t)$ $K_2 = 0.3$	$E_p(t)$ $K_2 = 0.4$	$E_p(t)$ $K_2 = 0.5$	$E_p(t)$ $K_2 = 0.6$
0	−0.00003	−0.00003	−0.00003	−0.00003	−0.00003	−0.00003
1	0.89196	0.79196	0.69196	0.59196	0.49196	0.39196
2	1.78936	1.58936	1.38936	1.18936	0.98936	0.78936
3	2.70468	2.40468	2.10468	1.80468	1.50468	1.20468
4	3.64128	3.24128	2.84128	2.44128	2.04128	1.64128
5	4.60054	4.10054	3.60054	3.10054	2.60054	2.10054
6	5.58328	4.98328	4.38328	3.78328	3.18328	2.58328
7	6.59014	5.89014	5.19014	4.49014	3.79014	3.09014
8	7.62170	6.82170	6.02170	5.22170	4.42170	3.62170
9	8.67852	7.77852	6.87852	5.97852	5.07852	4.17852

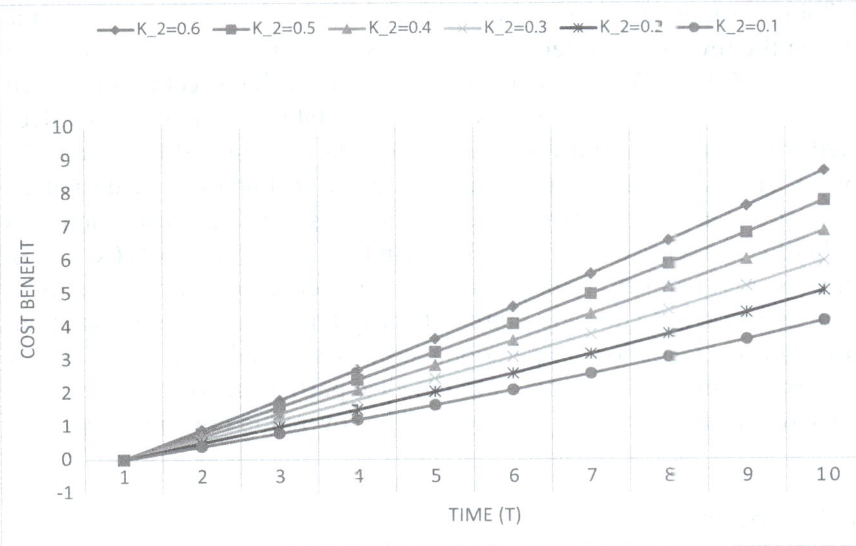

Figure 5.6 Cost-benefit against time.

5.5 Conclusions Through Result Discussion

Table 5.1 and Figure 5.2 show clearly how the availability of the repairable series system changes with time t when failure rates are fixed at different values. When failure rates are low, $\beta_1 = 0.01$, $\beta_2 = 0.02$, $\beta_3 = 0.03$, and $\beta_4 = 0.04$, the system's availability decreases gradually over time as the value of t increases, eventually reaching zero after a long interval of time. As a result, for any given set of parametric values, the next step of the repairable system can be reliably predicted at any stage. This is clear from the model's graphical construction.

It is relatively good that when repair is performed, the system performance is far superior to when the repair is not performed. Tables 5.2 and 5.3 show that the corresponding availability values are greater than the corresponding reliability values. To improve system performance, this mathematical analysis suggests that regular repairs are expected to be performed

Furthermore, Table 5.4 and Figure 5.5 yield the MTTF of the system with respect to variation in β_1, β_2, β_3, and β_4, respectively, when other parameters are kept constant. The variation in MTTF corresponding to β and β_2 is almost close but the variation in MTTF corresponding to β_3 and β_4 is higher than β_1 and β_2.

Table 5.5 and Figure 5.6 show the result of the cost function against time t when the revenue cost per unit time K_2 is fixed at 1, and the service cost $K_2 = 0.6, 0.5, 0.4, 0.3, 0.2, 0.1$. The cost increases with respect to time when the service cost K_2 decreases, as shown in the table and figure. The calculated cost in the table shows that $K_2 = 0.1$ is the highest and $K_2 = 0.6$ is the lowest. Finally, it has been observed that as the cost of a service decreases, the cost increases with the passage of time. In general, the cost function is higher for low service costs ($K_2 = 0.1$) than for high service costs ($K_2 = 0.6$). This research will help in the reliability of the engineers and designers of computer system to develop more critical systems in order to bring about more efficiency and lower operational costs. As a result, the current study will include maintenance and replacement at partial and total failure under a free renewing warranty. This topic will be covered in our future work.

References

1. Garg, H., Predicting uncertain behaviour in critical engineering systems under vague environment. *J. Mult.-Valued Log. Soft Comput.*, 25, 1, 1–21, 2015.
2. Garg, H., An approach for analysing the reliability of industrial system using fuzzy Kolmogrov's differential equations. *Arab. J. Sci. Eng.*, 40, 3, 975–987, 2016a.
3. Garg, H., A novel approach for analysing the reliability of series-parallel system using credibility theory and different types of intuitionistic fuzzy numbers. *J. Braz. Soc. Mech. Sci. Eng.*, 38, 3, 1021–1035, 2016b.
4. Isa, M.S., Ali, U.A., Yusuf, I., Yusuf, B., Cost-benefit analysis of three different series-parallel dynamo configurations. *Life Cycle Reliability Saf. Eng.* 9, 10, 413–423, 2020. https://doi.org/10.1007/s41872-020-00141-0.
5. Kumar, A., Saini, M., Malik, S.C., Performance analysis of a computer system with imperfect fault detection of hardware. *Procedia Comput. Sci., International Conference on Advanced Computing Technologies and Applications*, 45, 602–610, 2015.
6. Lado, A., Singh, V.V., Ismail, K.H., Yusuf., I., Performance and cost assessment of repairable complex system with two subsystems connected in series configuration. *Int. J. Reliab. Appl.*, 19, 1, 27–42, 2018.
7. Malik, S.C., Reliability modeling of a computer system with preventive maintenance and priority subject to maximum operation and repair times. *Int. J. Syst. Assur. Eng. Manage.*, 4, 1, 94–100, 2013.
8. Gahlot, M., Singh, V.V., Ayagi, H., II, Goel, C.K., Performance assessment of repairable system in series configuration under different types of failure and repair policies using copula linguistics. *Int. J. Reliab. Saf.*, 12, 4, 348–374, 2018.

9. Gahlot, M., Singh, V.V., Ayagi, H., II, Abdullahi, I., Stochastic analysis of a two units' complex repairable system with switch and human failure using copula approach. *Life Cycle Reliability Saf. Eng.*, 9, 1, 1–1_, 2019.
10. Niwas, R. and Garg, H., An approach for analyzing the reliability and profit of an industrial system based on the cost free warranty policy. *J. Braz. Soc. Mech. Sci. Eng.*, 40, 265, 2018.
11. Wu, Y., Modeling of distributed file systems for practical performance analysis. *IEEE Trans. Parallel Distrib. Syst.*, 25, 1, 156–166, 2014.
12. Yusuf, I., Reliability modeling of a parallel system with a supporting device and two types preventive maintenance. *Int. J. Oper. Res.*, 25, 3, 269–287, 2016.
13. Yusuf, I., Sani, B., Yusuf, B., Profit analysis of a serial-parallel system under partial and complete failures. *J. Appl. Sci.*, 19, 565–574, 2019.

6

Applications of Artificial Intelligence in Sustainable Energy Development and Utilization

Aditya Kolakoti[1]*, Prasadarao Bobbili[2], Satyanarayana Katakam[3], Satish Geeri[4] and Wasim Ghder Soliman[5]

[1]Department of Mechanical Engineering, Raghu Engineering College, Visakhapatnam, India
[2]Department of Electrical & Electronics Engineering, Vignan's Institute of Information Technology, Visakhapatnam, India
[3]Department of Mechanical Engineering, ANITS Engineering College, Visakhapatnam, India
[4]Department of Mechanical Engineering, Pragati Engineering College, Kakinada, India
[5]Department of Industrial Automation Engineering, Faculty of Technical Engineering, Tartous University, Tartus, Syria

Abstract
Clean energy and the environment are considered as the most influential parameters to promote sustainable development. However, with the increase in the global population, the consumption of energy sources from the nonrenewable category is increasing rapidly. On the other hand, with the utilization of nonrenewable energy sources like petro-diesels, environmental air pollution is also increasing. To combat the energy and environmental crises, clean and renewable fuels like biofuels are popular as a petrodiesel replacement fuels. Biofuels can be obtained from different feedstocks, and they are successfully tested in diesel engines. However, during their production and engine testing, several parameters influence their output results. The accurate prediction of end results is considered a challenge with the traditional techniques. Therefore, artificial intelligence (AI) techniques emerged as the most successful in solving nonlinear problems and achieve a high

**Corresponding author*: aditya.kolakoti@gmail.com

S. C. Malik, Deepak Sinwar, Ashish Kumar, S. R. Gadde, Prasenjit Chatterjee and Bui Thanh Hung (eds.) Computational Intelligence in Sustainable Reliability Engineering, (129–144) © 2023 Scrivener Publishing LLC

success rate in prediction. In this chapter, different AI techniques that are successfully used in finding a feasible solution for complex problems in biodiesel production and engine testing are discussed in detail.

***Keywords*:** Artificial intelligence (AI), machine learning (ML), renewable energy, biofuels, Industry 4.0

6.1 Energy and Environment

The demand for the energy sources that are obtained from fossil fuels is increasing day today. Approximately 80% of the world's energy utilizations rely on fossil fuels, and hence these fossil fuels are well known for powering the economy and development of the nations [1,2]. On the other hand, fossil fuels like coal and petro-diesel burning increase environmental air pollution by releasing greenhouse gas emissions into the atmosphere [3]. The greenhouse gases increase the atmospheric temperatures and result in sudden climatic changes. The situation is currently witnessed in most countries with the sudden changes in the climatic conditions with increased environmental air pollution. In general, naturally available greenhouse gases like carbon dioxide (CO_2) and methane (CH_4) in the atmosphere helps in maintaining the earth's temperature optimum by utilizing the sun heat. With the rapid utilization of fossil fuels, excess greenhouse gases pumped into the atmosphere increase global temperatures [4]. Furthermore, forest fires and clearing the forestlands by burning the trees will also promote the release of CO_2 into the atmosphere and contribute to the greenhouse effect. Now there is an urgent need to overcome the energy and environmental challenges. As per the environmental scientist, with the implementation of clean and sustainable energy in place of fossil fuels is the only possible way to mitigate the energy and environmental challenges.

6.2 Sustainable Energy

Efficient production and clean utilization of green energy sources are the need of the century and energy sources play a predominant role in shaping the economy of the nations. Currently, sustainable energy production is gaining wide popularity among nations as it meets the present and future needs without putting the available sources in danger of depletion [5]. With the increase in global population, the consumption rate of the energy sources has increased rapidly rather than its production. Many nations are inclined towards sustainable energy production to bridge the gap between production and demand. For instance, alternative fuels like biodiesels are successfully utilized in place of petro-diesels in many developed and developing nations [6]. Similarly, the power production from solar and wind are gradually replacing the dependency on nonrenewable sources of power production like thermal power plants. Utilizing renewable and abundantly available energy sources to meet the current demand and future needs can reduce the reliance on imports and, at the same time saving a huge currency in the form of foreign exchange. Furthermore, the challenge of achieving a clean and green environment is possible with sustainable energy utilization. Sustainable energy includes energy production from solar, wind, biomass, wave, tidal, etc. In this chapter, special attention was paid to one of the sustainable energy products, so-called biodiesel, which is considered as a promising alternative to the petro-diesel due to carbon balance and often more environmentally friendly fuel. Furthermore, the role of artificial intelligence (AI) in improving biodiesel production yield and its performance and emissions is discussed.

6.3 Artificial Intelligence in Industry 4.0

The progressive stages of the industrial revolution (IR), from 1.0, which started in the 18th century, to the latest IR 4.0 (Figure 6.1), show the success of humanity in mitigating the existing challenges from time to time. This transformation has been witnessed clearly from IR 3.0, where computer technology replaces human interference with the automation of the production process. With that initiative, computer technology further expands its networks to the digital twin modes where smart technology comes into existence and is currently called IR 4.0 [7]. With IR 4.0, there is a great potential to raise the global income and at the same time improve the living standards of the people. In IR 4.0, special attention was given to intelligent networking of machines and processes where artificial intelligence (AI), machine learning (ML), internet of things (IoT), cloud computing, cognitive computing, etc., play a significant role [8]. With the advent of the above-said tools, the concept of a smart factory came into reality, and the ultimate results are exceptional. Since the smart machines can access to huge data and the machines become more efficient and have high productivity with less wastage. Hence these tools are gaining wide popularity and are successfully utilized in most engineering applications. Among the mentioned tools, AI and ML has a predominant role in IR 4.0 due to its high accuracy in estimating the results and convergence in solving complex and nonlinear problems with insignificant errors. AI and ML

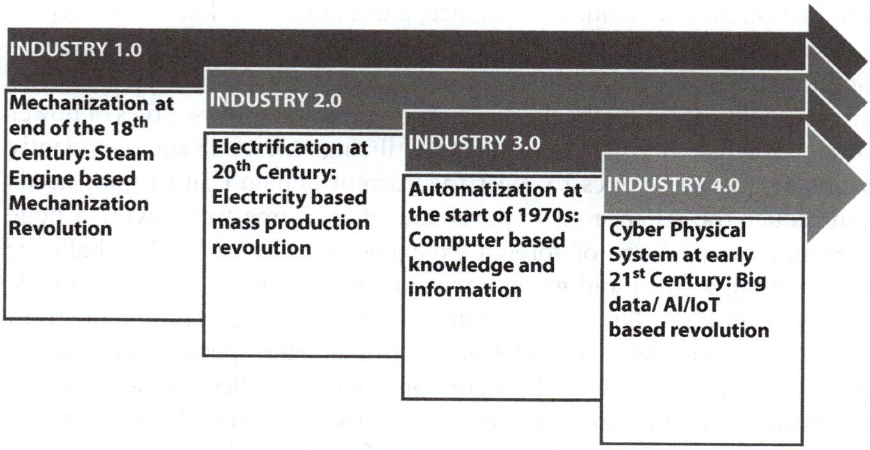

Figure 6.1 Stages of Industrial Revolution from industry 1.0 to 4.0.

find many applications in engineering, science and technology; however, in this chapter, AI applications are devoted to biofuel production and diesel engine experimental performance and emission studies.

6.4 Introduction to AI and its Working Mechanism

Artificial Intelligence (AI) tools have emerged as the most intelligent and high accuracy in solving complex nonlinear problems. They are also known for solving complex problems from a set of examples with high accuracy and in a short time without knowing prior information on the mechanisms or principles behind them. Artificial learning (AL) tools include artificial neural networks (ANN), genetic algorithms (GA), random forest (RF), support vector machines (SVM), etc. are gaining wide popularity in solving complex dynamic problems [9]. Among the aforesaid AI tools, ANN is widely used in the biodiesel production process as a modeling tool to predict maximum biodiesel yield, even with uncertain and noisy data patterns [10, 11].

The working mechanism of the ANN model is similar to the human nervous system, which is formed by the interconnections of billions of neurons. A typical biological neuron consists of four essential components, i.e., cell body/soma, dendrites, axon, and synapses, as shown in Figure 6.2. Each component has a unique function in the successful transformation of memory or signals to the destinations. The dendrites receive the input

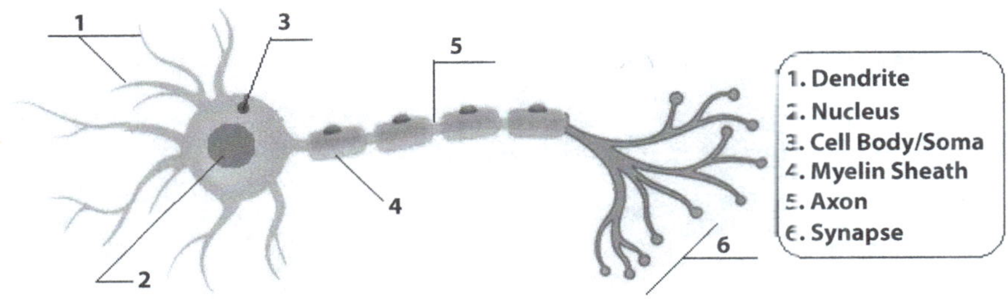

Figure 6.2 Stricture of a biological neuron.

signals from the other neurons with the support of fiber branches attached to the dendrites. The cell body is also called soma, contains the nucleus, and controls all the functions of the cell. The axon acts as a bridge between the cell body and synapse where the information from soma is passed to the synapse, which are located near neighboring neuron. During this transformation of biological information, the received neuron generates an electrical signal, which is carried forward to the next neighboring neuron. In this way, the information is processed and shared in a biological neural network.

Similar to the biological neural network model, an artificial neural network (ANN) will also work. However, ANN consists of analog signal processing units called artificial neurons, which are interconnected with synaptic weights in unidirectional, as shown in Figure 6.3. Each neuron will connect to at least one other neuron using the synaptic weight coefficient and the artificial neuron can receive the input signals, process it for the required information and generate the desired output signal. From Figure 6.3, it was evident that the input to artificial neuron is X_1, X_2, X_3, … X_n and the input data were multiplied with the respective synaptic weights (W_1, W_2, W_3, …W_n). A bias input is also utilized in addition to the input data and both the weight input and bias input are added at the summing junction (Eq. 6.1). The activation function (Eq. 6.2) was used to reduce the intensity of input signals for better and error-free results. The commonly used activation functions are tangent, hyperbolic tangent, sigmoid, linear, Gaussian, etc.

$$\varphi_k = \sum_{j=1}^{n} W_j X_j + b_k \tag{6.1}$$

$$y_k = f(\varphi_k) \tag{6.2}$$

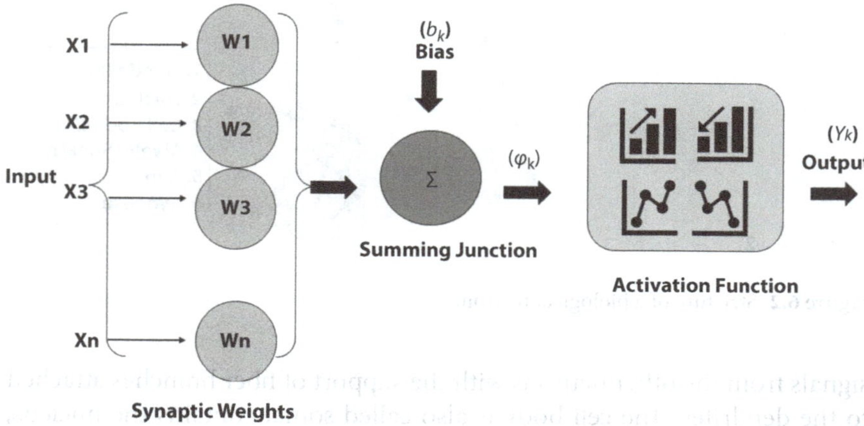

Figure 6.3 Stricture of artificial neuron.

$$R = \frac{\sum_{i=1}^{n}(X_{p,i} - X_{p,avg})(X_{a,i} - X_{a,avg})}{\sqrt{\left[\sum_{i=1}^{n}(X_{p,i} - X_{p,avg})^2\right]\left[\sum_{i=1}^{n}(X_{a,i} - X_{a,avg})^2\right]}} \tag{6.3}$$

$$R^2 = 1 - \frac{\sum_{i=1}^{n}(X_{a,i} - X_{p,i})^2}{\sum_{i=1}^{n}(X_{p,i} - X_{a,avg})^2} \tag{6.4}$$

$$MSE = \frac{1}{n}\sum_{i=1}^{n}(X_{p,i} - X_{a,i})^2 \tag{6.5}$$

$$RMSE = \sqrt{\frac{1}{n}\sum_{i=1}^{n}(X_{p,i} - X_{a,i})^2} \tag{6.6}$$

The network models are mainly classified into two categories based on their learning supervised/unsupervised, based on their stricture feed-forward or feedback [12]. In supervised learning, both input and output data are used for training the network, whereas in unsupervised learning, only input data are utilized for training. Similarly, for the feedforward network model, use the unidirectional information where the information is permitted to transfer from inputs to outputs only. On the other hand, it is bidirectional information flow, where a feedback network is used. The input data for the network can be obtained from any of the optimization algorithms, such as response surface optimization (RSM), Taguchi, etc [13, 14]. The accuracy of the network model can be defined based on the regression values (R and R^2) and low mean square error (RSM) values as shown in equations 6.3-6.6.

6.5 Biodiesel

Biodiesels are derived from plant and animal oils of the edible and nonedible category, and they are popularly known for low or no sulfur fuels [15]. The important fuel properties of biodiesel are in line with conventional diesel fuel, and hence, they are widely used as diesel fuel replacement in internal combustion (IC) engines. Like the industrial revolution (from 1.0 to 4.0), the transformation of biofuel has happened into four generations. The first-generation biodiesels are obtained from food crop oils, like peanut, soybean, palm, etc. In fact, the first experimental diesel engine runs with neat peanut oil in the 1900s by Dr. Rudolf Diesel, a German inventor. Due to the high cost of first-generation raw oils and food versus fuel, conflicts pay away to the second-generation biodiesels. The second-generation biodiesels are mostly obtained from nonedible oils, and these are divided into three categories [16]. Homogeneous (oils obtained from white wood chips), quasi-homogeneous (oil obtained from agricultural and forest residues) and nonhomogeneous (municipal solid wastes). With the utilization of nonedible feedstocks for biodiesel production, the cost has been reduced. Waste cooking oils (WCO) are also considered as one of the potential low-cost feedstocks for biodiesel production due to their easy availability. However, the filtering and process of WCO are costly affairs. The third-generation biodiesels are obtained from the algal biomass, which is grown anywhere on the earth and has a high oil yield compared with other generation oils [17]. The fourth generation belongs to genetically modified algae, where the research is under progress. Figure 6.4 represents the selected raw oils for the production of biodiesel like palm, waste cooking oil, and Jatropha oils. From the above discussions, it was evident that the feedstock for the biodiesel production process can be obtained from any of the fourth-generation oils. However, low-cost and nonedible oils are attracted as the most suitable feedstocks for biodiesel production. In general, raw feedstocks contain high free fatty acids (FFA) and triglyceride, due to which, their viscosity is observed very high at atmospheric

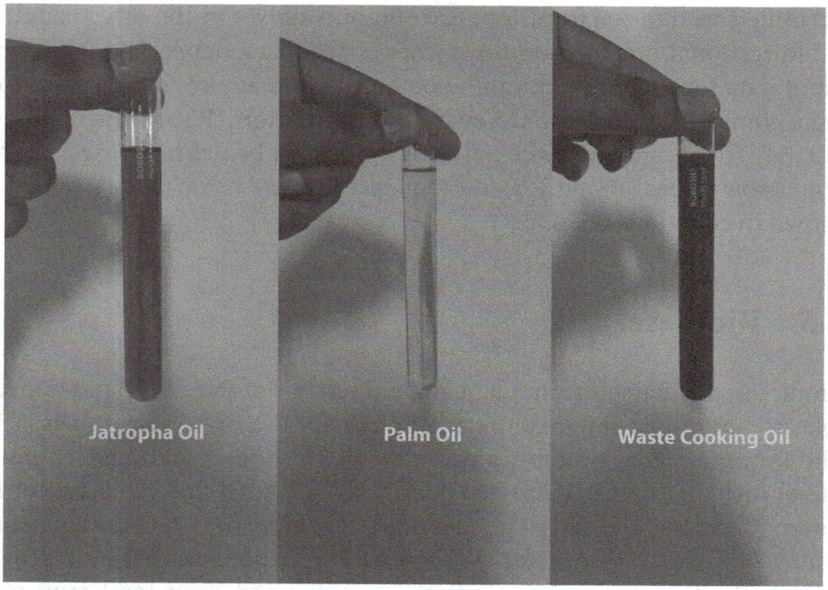

Figure 6.4 Different feedstocks for biodiesel production.

conditions. High viscous fuel leads to incomplete combustion and an increase in harmful exhaust emissions. Hence, the high viscosity in raw oils has to be regulated with the aid of a chemical treatment process.

6.6 Transesterification Process

Kinematic viscosity is considered as one of the important fuel properties for biodiesels, and it is always preferred to use low viscous fuels (< 5 cSt) in diesel engines [18]. The low viscosity of fuel in diesel engines improves the swirl and atomization, thereby achieving the total combustion of fuel inside the combustion chamber. Raw oils possess high viscosity (> 20 cSt) at atmospheric conditions, and hence, they are not recommended to use directly in the diesel engine. High viscosity results in injector clogging with poor atomization of fuel inside the combustion chamber [18]. Therefore, high viscosity in raw oils has to be regulated. Different methods are available to lower the high kinematic viscosity of raw oils, and the transesterification process is widely in use [19]. The transesterification utilizes alcohol (methanol or ethanol) and catalyst (homogeneous or heterogeneous) to convert the high free fatty acids (FFA) present in the oil to low fatty acid methyl or ethyl esters (FAME/FAEE) [20], as shown in Figure 6.5 and the stages of the transesterification process are presented in Figure 6.6.

During the transesterification, process, catalysts play a significant role in converting the triglyceride into diglycerides, monoglycerides, and glycerol. The commonly used catalyst belongs to the homogeneous or heterogeneous groups. Sodium hydroxide (NaOH) and potassium hydroxide (KoH) are commonly used homogeneous catalysts for the biodiesel production process due to their easy availability and high biodiesel yield conversion [20]. However, the catalyst cannot be reused, and the consumption of water is very high during the washing of biodiesel. To mitigate the challenge, heterogeneous catalysts are used in place of homogeneous catalysts [21]. The

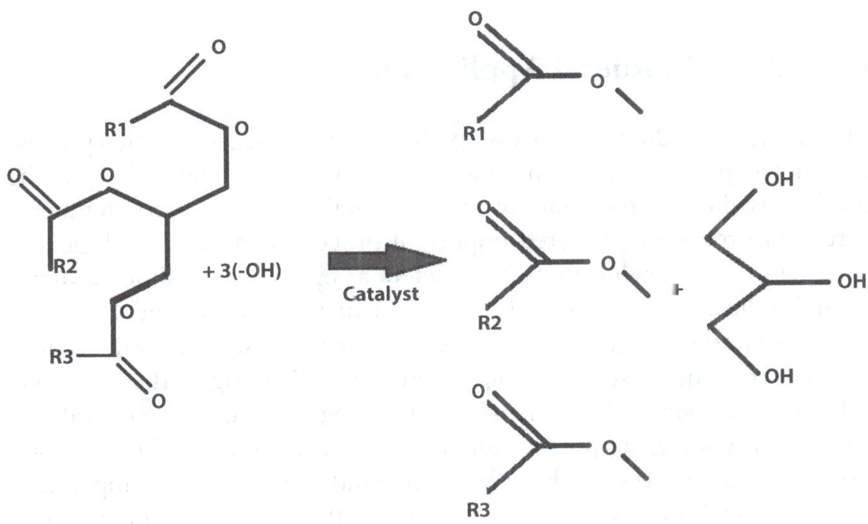

Figure 6.5 Transesterification reaction process.

Figure 6.6 Biodiesel production stages.

heterogeneous catalysts are obtained from natural products by calcination techniques, for instance, waste eggshells of birds, potassium-rich leaves, banana and orange peel, etc. Apart from the catalyst, alcohol and reaction time and temperatures are also play a vital role in biodiesel production. By adjusting the four reaction parameters, the optimum biodiesel yield can be achieved with less waste of time and material. The reaction parameters are observed to be nonlinear and difficult to maintain the optimum combinations to achieve maximum biodiesel yield with traditional statistical tools. Therefore, optimization techniques like Taguchi, response surface methods, screening, regression, etc., are used [22, 23]. In recent times, AI tools dominate the optimization models with relatively high prediction accuracy and low error. Therefore, the role of AI techniques in biodiesel production and its performance evaluation are highlighted in this chapter.

6.7 AI in Biodiesel Applications

The biodiesel production process involves different stages in which process parameters play a vital role in achieving the maximum biodiesel yield. The alcohol used either methanol or ethanol, catalyst type, reaction temperature, and time are considered important process parameters in biodiesel production. By varying these parameters at optimum conditions, maximum biodiesel yield can be achieved. To achieve the exact optimum conditions requires a huge number of experimental trials, which will account for high cost and wastage of material and time. To mitigate the challenge, artificial intelligence (AI) and machine learning (ML) tools like neural networks, fuzzy logic, support vector machines, genetic algorithms, etc. are popular in usage. The available literature studies reveal that compared to the traditional tool like response surface method (RSM) or Taguchi [22, 23], AI tools predict the biodiesel yield more accurately with a high coefficient of correlation values (R^2). Aditya *et al.* [24] used RSM and genetic algorithm (GA) tools to predict the mahua oil biodiesel yield. The experimental results proved that GA modeling improves the biodiesel yield by 4.96% compared to the RSM technique. In another study [25], artificial neural network (ANN) and RSM techniques were used to predict the low-grade waste cooking oil biodiesel yield using waste chicken eggshells as heterogeneous catalysts. The experimental results showed that both RSM and ANN predict the biodiesel yield of 91% with low mean square error (MSE) and a high coefficient of correlation (R^2) were acclaimed for the ANN technique. Similarly, in another study [26, 27], waste and discarded cooking oils of palm, sunflower, groundnut and rice brane are collected in a local restaurant for biodiesel production and 18 experiments were conducted based on RSM design and ANN was used as a modeling tool. The results reported that 1.70% high biodiesel yield was achieved with ANN modeling than RSM optimization. Hariram *et al.* [28] used the Bayesian Regularized Neural Network (BRNN) coupled with GA to optimize the significant reaction parameters during the biodiesel production from calophyllum inophyllum oil. The results showed that the developed BRNN-GA successfully estimated the optimum reaction parameters. Furthermore, the AI techniques are successfully applied in the transesterification of biodiesels from different edible and nonedible oils by mechanically assisted [29], ultrasound-assisted [30], microwave-assisted [31], supercritical alcohol [32], and enzyme-catalyzed [33] transesterification process. The reported results show an improvement in biodiesel yield compared to the traditional techniques. Converting the raw oils into biodiesels involves a complex process; therefore, controlling the various chemical reaction processes in the transesterification process requires high accuracy tools capable of handling the uncertainty, dynamic and complex problems. AI techniques have emerged as an alternative to the conventional methods, and they are successfully utilized in the prediction of heterogeneous combustion in diesel engines. Kolakoti [34] tested three different neat biodiesels in a compression ignition (CI) engine to evaluate the performance, combustion and exhaust emissions characteristics. The best performance with low exhaust emission biodiesel among the three has to identify. Due to the involvement of several nonlinear parameters in CI engine combustion, identifying the best biodiesel has become a difficult task. Hence, ANN modeling was performed by the author. The training, testing, and validation results from the ANN modeling show a good agreement with the experimental results. Similarly, different authors reported that the CI engine performance with diesel and biodiesel blends are successfully predicted by AI tools [35, 36]. From the above discussions, it was evident that biodiesel production and its applications in CI engines have a great potential to combat energy and environmental issues. As biodiesels are known for clean burning and ecofriendly fuels, with their utilization as a diesel fuel replacement, the exhaust emissions from the CI engines can be regulated. Furthermore, biodiesel production is currently restricted to lab-scale due to the selection of raw oils for production and the complexity of the chemical treatment process. To mitigate the challenge, AI tools can be successfully utilized in biodiesel production and its commercialization.

6.8 Conclusion

Biodiesels are considered one of the promising and sustainable alternative fuels for diesel engines to combat the energy and environmental air pollution challenges. Due to the significant benefits of utilizing biodiesel as an alternative fuel for diesel engines, biodiesel research has huge popularity. In this chapter, the role of artificial intelligence (AI) in biodiesel applications is enlightened. Based on the observations from the available literature, it was concluded that AI tools have great potential to solve dynamic, complex, and nonlinear problems. The same is evident from the above discussions that the AI tools are successfully implemented in biodiesel production, its combustion and performance estimations.

References

1. Kayalvizhi, S.B., Thenmalar, K., Raja, T., Gopal, S.B., Nanthiya, D., Sreemanjari, S., A campus energy auditing in energy utilization and conservation using renewable source. *2021 Third International Conference on Intelligent Communication Technologies and Virtual Mobile Networks (ICICV)*, pp. 535–539, 2021.
2. Babalola, J.O. and Akinoso, R., Energy utilization and conservation: An approach towards cashew nut mill, in: *Cutting-Edge Research in Agricultural Sciences*, vol. 12, pp. 6–16, 2021, https://doi.org/10.9734/bpi/cras/v12/10069D.
3. Vohra, K., Vodonos, A., Schwartz, J., Marais, E.A., Sulprizio, M.P., Mickley., L.J., Global mortality from outdoor fine particle pollution generated by fossil fuel combustion: Results from GEOS-Chem. *Environ. Res.*, 195, 110754, 2021. https://doi.org/10.1016/j.envres.2021.110754.
4. Taghizadeh-Hesary, F. and Taghizadeh-Hesary, F., The impacts of air pollution on health and economy in Southeast Asia. *Energies*, 13, 1812, 2020. https://doi.org/10.3390/en13071812.
5. Neofytou, H., Nikas, A., Doukas, H., Sustainable energy transition readiness: A multicriteria assessment index. *Renew. Sust. Energ. Rev.*, 131, 109988, 2020. https://doi.org/10.1016/j.rser.2020.109988.
6. Verma, P. and Sharma, M.P., Review of process parameters for biodiesel production from different feedstocks. *Renew. Sust. Energ. Rev.*, 62, 1063–1071, 2016.
7. Setiyo, M., Purnomo, T.A., Yuvenda, D., Biddinika, M.K., Sidik, N.A.C., Samuel, O.D., Kolakoti, A., Calam, A., Industry 4.0: Challenges of mechanical engineering for society and industry. *Mechanical Eng. Soc. Industry*, 1, 1, 3–6, 2021.

8. Becue, A., Praca, I., Gama, J., Artificial intelligence, cyber-threats and industry 4.0: Challenges and opportunities. *Artif. Intell. Rev.*, 54, 3849–3886, 2021. https://doi.org/10.1007/s10462-020-09942-2.
9. Zhao, S., Blaabjerg, F., Wang, H., An overview of artificial intelligence applications for power electronics. *IEEE Trans. Power Electron.*, 36, 4, 4633–4658, 2021.
10. Marinakis, V., Koutsellis, T., Nikas, A., Doukas, H., AI and data democratisation for intelligent energy management. *Energies*, 14, 4341, 2021. https://doi.org/10.3390/en14144341.
11. Kumari, A., Gupta, R., Tanwar, S., Kumar, N., Blockchain and AI amalgamation for energy cloud management: Challenges, solutions, and future directions. *J. Parallel Distrib. Comput.*, 143, 148–166, 2020. https://doi.org/10.1016/j.jpdc.2020.05.004.
12. Baba, A., Advanced AI-based techniques to predict daily energy consumption: A case study. *Expert Syst. Appl.*, 184, 115508, 2021. https://doi.org/10.1016/j.eswa.2021.115508.
13. Manojkumar, N., Muthukumaran, C., Sharmila, G., A comprehensive review on the application of response surface methodology for optimization of biodiesel production using different oil sources. *J. King Saud Univ.-Eng. Sci.*, 34, 198–208, 2020. https://doi.org/10.1016/j.jksues.2020.09.012.
14. Naveenkumar, R. and Baskar, G., Optimization and techno-economic analysis of biodiesel production from Calophyllum inophyllum oil using heterogeneous nanocatalyst. *Bioresour. Technol.*, 315, 123852, 2020. https://doi.org/10.1016/j.biortech.2020.123852.
15. Kolakoti, A. and Rao, B.V.A., A comprehensive review of biodiesel application in IDI engines with property improving additives. *I-Manager's J. Mech. Eng.*, 5, 4, 35–45, 2015.
16. Nabi, M.N. and Rasul., M.G., Influence of second generation biodiesel on engine performance, emissions, energy and exergy parameters. *Energy Convers. Manage.*, 169, 326–333, 2018. https://doi.org/10.1016/j.enconman.2018.05.066.
17. Sakthivel, R., Ramesh, K., Purnachandran, R., Shameer, P.M., A review on the properties, performance and emission aspects of the third generation biodiesels. *Renew. Sust. Energ. Rev.*, 82, 2970–2992, 2018. https://doi.org/10.1016/j.rser.2017.10.037.
18. Kolakoti, A. and Rao, B.V.A., Relative testing of neat jatropha methyl ester by preheating to viscosity saturation in IDI engine-an optimisation approach. *Int. J. Automot. Mech. Eng.*, 17, 2, 8052–8066, 2020. https://doi.org/10.15282/ijame.17.2.2020.23.0604.
19. Babji, P.S., Rao, B.V.A., Kolakoti, A., Optimization of process variables for the production of methyl esters from waste cooking oil. *Int. J. Sci. Res.*, 6, 10 1201–1204, 2017.
20. Kolakoti, A. and Rao, B.V.A., Effect of fatty acid composition on the performance and emission characteristics of an IDI supercharged engine using

neat palm biodiesel and coconut biodiesel as an additive. *Biofuels*, 10, 5, 591–605, 2019.

21. Bharti, M.K., Chalia, S., Thakur, P., Sridhara, S.N., Thakur, A., Sharma, P.B., Nanoferrites heterogeneous catalysts for biodiesel production from soybean and canola oil: A review. *Environ. Chem. Lett.*, 19, 3727–3746, 2021. https://doi.org/10.1007/s10311-021-01247-2.
22. Rajesh, Y., Kolakoti, A., Sheakar, B.G.C., Bhargavi, J., Optimization of biodiesel production from waste frying palm oil using definitive screening design. *Int. J. Eng. Sci. Technol.*, 11, 2, 48–57, 2019.
23. Satya, S., Kolakoti, A., Naga Raju, B., Shyam Sundar, R., Rao, R., Optimization of palm methyl ester and its effect on fatty acid compositions and cetane number. *Math. Models Eng.*, 5, 1, 25–34, 2019. https://doi.org/10.21595/mme.2019.20469.
24. Kolakoti, A., Jha, P., Mosa, P.R., Mahapatro, M., Kotaru, T.G., Optimization and modelling of mahua oil biodiesel using RSM and genetic algorithm techniques. *Math. Models Eng.*, 6, 2, 134–146, 2020. https://doi.org/10.21595/mme.2020.21357.
25. Kolakoti, A. and Satish, G., Biodiesel production from low-grade oil using heterogeneous catalyst: An optimisation and ANN modelling. *Aust. J. Mech. Eng.*, 2020. https://doi.org/10.1080/14484846.2020.1842298.
26. Kolakoti, A., Optimization of biodiesel production from waste cooking sunflower oil by taguchi and ANN techniques. *J. Therm. Eng.*, 6, 5, 712–723, 2020.
27. Kolakoti, A., Setiyo, M., Waluyo, B., Biodiesel production from waste cooking oil: Characterization, modeling and optimization. *Mechanical Eng. Soc. Industry*, 1, 1, 22–30, 2021.
28. Hariram, V., Seralathan, S., Micha Premkumar, T., Performance assessment of artificial neural network on the prediction of calophyllum inophyllum biodiesel through two-stage transesterification. *Energ. Source. Part A*, 43, 1060–1072, 2019.
29. Saeidi, S., Jouybanpour, P., Mirvakilli, A., Iranshahi, D., Klemes, J.J., A comparative study between modified data envelopment analysis and response surface methodology for optimisation of heterogeneous biodiesel production from waste cooking palm oil. *J. Clean. Prod.*, 136, 23–30, 2016.
30. Aghbashlo, M., Tabatabaei, M., Amid, S., Hosseinzadeh-Bandbafha, H., Khoshnevisan, B., Kianian, G., Life cycle assessment analysis of an ultrasound-assisted system converting waste cooking oil into biodiesel. *Renew. Energy*, 151, 352–1364, 2020.
31. Ma, L., Lv, E., Du, L., Lu, J., Ding, J., Statistical modeling/optimization and process intensification of microwave-assisted acidified oil esterification. *Energy Convers. Manage.*, 122, 411–418, 2016. https://doi.org/10.1016/j.enconman.2016.06.001.

32. Farobie, O., Hasanah, N., Matsumura, Y., Artificial neural network modeling to predict biodiesel production in supercritical methanol and ethanol using spiral reactor. *Procedia Environ. Sci.*, 28, 214–23, 2015.
33. Bajaj, A., Lohan, P., Jha, P.N., Mehrotra, R., Biodiesel production through lipase catalyzed transesterification: An overview. *J. Mol. Catal. B Enzym.*, 62, 1, 9–14, 2010. https://doi.org/10.1016/j.molcatb.2009.09.018.
34. Kolakoti, A., An experimental based artificial neural network modeling in prediction of optimum combustion, performance, and emission from diesel engine operated with three biodiesels. *World J. Eng.*, 18, 5, 805–814, 2021. https://doi.org/10.1108/WJE-01-2021-0010.
35. Dey, S., Reang, N.M., Majumder, A., Deb, M., Das, P.K., A hybrid ANN-fuzzy approach for optimization of engine operating parameters of a CI engine fueled with diesel-palm biodiesel-ethanol blend. *Energy*, 202, 1, 117813, 2020.
36. Joshi, M.P. and Thipse, S.S., Combustion analysis of a compression-ignition engine fuelled with an algae biofuel blend and diethyl ether as an additive by using an artificial neural network. *Biofuels*, 12, 429–438, 2019.

7

On New Joint Importance Measures for Multistate Reliability Systems

Chacko V. M.

St. Thomas College (Autonomous), Thrissur, Kerala, India

Abstract
In system reliability engineering, the use of importance and joint importance measures to identify the weak areas of a system and signify the roles of components in either causing or contributing to proper functioning of the system is addressed by several researchers. But a few papers are available in the literature for finding joint importance measures for more than two components. This chapter introduces new Joint Reliability Achievement Worth (JRAW), Joint Reliability Reduction Worth (JRRW), and Joint Reliability Fussell-Vesely measure (JRFV) for three multistate components of a multistate system. This is a new approach to find out the joint effect of group of components in improving system reliability. The differencing technique is used in the proposed measures. A steady-state performance level distribution with restriction to the component's states is used to evaluate the proposed measures. Universal generating function (UGF) technique is applied for the evaluation of proposed joint importance measures. An illustrative example is provided.

Keywords: Multistate system, reliability, joint importance measure, universal generating function

7.1 Introduction

The role of group of components in improving system reliability is inevitable in system engineering with regard to the effect of interaction of the group of components on performance of system.

Email: chackovm@gmail.com

S. C. Malik, Deepak Sinwar, Ashish Kumar, S. R. Gadde, Prasenjit Chatterjee and Bui Thanh Hung (eds.) Computational Intelligence in Sustainable Reliability Engineering, (145–158) © 2023 Scrivener Publishing LLC

Various joint importance measures provide useful information to understand the system and apply reliability improvement activities on systems [1–3]. Interaction importance of groups of components, with respect to output performance measure (OPM)s is more useful to the designers, engineers and managers in design perspective [4]. The importance measures will be helpful in discriminating the most important group of components with respect to the overall system reliability [10]. Group of system components can be ranked according to the influence on system reliability based on a given joint importance measure.

Apart from binary state systems, some systems have more complicated behavior. Some components of the system may be operating in a degraded state. It may be the reason for the system to provide service at less than full capacity, the system may still be providing an acceptable level of service, or perhaps, a partial level of service. Multistate reliability models have been available to describe such systems [6]. Component importance and joint importance measures are available in literature to address the problems of identification of most important component or group of components [7]. The joint importance measures of two components for MSS with respect to reliability and expected output performance are discussed in literature in the Birnbaum sense [5]. Joint Reliability Achievement Worth (JRAW), Joint Reliability Reduction Worth (JRRW), and Joint Reliability Fussel-Vesely (JRFV) importance measures by considering groups with two components are useful while considering reliability as performance measure [7]. Measuring the role of interaction of components in a group consisting of two components, in performance measure achievement, reduction, and fractional contribution sense, has to be investigated while another component is fully available or changes from available states to unavailable states. If there is any change in JRAW, JRRW, and JRFV of two components while performance of another component changes from high level to low level, one can say some interaction effect exist between three components. This interaction effect can be compared, to prioritize reliability improvement activities. In this paper, for three components of binary and MSS, change in JRAW measures the change in reliability achievement worth of two components when another components changes from higher level to lower level, JRRW measures the change in reliability reduction worth of two components of the system when other component changes from higher level to lower level and JRFV measures the change in fractional contribution of interaction effect of two components of the system when another component changes from higher level to lower level. These measures are generalized to the expected output performance.

A steady-state performance level distribution for the system is considered [6]. The information derived by these joint importance measures will be helpful to assess the interaction effect of three components for system OPM improvement.

The components *i*, *j* and *q* are restricted with respect to performance thresholds α, β, and δ respectively. Let $OPM_{i,j}^{\le\alpha,\le\beta}$, $OPM_{i,j}^{>\alpha,\le\beta}$, $OPM_{i,j}^{\le\alpha,>\beta}$, $OPM_{i,j}^{>\alpha,>\beta}$, $OPM_{i,j,q}^{>\alpha,>\beta,>\delta}$, $OPM_{i,j,q}^{>\alpha,>\beta,<\delta}$, $OPM_{i,j,q}^{>\alpha,<\beta,>\delta}$, $OPM_{i,j,q}^{>\alpha,<\beta,<\delta}$, $OPM_{i,j,q}^{<\alpha,>\beta,>\delta}$, $OPM_{i,j,q}^{<\alpha,>\beta,<\delta}$, $OPM_{i,j,q}^{<\alpha,<\beta,>\delta}$, and $OPM_{i,j,q}^{<\alpha,<\beta,<\delta}$ are state space restricted OPMs, in which the restrictions to the performance of the components are indicated in power. UGF method can be used to evaluate system performance measures with restrictions.

The chapter is arranged as follows. The performance measures of the MSS are reviewed and new joint importance measures of three components of the binary and MSS are introduced in section 7.2. Discussion is given in section 7.3. Illustrative example is given in section 7.4. Conclusion is given in section 7.5.

7.2 New Joint Importance Measures

Reliability, availability, expected system performance, risk, unavailability, etc. are commonly used performance measures in system engineering. The performance measures used for the present study are Reliability and expected output performance measure. Joint importance measures from a differential perspective is addressed in this chapter. This approach is useful for the identification of group of components, which influence the system performance through their interaction.

1. Performance Measures of a Multistate system

Most of the engineering systems are either binary state system or multi-state system. Binary state system can be considered as special case of multistate system. Let the structure function of a MSS at time *t* be denoted by $\varphi(X(t)) = i, i \in \{0, 1, 2, \ldots, M\}$, where $X(t) = (X_1(t), X_2(t), \ldots, X_n(t))$, $X_i(t) \in \{0, 1, 2, \ldots, M_i\}$, and $M = \max_{1\le i\le n}\{M_i\}$. The system state corresponds to the output performance measure.

Let the output performance of the MSS at time *t*, $W(t)$ with $p_i = \lim_{t\to\infty} \Pr\{W(t) = w_i\} = \lim_{t\to\infty} \Pr\{\varphi(X(t)) = i\}$, where $W(t) = w_i$, is equivalent to $\varphi(X(t)) = i, 0 \le i \le M$.

The steady-state performance distribution of the output performance of system, W = {w_i, $0 \le i \le M$} is represented by P = {p_i, $0 \le i \le M$}. Steady state expected performance is

$$E(W) = \sum_{i=0}^{M} p_i w_i \tag{7.1}$$

and expected system state is

$$E_s(\varphi(X)) = \sum_{i=0}^{M} i p_i. \tag{7.2}$$

For constant demand D_k, to state k of the multistate system, reliability is

$$R\,(t) = Pr\,\{W\,(t) \ge D_k\} = Pr\,\{\varphi\,(t) \ge k\}. \tag{7.3}$$

The stationary reliability is

$$R(D_k) = \sum_{i=0}^{M} p_i 1(w_i - D_k) \tag{7.4}$$

These performance measures are commonly used for reliability importance analysis [6].

2. New Joint Importance Measures for two components in the MSS

Suppose now the components are statistically independent and reliabilities are known. The measures to analyze the interaction effect of three components in reliability achievement, reliability reduction and fractional contribution to reliability improvement are proposed.

7.2.1 Multistate Differential Joint Reliability Achievement Worth (MDJRAW)

Joint reliability achievement worth (JRAW) of two components is used to measure the improvement in system reliability with reference to the interaction [7]. The groups having highest joint reliability achievement worth needs more care. In order to assess the change in JRAW of two components by the presence or functioning or switching to functioning states of another component, the Differential Joint Reliability Achievement Worth has to be measured.

The role of interaction of components in a group consisting of two or three components, in increasing reliability of a system, defines the following:

Let

$R_{\{i+,j+\}} = P(\varphi(X(t)) = 1, X_i(t) = 1, X_j(t) = 1)$, $R_{\{i-,j+\}} = P(\varphi(X(t)) = 1, X_i(t) = 0, X_j(t) = 1)$,

$R_{\{i+,j-\}} = P(\varphi(X(t)) = 1, X_i(t) = 1, X_j(t) = 1)$, $R_{\{i-,j-\}} = P(\varphi(X(t)) = 1, X_i(t) = 0, X_j(t) = 0)$.

$R_{\{i+,j+,q+\}} = P(\varphi(X(t)) = 1, X_i(t) = 1, X_j(t) = 1, X_q(t) = 1)$,
$R_{\{i-,j+,q+\}} = P(\varphi(X(t)) = 1, X_i(t) = 0, X_j(t) = 1, X_q(t) = 1)$,
$R_{\{i+,j-,q+\}} = P(\varphi(X(t)) = 1, X_i(t) = 1, X_j(t) = 0, X_q(t) = 1)$,
$R_{\{i-,j-,q+\}} = P(\varphi(X(t)) = 1, X_i(t) = 0, X_j(t) = 0, X_q(t) = 1)$.
$R_{\{i+,j+,q-\}} = P(\varphi(X(t)) = 1, X_i(t) = 1, X_j(t) = 1, X_q(t) = 0)$,
$R_{\{i-,j+,q-\}} = P(\varphi(X(t)) = 1, X_i(t) = 0, X_j(t) = 1, X_q(t) = 0)$,
$R_{\{i+,j-,q-\}} = P(\varphi(X(t)) = 1, X_i(t) = 1, X_j(t) = 0, X_q(t) = 0)$, and
$R_{\{i-,j-,q-\}} = P(\varphi(X(t)) = 1, X_i(t) = 0, X_j(t) = 0, X_q(t) = 0)$.

Let

$$JRAW = \frac{\text{Maximum Reliability due to high level interaction effect of two components}}{\text{The present reliability level.}}$$

$JRAW_{i,j} = \frac{[R_{\{i+,j+\}} - R_{\{i-,j+\}}]}{R}, i, j = 1, 2, \ldots, n$ which quantifies the maximum possible achievement of reliability due to interaction effect of component i and j, [7]. Now let us define JRAW when another component, say q, is restricted to above and below some prespecified performance level δ.

Let $JRAW_{i,j,q+} = \frac{[R_{\{i+,j+,q+\}} - R_{\{i-,j+,q+\}}]}{R}$ and $JRAW_{i,j,q-} = \frac{[R_{\{i+,j+,q-\}} - R_{\{i-,j+,q-\}}]}{R}, i, j, q = 1, 2, \ldots, n$. The new differential joint reliability achievement worth (DJRAW) of three components can be defined as

$$DJRAW = JRAW_{i,j,q+} - JRAW_{i,j,q-}, \; i, j, q = 1, 2, \ldots, n$$

For a constant demand D_k, to define Multistate Differential Joint Reliability Achievement Worth (MDJRAW) of components i, j and q, let,

$$R_{\{i^{\geq\alpha},j^{\geq\beta}\}} = P\big(\varphi(X(t)) \geq k, X_i(t) \geq x_{ik_{i\alpha}}, X_j(t) \geq x_{jk_{j\beta}}\big),$$
$$R_{\{i^{<\alpha},j^{\geq\beta}\}} = P\big(\varphi(X(t)) \geq k,\ X_i(t) < x_{ik_{i\alpha}}, X_j(t) \geq x_{jk_{j\beta}}\big),$$
$$R_{\{i^{\geq\alpha},j^{<\beta}\}} = P\big(\varphi(X(t)) \geq k, X_i(t) \geq x_{ik_{i\alpha}}, X_j(t) < x_{jk_{j\beta}}\big), \text{and}$$
$$R_{\{i^{<\alpha},j^{<\beta}\}} = P\big(\varphi(X(t)) \geq k, X_i(t) < x_{ik_{i\alpha}}, X_j(t) < x_{jk_{j\beta}}\big)$$
$$R_{\{i^{\geq\alpha},j^{\geq\beta},q^{\geq\delta}\}} = P\big(\varphi(X(t)) \geq k, X_i(t) \geq x_{ik_{i\alpha}}, X_j(t) \geq x_{jk_{j\beta}}, X_q(t) \geq x_{qk_{q\delta}}\big),$$
$$R_{\{i^{<\alpha},j^{\geq\beta},q^{\geq\delta}\}} = P\big(\varphi(X(t)) \geq k, X_i(t) < x_{ik_{i\alpha}}, X_j(t) \geq x_{jk_{j\beta}}, X_q(t) \geq x_{qk_{q\delta}}\big),$$
$$R_{\{i^{\geq\alpha},j^{<\beta},q^{\geq\delta}\}} = P\big(\varphi(X(t)) \geq k, X_i(t) \geq x_{ik_{i\alpha}}, X_j(t) < x_{jk_{j\beta}}, X_q(t) \geq x_{qk_{q\delta}}\big),$$
$$R_{\{i^{\geq\alpha},j^{\geq\beta},q^{<\delta}\}} = P\big(\varphi(X(t)) \geq k, X_i(t) \geq x_{ik_{i\alpha}}, X_j(t) \geq x_{jk_{j\beta}}, X_q(t) < x_{qk_{q\delta}}\big),$$
$$R_{\{i^{<\alpha},j^{\geq\beta},q^{<\delta}\}} = P\big(\varphi(X(t)) \geq k, X_i(t) < x_{ik_{i\alpha}}, X_j(t) \geq x_{jk_{j\beta}}, X_q(t) < x_{qk_{q\delta}}\big),$$
$$R_{\{i^{\geq\alpha},j^{<\beta},q^{<\delta}\}} = P\big(\varphi(X(t)) \geq k, X_i(t) \geq x_{ik_{i\alpha}}, X_j(t) < x_{jk_{j\beta}}, X_q(t) < x_{qk_{q\delta}}\big),$$
$$R_{\{i^{<\alpha},j^{<\beta},q^{\geq\delta}\}} = P\big(\varphi(X(t)) \geq k, X_i(t) < x_{ik_{i\alpha}}, X_j(t) < x_{jk_{j\beta}}, X_q(t) \geq x_{qk_{q\delta}}\big)$$
$$\text{and } R_{\{i^{<\alpha},j^{<\beta},q^{<\delta}\}} = P\big(\varphi(X(t)) \geq k, X_i(t) < x_{ik_{i\alpha}}, X_j(t) < x_{jk_{j\beta}}, X_q(t) < x_{qk_{q\delta}}\big),$$

where α s the performance threshold corresponding to performance $x_{ik_{i\alpha}}$ of component i in state $k_{i\alpha}$, β is the performance threshold corresponding to performance $x_{jk_{j\beta}}$ of component j in the state $k_{j\beta}$, δ is the performance threshold corresponding to performance $x_{qk_{q\delta}}$ of component q in the state $k_{q\delta}$, $i, j, q=1,2,\ldots,n$.

MJRAW of two components i and j is, $MJRAW_{i,j} = \dfrac{\left[R_{\{i^{\geq\alpha},j^{\geq\beta}\}} - R_{\{i^{<\alpha},j^{\geq\beta}\}}\right]}{R}$, the reliability achievement worth of interaction effect of two components.

Let $MJRAW_{i,j,q+} = \dfrac{\left[R_{\{i+,j+,q+\}} - R_{\{i-,j+,q+\}}\right]}{R}$ and $MJRAW_{i,j,q-} = \dfrac{\left[R_{\{i+,j+,q-\}} - R_{\{i-,j+,q-\}}\right]}{R}$, $i, j, q = 1, 2, \ldots, n$. The new multistate differential joint reliability achievement worth (MDJRAW) of three components can be defined as $MDJRAW = MJRAW_{i,j,q+} - MJRAW_{i,j,q-}$, $i, j, q = 1, 2, \ldots, n$.

7.2.2 Multistate Differential Joint Reliability Reduction Worth (MDJRRW)

Joint Reliability Reduction Worth (JRRW) is used to measure the role of interaction effect of two components in reducing the present reliability [7].

To examine change in JRRW of two components while another component changes from high level to low level, differential joint reliability reduction worth has to be introduced.

Let R_G^- = The decreased reliability level by the low level interaction of two omponents and R_0 = Present reliability level.

The JRRW of two components is defined as:

$$JRRW = \frac{R_0}{R_G^-}$$

JRRW of two binary components i and j is

$$\text{JRRW} = \frac{\text{Present Reliability Level}}{\text{Reliability when interaction of two components is at low level}}$$

$\text{JRRW}_{i,j} = \frac{R}{\left[R_{\{i+,j-\}} - R_{\{i-,j-\}}\right]}, i, j = 1,2,\ldots,n, which$ quantifies the maximum possible reduction of reliability due to low level of interaction effect of component i and j.

Let $\text{JRRW}_{i,j,q+} = \frac{R}{\left[R_{\{i+,j-,q+\}} - R_{\{i-,j-,q+\}}\right]}$ and $\text{JRRW}_{i,j,q-} = \frac{R}{\left[R_{\{i+,j-,q-\}} - R_{\{i-,j-,q-\}}\right]}, i, j, q = 1,2,\ldots,n$

The new differential joint reliability reduction worth (DJRRW) of three components can be defined as $DJRRW = JRRW_{i,j,q+} - JRRW_{i,j,q-}$, $i, j, q = 1, 2, \ldots, n$.

For a constant demand D_k, Multistate Joint Reliability Reduction Worth (MJRRW) of a module consisting of two components i and j is,

$$MJRRW_{i,j} = \frac{R}{\left[R_{\{i^{\geq\alpha}, j^{<\beta}\}} - R_{\{i^{<\alpha}, j^{<\beta}\}}\right]}, i, j = 1,2,\ldots,n.$$

Let $MJRRW_{i,j,q+} = \frac{R}{\left[R_{\{i^{\geq\alpha}, j^{<\beta}, q^{\geq\delta}\}} - R_{\{i^{<\alpha}, j^{<\beta}, q^{\geq\delta}\}}\right]}$ and $MJRRW_{i,j,q-} = \frac{R}{\left[R_{\{i^{\geq\alpha}, j^{<\beta}, q^{<\delta}\}} - R_{\{i^{<\alpha}, j^{<\beta}, q^{<\delta}\}}\right]}, i, j, q = 1,2,\ldots,n.$

MJRRW measures the reliability reduction worth of interaction effect of two components i and j. The new multistate differential joint reliability reduction worth (MDJRRW) of three components can be defined as $MDJRRW = MJRRW_{i,j,k+} - MJRRW_{i,j,k-}$, $i, j, q = 1, 2, \ldots, n$

7.2.3 Multistate Differential Joint Reliability Fussel-Vesely (MDJRFV) Measure

Joint Reliability Fussel-Vesly (JRFV) measure the fractional contribution of interaction effect of two components to the increase of reliability, [7]. JRFV measure can be expressed as, $JRFV = \frac{R_0 - R_{\bar{G}}}{R_0}$.

$$JRFV = \frac{\text{Present Reliability Level} - \text{Reliability when interaction oftwo componentsis in low level}}{\text{Present Reliability Level}}$$

$JRFV_{i,j} = \frac{R - \left[R_{\{i+,j-,\}} - R_{\{i-,j-\}}\right]}{R}, i,j=1,2,\ldots,n$ which quantifies the maximum fractional contribution of reliability due to high level of interaction effect of component *i* and *j*.

Let $JRFV_{i,j.q+} = \frac{R - \left[R_{\{i+,j-,q+\}} - R_{\{i-,j-,q+\}}\right]}{R}$ and $JRFV_{i,j.q-} = \frac{R - \left[R_{\{i+,j-,q-\}} - R_{\{i-,j-,q-\}}\right]}{R}, i,j,q=1,2,\ldots,n.$

The new multistate differential joint reliability reduction worth (MDJRRW) of three components can be defined as $DJRFV = JRFV_{i,\,j,\,q+} - JRFV_{i,\,j,\,q-}$, $i, j, q = 1, 2, \ldots, n$.

For a constant demand D_k, Multistate Joint Fussel-Vesly (MJRFV) of two components *i* and *j* is defined as,

$$MJRFV_{i,j} = \frac{R - \left[R_{\{i^{\geq\alpha},j^{<\beta}\}} - R_{\{i^{<\alpha},j^{<\beta}\}}\right]}{R},$$

measures the reliability FV of a module consisting of two components.

Let $MJRFV_{i,j,q+} = \frac{R - \left[R_{\{i^{\geq\alpha},j^{<\beta},q^{\geq\delta}\}} - R_{\{i^{<\alpha},j^{<\beta},q^{\geq\delta}\}}\right]}{R}$ and $MJRFV_{i,j,q-} = \frac{R - \left[R_{\{i^{\geq\alpha},j^{<\beta},q^{<\delta}\}} - R_{\{i^{<\alpha},j^{<\beta},q^{<\delta}\}}\right]}{R}, i,\ j,\ q=1,2,\ldots,n.$

The new multistate differential joint reliability reduction worth (MDJRRW) of three components can be defined as $MDJRFV = MJRFV_{i,\,j,\,q+} - MJRFV_{i,\,j,\,q-}$, $i, j, q = 1, 2, \ldots, n$.

Using expected output performance measure, define Multistate Differential Joint Output Performance Measure Achievement Worth (MDJOPMAW), Multistate Differential Joint Output Performance Measure Reduction Worth (MDJOPMRW) and Multistate Differential Joint Output

Performance Measure Fussel-Vesely (MDJOPFV) measures as below. For three components *i*, *j* and *q*, let

$$MJOPMAW_{i,j,q\geq} = \frac{\left[OPM_{\{i\geq,j\geq,q^{\geq}\}} - OPM_{\{i<,j\geq,q^{\geq}\}}\right]}{OPM}$$

$$MJOPMAW_{i,j,q<} = \frac{\left[OPM_{\{i\geq,j\geq,q<\}} - OPM_{\{i<,j\geq,q<\}}\right]}{OPM}$$

$$MJOPMRW_{i,j,q\geq} = \frac{OPM}{\left[OPM_{\{i^{\geq\alpha},j^{<\beta},q^{\geq}\}} - OPM_{\{i^{<\alpha},j^{<\beta},q^{\geq}\}}\right]}$$

$$MJOPMRW_{i,j,q<} = \frac{OPM}{\left[OPM_{\{i^{\geq\alpha},j^{<\beta},q^{<}\}} - OPM_{\{i^{<\alpha},j^{<\beta},q^{<}\}}\right]}$$

$$MJOPMFV_{i,j,q\geq} = \frac{OPM - \left[OPM_{\{i^{\geq\alpha},j^{<\beta},q^{\geq}\}} - OPM_{\{i^{<\alpha},j^{<\beta},q^{\geq}\}}\right]}{OPM}, \text{ and}$$

$$MJOPMFV_{i,j,q^{<}} = \frac{OPM - \left[OPM_{\{i^{\geq\alpha},j^{<\beta},q^{<}\}} - OPM_{\{i^{<\alpha},j^{<\beta},q^{<}\}}\right]}{OPM} i,j,q = 1,2,\ldots,n.$$

$$\text{MDJOPMAW} = \text{MJOPMAW}_{i,j,q\geq} - \text{MJOPMAW}_{i,j,q<}$$

$$\text{MDJOPMRW} = \text{MJOPMRW}_{i,j,q\geq} - \text{MJOPMRW}_{i,j,q<}$$

$$\text{MDJOPMFV} = \text{MJOPMRFV}_{i,j,q\geq} - \text{MJOPMRFV}_{i,j,q<}$$

Performance restriction approach is useful for the computation of the joint importance measures and evaluation of proposed measures can be carried out using UGF [6–9].

7.3 Discussion

In binary and multistate context, the proposed measures quantify the change in JRAW, JRRW, and JRFV measures of two components which measure the interaction effect of three components, in a differential sense. For three components, DMJRW measures the change in reliability achievement due to interaction effect of two components while performance of another component changes from higher level to lower level, DMJRRW measures the change in reliability reduction due to interaction effect of two components while performance of another component changes from higher level to lower level and DMJRFV measures the change in fractional contribution of interaction effect of two components, while performance of another component changes from higher level to lower level. Using the information of DMJRAW, it is easy to understand and identify the group of three components with highest contribution to system reliability improvement. DMJRRW measure the lowest reduction in system reliability with lower level of performance of group of three components. The change in fractional contribution in reliability improvement of a group of components can be measured using DMJRFV. While considering expected output performance measure for the evaluation of joint importance measures, DMJOPMAW, DMJOPMRW, and DMJOMPFV measures are more plausible.

7.4 Illustrative Example

In this chapter, a system made up of n = 4 multistate components in series logic, is considered (see Figure 7.1). Each component has M_j = 4 nonzero states, 0, 1, 2, 3, and 4. The performance of the components are $x_{j0}=0$, $x_{j1}=50$, $x_{j2}=100$, $x_{j3}=200$, $x_{j4}=500$, j=1, 2, 3,4.

The probability distribution of each component is given in Table 7.1. p_{jk} represent the probability of component j of being in state k. Let the states 0, 1, and 2 represent the unreliable states corresponding to performances $<\alpha$ or $<\beta$ or $<\delta$ and states 3 and 4 represent reliable states corresponding to performance $\geq\alpha$ *or* $\geq\beta$ *or* $\geq\delta$, of the components. The multistate expected output performance measure with restriction to reliable or unreliable states are obtained, for group of components 1, 2, and 3 and 2, 3, and 4, and given in Table 7.2 and Table 7.3, respectively.

Multistate joint importance measures of components 1, 2, and 3 are given in Table 7.2 and plotted in Figure 7.2.

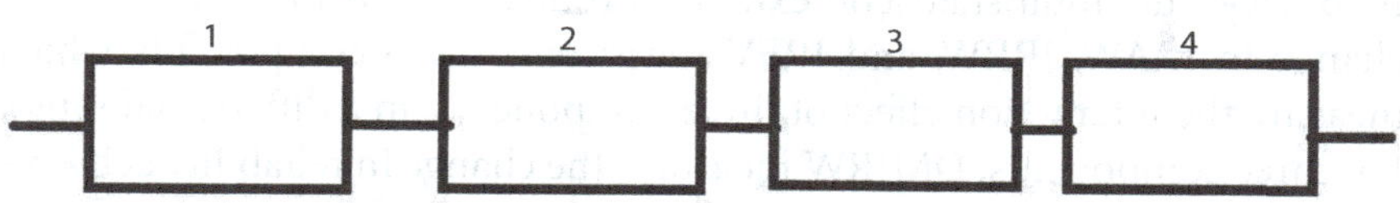

Figure 7.1 Series system.

Table 7.1 Probability distributions of components 1, 2, 3, and 4.

	Component Number							
Probability distribution	**1**		**2**		**3**		**4**	
P(Xi0=0)	p10	=0.5	p20	=0.45	p30	=0.4	p40	=0.45
P(Xi1=25)	p11	=0	p21	=0	p31	=0.1	p41	=0
P(Xi2=50)	p12	=0	p22	=0.1	p32	=0	p42	=0.1
P(Xi3=75)	p13	=0	p23	=0	p33	=0.1	p43	=0
P(Xi4=100)	p14	=0.5	p24	=0.45	p34	=0.4	p44	=0.45

Table 7.2 Multistate joint importance measures (1, 2, 3).

MDJOPMARW	−7.60907
MDJOPMAW	0.039628
MDJOPMFV	−0.03092

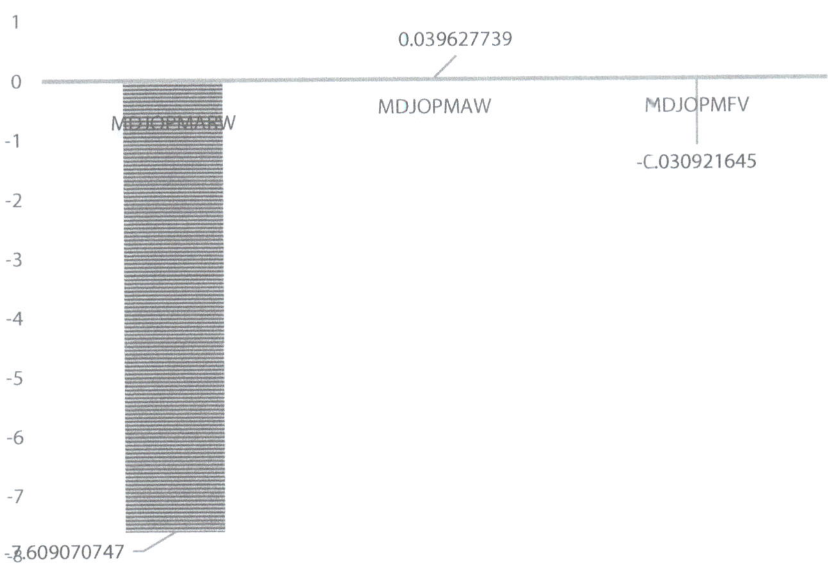

Figure 7.2 Multistate joint importance measures (1, 2, 3).

Multistate joint importance measures of components 2, 3, and 4 are given in Table 7.3 and plotted in Figure 7.3.

Multistate joint importance measures for group 1, 2, and 3 and 2, 4, and 4 are given in Table 7.2 and Table 7.3, and plotted in Figure 7.2 and Figure 7.3, respectively. The sign and size of the value of joint importance measure give more information about the impact of interaction effect of group of components on expected system output performance. The positive value indicates the increase in JRAW, JRRW and JFV of group of two components with reference to the switching of third component from lower level to higher level. So, the most influencing group and components can be identified with the proposed measures.

Consider two groups, group 1 with components 1 and 2 and group 2 with components 2 and 3. The variation in MDJOPMAW, MDJOPMRW and MDJOPMFV is measured for group of three components 1, 2, and 3

Table 7.3 Multistate joint importance measures (2, 3, 4).

MDJOPMARW	0.204709
MDJOPMAW	−2.06978
MDJOPMFV	0.09574

Figure 7.3 Multistate joint importance measures (1, 2, 3).

and 2, 3, and 4. Highest values of joint importance measures give insight regarding the need of reliability improvement activities. The variation in system reliability with reference to the interaction effect of three components can be assessed by the proposed measures in some differential sense.

7.5 Conclusion

With reference to the OPMs reliability and expected system output performance, three new joint importance measures are proposed. The joint importance measures MDJRAW, MDJRRW, and MDJRFV for three components are introduced while using reliability as output performance measure. Apart from using reliability, expected output performance measure is considered for the evaluation of joint importance measure. The new joint importance measures are useful for giving priority for reliability improvement activities. The UGF method is used to evaluate the proposed joint importance measures. This method is useful when system performance is measured in terms of productivity or capacity. Joint importance measure is helpful for reliability improvement activities. A data analysis is given to illustrate the proposed measures. The sign and size of the importance measure can be used to make a priority to give reliability importance activities like redundancy or repair or maintenance.

References

1. Barlow, R.E. and Proschan, F., *Statistical Theory of Reliability and Life Testing*, Holt, Rinheart & Winston, New York, 1975.
2. Barlow, R.E. and Wu, A., Coherent system with multistate components. *Math. Oper. Res.*, 3, 275–281, 1978.
3. Bueno, V.C., On the importance of components for multistate monotone systems. *Stat. Probab. Lett.*, 7, 51–59, 1989.
4. Chacko, V.M. and Manoharan, M., Joint Importance measures for multistate reliability system. *Opsearch*, 48, 257–278, 2011.
5. Chacko, V.M., New joint importance measures for multistate systems. *IJSRE*, 7, 140–148, 2020.
6. Chacko, V.M., On birnbaum type joint importance measures for multistate reliability systems. *Commun. Stat. Theory Methods*, 2021. doi.10.1080/03610926.2021.1961000.
7. Chacko, V.M., On joint importance measures for multistate systems. *Reliab. Theory Appl.*, 16, 4, 286–293, 2021.

8. Chacko, V.M., On joint importance measures for multistate system's, in: *Reliability, Operations Research: Methods, Techniques, and Advancements*, CRC Press, Taylor & Francis Group, Chapter 9, pp. 147–165, 2022.
9. Ushakov, I., A universal generating function. *Soviet J. Comp. Syst. Sci.*, 24, 37–39, 1986.
10. Wu, S., Joint importance measures of multistate systems. *Comput. Ind. Eng.*, 49, 63–75, 2005.

8

Inferences for Two Inverse Rayleigh Populations Based on Joint Progressively Type-II Censored Data

Kapil Kumar* **and Anita Kumari**

Department of Statistics, Central University of Haryana, Mahendergarh, India

Abstract

The rising demand for reliable products made comparative studies a must for the products manufactured from two different lines. For the comparative studies, researchers use the joint progressive type-II censoring scheme in order to minimize the capital and time involved in the experiment. In this article, we draw some inferences for inverse Rayleigh distribution based on joint progressive type-II censoring. The maximum likelihood estimation and the corresponding asymptotic confidence interval estimation are used as the classical estimation methods. The Bayes estimates are calculated under the squared error loss function (SELF) using Tierney-Kadane's approximation and Metropolis-Hastings algorithm along with the Bayes estimates highest posterior density credible intervals are also constructed. A Markov Chain Monte Carlo simulation study is carried out to compare different estimation methods. A real life problem is discussed for illustrative purposes.

Keywords: Inverse Rayleigh distribution, joint progressive type-II censoring, maximum likelihood estimation, bayesian estimation, metropolis-hastings algorithm, Markov chain Monte Carlo simulation

8.1 Introduction

There is a rise in the demand for more reliable products in the market nowadays. To provide better and more reliable products to the consumer, the producers are conducting comparative studies for the products

**Corresponding author*: kapilstats@gmail.com

S. C. Malik, Deepak Sinwar, Ashish Kumar, S. R. Gadde, Prasenjit Chatterjee and Bui Thanh Hung (eds.) Computational Intelligence in Sustainable Reliability Engineering, (159–180)

coming from different lines of the production units. In view of the time and money involved in the experiment, censoring schemes are used. Type I and type II censoring schemes are known as the most conventional censoring schemes. The removals of surviving units are made at the termination point while applying these censoring schemes. In many scenarios, it is needed to remove the surviving units other than the termination point. For such cases, experimenter uses a progressive censoring scheme. This censoring scheme was first introduced by Cohen [8]. One can describe progressive censoring as let n units are subject to test in a life testing experiment, and the required number of failures are prefixed as m and $(S_1, S_2, \ldots, S_m)$ be prefixed positive integers. Now, failure of a unit is considered as an event, and at the time when first event takes place, experimenter withdraws S_1 live units from $n - 1$ surviving units. Similarly, when the second event occurs, withdrawal of S_2 live units from $n - 1 - R_1$ surviving units takes place, and so on, until the mth event occurs, and all the remaining S_r live units are removed, and the experiment is stopped. Balakrishnan and Aggarwala [2] provide the extensive work done on the progressive type-II censoring scheme. More details about progressive censoring can be found in Krishna and Kumar [6, 15–17] and the references cited therein.

Most of the conventional censoring schemes work for the one sample problem, whereas in real life, there are many situations where comparative study is needed. A joint censoring scheme was introduced to deal with such problems. The exact likelihood inferences for two and k-exponential populations are investigated by Balakrishnan and Rasouli [3] and Balakrishnan and Su [4], respectively, in case of joint type-II censoring. Rasouli and Balakrishnan [23] introduced joint progressive type-II (JPTII) censoring, which we will discuss in this article. Let us run an experiment with two independent samples of sizes m_1 and m_2. Both samples were taken in the same environment. To save the money and time indulging in the experiment JPTII censoring is applied. The experiment under JPTII censoring is terminated at the prespecified number of failures. Several researchers have generated literature inferences on the joint censoring scheme. Rao *et al.* [22] investigated contributions to the theory of rank order statistics for two exponential populations with two samples. The challenge of testing equality under type-II censoring was addressed by Bhattacharyya and Mehrotra [5]. Parsi and Bairamov [21] describe the estimated number of failures for two populations under joint type-II censoring. The exact likelihood inference for two exponential populations is discussed by Rasouli and Balakrishnan [23] with joint progressively type-II censored data. Doostparast *et al.* [10] investigate Bayesian estimation in a joint progressive type-II censored sample using the Linex loss function and general distribution. Krishna and Goel [14] have investigated conclusions for two Lindley populations using joint progressively type-II censored data. When the experimental units follow a two-parameter generalized exponential distribution, Mondal and Kundu [20] looked at the joint progressively censored data.

The inverse Rayleigh (IR) distribution possesses various properties in reliability studies. To approximate the distribution of experimental units, Voda [28] suggested IRD. The Bayesian estimates of the parameters of Rayleigh distribution are obtained by El-Helbawy [11] using four different types of loss functions. Soliman *et al.* [26] and Shawky, A. and Badr [24] worked on the estimation and prediction from the IR distribution based on lower record values and lower record statistics, respectively. Dey [9] estimated the unknown parameters and reliability of the inverse Rayleigh distribution in the Bayesian paradigm. For the recent advances in IR distribution one can refer to Sindhu *et al.* [25] and Athirakrishnan, R. and Abdul-Sathar [1].

The remaining article is organized in such a manner that section 8.2 contains the model description and section 8.3 contains the information on classical estimation of the parameters under JPTII censoring. For classical estimation, maximum likelihood estimates of the parameters are calculated along with their respective asymptotic confidence intervals. Section 8.4, deals with the Bayesian estimation of the unknown parameters using Tierney-Kadane's approximation and Metropolis-Hastings algorithm under the squared error loss function. The highest posterior density credible intervals are also constructed in section 8.4. To compare different estimation methods followed during the study a simulation study is carried out in section 8.5. In section 8.6, a real data set is analyzed in view of the applicability of the study. Lastly, some conclusions are made on the basis of the whole study in section 8.7.

8.2 Model Description

Let $u_1, u_2, u_3, \ldots, u_{m_1}$ be a random sample of size m_1 from product line A and $v_1, v_2, v_3, \ldots, v_{m_2}$ be another sample of size m_2 from product line B. Both the samples are independent and identically distributed (i.i.d.) random variables from IR distribution i.e. $u_1, u_2, u_3, \ldots, u_{m_1}$ are i.i.d. random variables from IR(θ_1) with probability density function $f(u)$ and distribution function $F(u)$ such as:

$$f(u)=\frac{2\theta_1^2}{u^3}e^{-\theta_1^2/u^2}, \quad u>0,\theta_1>0$$

$$F(u)=e^{-\theta_1^2/u^2}, \quad u>0,\theta_1>0.$$

Also, $v_1, v_2, v_3, \ldots, v_{m_2}$ are i.i.d. random variables from IR(θ_2) with pdf $g(v)$ and cdf $G(v)$ given as:

$$g(v)=\frac{2\theta_2^2}{v^3}e^{-\theta_2^2/v^2}, v>0,\theta_2>0$$

$$G(v)=e^{-\theta_2^2/v^2}, \quad v>0,\theta_2>0.$$

Now, we arrange both the samples as the order statistics $w_1, w_2, w_3, \ldots, w_M$ where $M = m_1 + m_2$ and $W=(w_1, w_2, w_3, \ldots, w_M)=(u_1, u_2, \ldots, u_{m_1}; v_1, v_2, \ldots, v_{m_2})$ and the JPTII censoring is applied as follows. At first, all the M test units are put on a life testing experiment and when the first event is recorded S_1 live test units (which can come from either U-sample or V-sample) are removed from $M - 1$ surviving units. Similarly, when the second event is recorded S_2 live test units (from either X-sample or Y-sample) are removed from $M - S_1 - 2$ surviving units. Now, this process continues until the mth failure where all the surviving units $S_m = M - m - S_1 - S_2 - \ldots - S_{m-1}$ are removed and the experiment is terminated. In JPTII censoring scheme the required number of failures i.e. m and the censoring scheme $(S_1, S_2, \ldots, S_m)$ are prefixed before the experiment is started. The decomposition of the Censoring scheme $(S_1, S_2, \ldots, S_m)$ is given as $S_i = T_i + D_i$; $i = 1,2, \ldots, m$ where T_i and D_i are random variables. Where, the number of units withdrawn when the ith failure is recorded which belongs to U-sample are denoted as T_i and which belongs to V-sample are denoted as D_i. Thus the joint progressively type-II censored data is denoted by (W, Δ, T). Where, $\mathbf{W} = (w_1, w_2, \ldots, w_m)$ with $1 \le m \le M$ and $\Delta = (\delta_1, \delta_2, \ldots, \delta_m)$ is defined as:

$$\delta_i = \begin{cases} 1 & ;\text{if } W_i \in U_i \\ 0 & ;\text{if } W_i \in U_i \end{cases} \tag{8.1}$$

i.e., δ_i takes value 1 when w_i is an failure from U-sample and it takes value 0 when w_i is an failure from V–sample. S_i's can be denoted as $S = T + D$

where $T = (T_1, T_2, \ldots, T_m)$ and $D = (D_1, D_2, \ldots, D_m)$. The joint type-II censoring becomes the particular case of JPTII censoring when $S_1 = S_2 = S_{m-1}$ are taken as 0 and $S_m = M - m$. Here, $m_{1m} = \sum_{i=1}^{m} \delta_i$ and $m_{2m} = \sum_{i=1}^{m} (1-\delta_i)$.

8.3 Classical Estimation

This section is dedicated to the classical estimation of the unknown parameters of the IR distribution. Here, we use the maximum likelihood (ML) estimation method for the point estimation and asymptotic confidence interval (ACI) estimation for interval estimation.

8.3.1 Maximum Likelihood Estimation

Now, in this subsection we obtain the ML estimates of θ_1 and θ_2. First, the likelihood function under JPTII censoring is given by

$$L(w,\delta,t) = C\prod_{i=1}^{m}\left[\{f(w_i)\}^{\delta_i}\{g(w_i)\}^{(1-\delta_i)}\{\bar{F}(w_i)\}^{t_i}\{\bar{G}(w_i)\}^{d_i}\right] 0 \le w_1 \le w_2 \le \cdots \le w_m \tag{8.2}$$

Where, $C = L_1 L_2$

$$L_1 = \prod_{j=1}^{m-1}\left[\left(m_1 - \sum_{i=1}^{j-1}\delta_i - \sum_{i=1}^{j-1} t_i\right)\delta_j + \left(m_2 - \sum_{i=1}^{j-1}(1-\delta_i) - \sum_{i=1}^{j-1}(S_i - t_i)\right)(1-\delta_j)\right]$$

$$L_2 = \prod_{j=1}^{m-1}\left[\frac{\binom{\left(m_1 - \sum_{i=1}^{j-1}\delta_i - \sum_{i=1}^{j-1} t_i\right)}{t_j}\binom{\left(m_2 - \sum_{i=1}^{j-1}(1-\delta_i) - \sum_{i=1}^{j-1}(S_i - t_i)\right)}{t_j}}{\binom{m_1 - m_2 - j - \sum_{i=1}^{j-1} s_i}{s_j}}\right]$$

The pdf and cdf of IR distribution are substituted in 8.2, and the likelihood is obtained in the following form

$$L = c\,\theta_1^{2m_{1m}}\theta_1^{2m_{2m}} e^{-\sum_{i=1}^{m}\left[\frac{\theta_1^2\delta_i+\theta_2^2(1-\delta_i)}{w_i^2}\right]} \prod_{i=1}^{m}\frac{1}{w_i^3}\left(1-e^{-\theta_1^2/w_i^2}\right)^{t_i}\left(1-e^{-\theta_2^2/w_i^2}\right)^{d_i} \tag{8.3}$$

The log likelihood function is obtained by taking the logarithm of 8.3. Hence, the log likelihood function is given as

$$\log L = \log C + 2m_{1m}\log\,\theta_1 + 2m_{2m}\log\,\theta_2 - \sum_{i=1}^{m}\left[\frac{\theta_1^2\delta_i+\theta_2^2(1-\delta_i)}{w_i^2}\right]$$
$$-3\sum_{i=1}^{m}\log w_i + \sum_{i=1}^{r} t_i \log\left(1-e^{-\theta_1^2/w_i^2}\right) + \sum_{i=1}^{r} d_i \log\left(1-e^{-\theta_2^2/w_i^2}\right).$$

The ML estimates of θ_1 and θ_2 are calculated by differentiating log likelihood equation with respect to θ_1 and θ_2 respectively and equating them to zero. Two normal equations are obtained as

$$\frac{\partial\, log\, L}{\partial\theta_1} = \frac{2m_{1m}}{\theta_1} - \sum_{i=1}^{m}\frac{2\theta_1\delta_i}{w_i^2} - \sum_{i=1}^{m} t_i \frac{2\theta_1 e^{-\theta_1^2/w_i^2}}{w_i^2\left(1-e^{-\theta_1^2/w_i^2}\right)} = 0$$
$$\frac{\partial\, log\, L}{\partial\theta_2} = \frac{2m_{2m}}{\theta_2} - \sum_{i=1}^{m}\frac{2\theta_2(1-\delta_i)}{w_i^2} - \sum_{i=1}^{m} d_i \frac{2\theta_2 e^{-\theta_2^2/w_i^2}}{w_i^2\left(1-e^{-\theta_2^2/w_i^2}\right)} = 0$$

From the above normal equations, it is clear that the closed form solutions for θ_1 and θ_2 cannot be obtained analytically. Hence, we use some iterative methods to get the ML estimates $\widehat{\theta_1}$ and $\widehat{\theta_2}$ of θ_1 and θ_2 respectively.

8.3.2 Asymptotic Confidence Interval

This subsection discusses construction of 100(1 – α)% asymptotic confidence interval (ACI) using the Fisher information matrix (FIM). FIM in case of JPTII censoring is given by

$$I(\widehat{\theta_1},\widehat{\theta_2})=\begin{bmatrix} -\dfrac{\partial^2 \log L}{\partial^2\theta_1} & -\dfrac{\partial^2 \log L}{\partial\theta_1\,\partial\theta_2} \\ -\dfrac{\partial^2 \log L}{\partial\theta_2\,\partial\theta_1} & -\dfrac{\partial^2 \log L}{\partial^2\,\partial\theta_2} \end{bmatrix}_{\theta_1=\widehat{\theta_1},\theta_2=\widehat{\theta_2}} \tag{8.4}$$

$$=-\begin{bmatrix} \hat{I}_{11} & \hat{I}_{12}(6) \\ \hat{I}_{21} & \hat{I}_{22} \end{bmatrix}$$

The elements of FIM are given by the following equations

$$\frac{\partial^2 \log L}{\partial\theta_1^2}=-\frac{2m_{1m}}{\theta_1^2}-\sum_{i=1}^{m}\frac{2\delta i}{w_i^2}-\sum_{i=1}^{m}t_i\frac{\left(1-e^{-\theta_1^2/w_i^2}\right)e^{-\theta_1^2/w_i^2}\left[\dfrac{2}{w_i^2}-\left(\dfrac{2\theta_1}{w_i^2}\right)^2\right]}{\left(1-e^{-\theta_1^2/w_i^2}\right)}$$

$$\frac{\partial^2 \log L}{\partial\theta_2^2}=-\frac{2m_{2m}}{\theta_2^2}-\sum_{i=1}^{m}\frac{2(1-\delta i)}{w_i^2}-\sum_{i=1}^{r}d_i\frac{\left(1-e^{-\theta_2^2/w_i^2}\right)e^{-\theta_2^2/w_i^2}\left[\dfrac{2}{w_i^2}-\left(\dfrac{2a_2}{w_i^2}\right)^2\right]}{\left(1-e^{-\theta_2^2/w_i^2}\right)^2}$$

$$\frac{\partial^2 \log L}{\partial\theta_1\theta_2}=\frac{\partial^2 \log L}{\partial\theta_2\theta_1}=0$$

The estimated variances of $\widehat{\theta_1}$ and $\widehat{\theta_2}$ are given by

$$\hat{V}ar(\hat{\theta}_1)=-\frac{1}{\hat{I}_{11}} \quad \& \ \hat{V}ar(\hat{\theta}_2)=-\frac{1}{\hat{I}_{22}} \tag{8.5}$$

Now, the $100(1-\alpha)\%$ asymptotic confidence intervals for (θ_1, θ_2) using the approximated standard normal distribution are given as

$$P\left[\left|\frac{(\theta_1-\hat{\theta}_1)}{\sqrt{\hat{V}ar(\hat{\theta}_1)}}\right|\le Z_{\frac{\alpha}{2}}\right] \quad \& \quad P\left[\left|\frac{(\theta_2-\hat{\theta}_2)}{\sqrt{\hat{V}ar(\hat{\theta}_2)}}\right|\le Z_{\frac{\alpha}{2}}\right] \tag{8.6}$$

Here, $Z_{\frac{\alpha}{2}}$ is the upper $\frac{\alpha}{2}$ percentile of standard normal distribution.

8.4 Bayesian Estimation

The Bayes estimates of unknown parameters are calculated in this section. The Bayes estimates are obtained using squared error loss function (SELF). The unknown parameters θ_1 and θ_2 are supposed to follow independent gamma priors with their respective hyper parameters (a, b) and (a', b'). The joint prior distribution for θ_1 and θ_2 is described as follows

$$g(\theta_1,\theta_2) \propto \theta_1^{a-1} e^{-b\theta_1} \theta_2^{a'-1} e^{-b'\theta_2} \;;\theta_1,\theta_2 > 0, a, b, a', b' \geq 0 \tag{8.7}$$

The joint posterior density function is given by

$$\pi(\theta_1,\theta_2|data) = K^{-1}\theta_1^{2m_{1m}+a-1}\theta_2^{2m_{2m}+a'-1} e^{-\left(\sum_{i=1}^{m}\left[\frac{\theta_1^2\delta i+\theta_2^2(1-\delta i)}{w_i^2}\right]+b\theta_1+b'\theta_2\right)}$$

$$\prod_{i=1}^{m}\frac{1}{w_i^3}\left(1-e^{-\theta_1^2/w_i^2}\right)^{t_i}\left(1-e^{-\theta_2^2/w_i^2}\right)^{d_i}$$

Here,

$$K^{-1} = K_1^{-1}K_2^{-1}$$

$$= \int\int \theta_1^{2m_{1m}+a-1}\theta_2^{2m_{2m}+a'-1} e^{-\left(\sum_{i=1}^{m}\left[\frac{\theta_1^2\delta_i+\theta_2^2(1-\delta_i)}{w_i^2}\right]+b\theta_1+b'\theta_2\right)}$$

$$\times\prod_{i=1}^{m}\frac{1}{w_i^3}\left(1-e^{-\theta_1^2/w_i^2}\right)^{t_i}\left(1-e^{-\theta_2^2/w_i^2}\right)^{d_i} d\theta_1 d\theta_2$$

Now, we calculate the marginal posterior densities of θ_1 and θ_2 respectively as

$$\pi(\theta_1|\theta_2,data) = \theta_1^{2m_{1m}+a-1} e^{-(\sum_{i=1}^{m}\theta_1^2\delta_i/w_i^2+b\theta_1)}\prod_{i=1}^{m}\left(1-e^{-\theta_1^2/w_i^2}\right)^{t_i}$$

$$\pi(\theta_2|\theta_1,data) = \theta_2^{2m_{2m}+a'-1} e^{-(\sum_{i=1}^{m}\theta_2^2(1-\delta_i)/w_i^2+b'\theta_2)}\prod_{i=1}^{m}\left(1-e^{-\theta_2^2/w_i^2}\right)^{d_i}$$

The posterior mean is the Bayes estimate under SELF. So, the Bayes estimates of θ_1 and θ_2 are given by

$$\tilde{\theta}_1 = E[\theta_1|data] = K^{-1}\int_0^\infty \theta_1^{2m_{1n}+a} e^{-(\sum_{i=1}^m \theta_1^2 \delta_i / w_i^2 + b\theta_1)} \prod_{i=1}^m \left(1-e^{-\theta_1^2/w_i^2}\right)^{t_i} d\theta_1$$

$$\tilde{\theta}_2 = E[\theta_2|data] = K^{-1}\int_0^\infty \theta_2^{2m_{2m}+a'} e^{-(\sum_{i=1}^m \theta_2^2 (1-\delta_i)/ w_i^2 + b'\theta_2)} \prod_{i=1}^m \left(1-e^{-\theta_2^2/w_i^2}\right)^{d_i} d\theta_2$$

From the above equations, we can get that one cannot calculate the Bayes estimates $\tilde{\theta}$ and $\tilde{\theta}$ analytically. Hence, we use some approximation techniques such as Tierney-Kadane's approximation and Metropolis-Hastings algorithm.

8.4.1 Tierney-Kadane's Approximation

The approximation method for posterior mean is given by Tierney and Kadane [27]. Now, according to Tierney-Kadane's (T-K) approximation the posterior mean of $u(\theta_1, \theta_2)$ is given by:

$$E[\phi(\theta_1,\theta_2|data] = \frac{\int\int exp[nl^*(\theta_1^*,\theta_2^*)d\theta_1 d\theta_2]}{\int\int exp[nl(\theta_1,\theta_2)d\theta_1 d\theta_2]}$$

where,

$$l(\theta_1,\theta_2) = \frac{log\ v(\theta_1,\theta_2) + l(\theta_1,\theta_2;w)}{n}$$

$$l^*(\theta_1,\theta_2) = l(\theta_1,\theta_2) + \frac{1}{n}\log\phi(\theta_1,\theta_2)$$

The Bayes estimate with respect to $u(\theta_1, \theta_2)$ comes out in the form

$$E[\phi(\theta_1,\theta_2|data] = \left[\frac{|\Sigma^*|}{|\Sigma|}\right]^{\frac{1}{2}} exp\{n[l^*(\theta_1^*,\theta_2^*) - l(\theta_1,\theta_2)]\}$$

Here, determinant of Σ^* and Σ is given as follows

$$|\Sigma| = \left(\frac{\partial^2 l}{\partial\theta_1^2}\frac{\partial^2 l}{\partial\theta_2^2} - \left(\frac{\partial^2 l}{\partial\theta_1\theta_2}\right)^2\right)^{-1} \text{ and } |\Sigma^*| = \left(\frac{\partial^2 l^*}{\partial\theta_1^2}\frac{\partial^2 l^*}{\partial\theta_2^2} - \left(\frac{\partial^2 l^*}{\partial\theta_1\theta_2}\right)^2\right)^{-1}$$

$$l(\theta_1,\theta_2) = \frac{1}{M}[(a-1)\log(\theta_1) - b\theta_1 + (a'-1)\log(\theta_2) - b'\theta_2 2m_{1m}\log\theta_1$$

$$+ 2m_{2m}\log\theta_2 - \left[\frac{\theta_1^2\delta_i + \theta_2^2(1-\delta_i)}{w_1^2}\right] - 3\sum_{i=1}^{m}\log w_i$$

$$+ \sum_{i=1}^{m} t_i \log\left(1-e^{-\theta_1^2/w_i^2}\right) + \sum_{i=1}^{m} d_i \log\left(1-e^{-\theta_2^2/w_i^2}\right)$$

$$l^*(\theta_1^*,\theta_2^*) = l(\theta_1,\theta_2) + \frac{1}{M}\log\phi(\theta_1,\theta_2)$$

For calculating the Bayes estimates of θ_1 and θ_2 we obtain the following expressions

$$l_{11} = \frac{\partial^2 l}{\partial^2\theta_1} = -\frac{(2m_{1m}+a-1)}{\theta_1^2} - \sum_{i=1}^{m}\frac{2\delta_i}{w_i^2} - \sum_{i=1}^{m} t_i \frac{\left(1-e^{-\theta_1^2/w_i^2}\right)e^{-\theta_1^2/w_i^2}\left[\frac{2}{w_i^2} - \left(\frac{2\theta_1}{w_i^2}\right)^2\right]}{\left(1-e^{-\theta_1^2/w_i^2}\right)^2}$$

$$l_{22} = \frac{\partial^2 l}{\partial^2\theta_2} = -\frac{(2m_{2m}+a'-1)}{\theta_2^2} - \sum_{i=1}^{m}\frac{2(1-\delta_i)}{w_i^2} - \sum_{i=1}^{m} d_i \frac{\left(1-e^{-\theta_2^2/w_i^2}\right)e^{-\theta_2^2/w_i^2}\left[\frac{2}{w_i^2} - \left(\frac{2\theta_2}{w_i^2}\right)^2\right]}{\left(1-e^{-\theta_2^2/w_i^2}\right)^2}$$

$$l_{12} = l_{21} = \frac{\partial^2 \log L}{\partial\theta_1\partial\theta_2} = \frac{\partial^2 \log L}{\partial\theta_2\partial\theta_1} = 0$$

Now, we substitute $\phi(\theta_1, \theta_2) = \phi_1(\theta_1, \theta_2) = \theta_1$ to compute the Bayes estimate of θ_1 under SELF hence, $l^*(\theta_1,\theta_2) = l_1^*(\theta_1,\theta_2) = l(\theta_1,\theta_2) + \frac{1}{n}\log\phi_1(\theta_1,\theta_2)$. So the Bayes estimates of θ_1 is given by

$$\widehat{\theta}_{1BE} = \sqrt{\frac{|\Sigma^*|}{|\Sigma|}} exp\{n[l_1^*(\theta_1^*,\theta_2^*) - l(\theta_1 - \theta_2)]\}$$

Similarly, Bayes estimate of θ_2 is obtained as

$$\widehat{\theta_2}_{BE} = \sqrt{\frac{|\Sigma^*|}{|\Sigma|}} exp\{n[l_2^*(\theta_1^*,\theta_2^*) - l(\theta_1 - \theta_2)]\}$$

8.4.2 Metropolis-Hastings Algorithm

Here, we use MCMC techniques to generate the random numbers from the marginal posterior distribution functions. To generate the random numbers from the marginal densities of θ_1 and θ_2 using MCMC techniques we use the Metropolis-Hastings (M-H) algorithm. A detailed discussion about MCMC and M-H algorithm can be found in Metropolis and Ulam [19], Hastings [13], and Gelman *et al.* [12]. The steps for the M-H algorithm are given below

Step 1. First, start with taking the initial guess as $\theta_1^{(0)}$.
Step 2. Generate a candidate point $\theta_{1c}^{(j)}$ from the proposal density $N(\theta_1^{(j-1)},1)$.
Step 3. Generate u from uniform (0,1).
Step 4. calculate $\phi(\theta_1^j|\theta_1^{(j-1)}) = min\left\{\frac{\pi_1(\theta_{1c}^{(j)}|\delta,w)q(\theta_1^{(j-1)}|\theta_{1c}^{(j)})}{\pi_1(\theta_1^{(j-1)}|\delta,w)q(\theta_{1c}^j|\theta_1^{(j-1)})},1\right\}$.
Step 5. IF $y \leq \phi$ set $\theta_1^{(j)} = \theta_{1c}^{(j)}$ with acceptance probability ϕ otherwise $\theta_1^{(j)} = \theta_1^{(j-1)}$.
Step 6. Now, repeat steps 2-5, for $j = 1,2, \ldots, N$ to obtain the sequence of the parameter θ_1 as $(\theta_{11}, \theta_{12}, \ldots, \theta_{1N})$.

To get the independent sample from the stationary distribution of Markov chain we discard the first N_0 values from $\theta_{1j}'s\ ; j = 1,2,\ldots,N$, here N_0 is known as the burn-in period.

Now, the Bayes estimators of the parameters θ_1 and θ_2 under SELF are given by

$$\hat{\theta}_{1MH} = \frac{1}{N-N_0}\sum_{j=N_0+1}^{N}\theta_{1j} \tag{8.8}$$

$$\hat{\theta}_{2MH} = \frac{1}{N-N_0}\sum_{j=N_0+1}^{N}\theta_{2j} \tag{8.9}$$

8.4.3 HPD Credible Interval

In this subsection, using the algorithm purposed by Chen and Shao [7], we develop HPD credible interval for θ_1 and θ_2. Let $\theta_{(11)},\theta_{(12)},\ldots,\theta_{1(N-N_0)}$ signify the ordered values of $\theta_{1(N_0+1)},\theta_{1(N_0+2)},\ldots,\theta_{1(N)}$, then the 100(1 – η)% HPD credible interval for θ_1 is calculated as

$$\left(\theta_{1(k)},\theta_{1(k+[(1-\eta)N])}\right),$$

here, k is chosen such that

$$\left(\theta_{1(k+[(1-\eta)(N-N_0)])}-\theta_{1k}\right)=\min_{1\le k\le \gamma N}\left(\theta_{1(k+[(1-\eta)(N-N_0)])}-\theta_{1(k)}\right),$$

$k = 1,2,\ldots,(N - N_0)$, where, $[x]$ is the integer part of x. Similarly, the 100(1 – η)% HPD credible interval for θ_2 is obtained as

$$\left(\theta_{2(k)},\theta_{2(k+[(1-\eta)N])}\right),$$

where k is chosen such that

$$\left(\theta_{2(k+[(1-\eta)(N-N_0)])}-\theta_{2k}\right)=\min_{1\le k\le \gamma N}\left(\theta_{2(k+[(1-\eta)(N-N_0)])}-\theta_{2(k)}\right),$$

$k = 1,2,\ldots,(N - N_0)$.

8.5 Simulation Study

A simulation study to compare various estimates obtained in above sections is carried out in this section. Under JPTII censoring, we calculate the average estimates (AEs) and mean squared errors (MSEs) of the ML and Bayes estimates of the unknown parameters from IR(θ_1) and IR(θ_2). Bayes estimates are calculated using two prior values: non-informative (prior A) and informative (prior B). The average length (AL) and coverage probability (CP) are calculated for the asymptotic confidence interval and HPD credible interval. Table 8.1 shows the various censoring schemes that were employed to conduct the simulation investigation. In the simulation study, two sets of the true values of the unknown parameters (θ_1, θ_2) =

Table 8.1 Censoring schemes used in simulation study.

CS	(m_1, m_2, m)	S	CS	(m_1, m_2, m)	S
[1]	(10,15,16)	(0*15,9)	[8]	(20,25,36)	(0*33,3*3)
[2]	(10,15,16)	(0*13,3*3)	[9]	(30,35,42)	(0*41,23)
[3]	(10,15,18)	(0*17,7)	[10]	(30,35,42)	(0*37,5*4,3)
[4]	(10,15,18)	(0*14,2*3,1)	[11]	(30,35,51)	(0*50,14)
[5]	(20,25,30)	(0*29,15)	[12]	(30,35,51)	(0*48,5*2,4)
[6]	(20,25,30)	(0*27,5*3)	[13]	(40,45,66)	(0*65,19)
[7]	(20,25,36)	(0*35,9)	[14]	(40,45,66)	(0*62,5*3,4)

(0.5,0.75) and $(\theta_1, \theta_2) = (0.75,1.5)$ are employed. For Bayesian estimation, the values of the hyper parameters for different prior densities are set as follows: $a = b = a' = b' = 0.0001$ for prior A; $a = 2$, $b = 4$, $a' = 3$, $b' = 4$ for prior B for set-1 and $a = 3$, $b = 4$, $a' = 3$, $b' = 2$ for set-2. The values of the hyper parameters are chosen in such a way that the mean of the prior density equals the true value of the parameter in the case of prior B, i.e. informative prior. We created 10,000 samples for the M-H algorithm, with a burn-in period of 4000.

The results of the simulation study are shown in Tables 8.2, 8.3, 8.4, and 8.5, respectively. The simulation Tables 8.2 and 8.3 provide that when we increase the sample size, MSEs of the parameters decreases. The Bayes estimates attain lesser MSEs than ML estimates. Hence Bayes estimates provide better estimates than the ML estimates. The Bayes estimates obtained in the case of informative prior provides better results than non-informative prior. The various censoring schemes used in the simulation study provides that the type-II censoring scheme and progressive type-II censoring schemes attains almost similar MSEs. Hence, we suggest using a progressive type-II censoring scheme instead of the type-II censoring scheme to reduce the involved time and cost in the experiment. From Tables 8.4 and 8.5, we suggest that the HPD credible intervals provide lesser AL than ACI. The HPD credible intervals constructed in the case of informative prior provides better interval estimates than asymptotic confidence intervals. The average length obtained in the case of type-II censoring and progressive type-II censoring has very slight differences. So, we suggest using progressive type-II censoring in place of type-II censoring to minimize

Table 8.2 The AE and MSE of θ_1 and θ_2 when $(\theta_1, \theta_2) = (0.5, 0.75)$.

	$\hat{\theta}_1$										$\hat{\theta}_2$									
			Bayes										Bayes							
			T-K				M-H						T-K				M-H			
	MLE		Prior A		Prior B		Prior A		Prior B		MLE		Prior A		Prior B		Prior A		Prior B	
CS	AE	MSE	AE	MSE	AE	MSE	AE	MSE	AE	MSE	AE	MSE	AE	MSE	AE	MSE	AE	MSE	AE	MSE
[1]	0.5232	0.0074	0.5143	0.0072	0.5132	0.0064	0.5144	0.0073	0.5133	0.0064	0.7642	0.0106	0.7602	0.0105	0.7593	0.0094	0.7601	0.0106	0.7592	0.0094
[2]	0.5195	0.0085	0.5234	0.0099	0.5216	0.0087	0.5234	0.0099	0.5217	0.0087	0.7983	0.0140	0.7939	0.0144	0.7911	0.0128	0.7939	0.0145	0.7911	0.0128
[3]	0.5223	0.0087	0.5143	0.0079	0.5131	0.0070	0.5143	0.0079	0.5133	0.0070	0.7681	0.0118	0.7553	0.0092	0.7547	0.0083	0.7555	0.0092	0.7548	0.0083
[4]	0.5245	0.0085	0.5173	0.0082	0.5158	0.0071	0.5174	0.0082	0.5160	0.0072	0.7880	0.0131	0.7886	0.0147	0.7896	0.0123	0.7888	0.0148	0.7896	0.0123
[5]	0.5116	0.0036	0.5090	0.0035	0.5051	0.0035	0.5090	0.0035	0.5050	0.0035	0.7563	0.0055	0.7596	0.0066	0.7535	0.0057	0.7598	0.0066	0.7534	0.0057
[6]	0.5140	0.0038	0.5095	0.0041	0.5091	0.0034	0.5095	0.0041	0.5090	0.0034	0.7835	0.0081	0.7791	0.0073	0.7764	0.0067	0.7792	0.0073	0.7767	0.0067
[7]	0.5122	0.0037	0.5068	0.0032	0.5058	0.0033	0.5068	0.0032	0.5058	0.0033	0.7580	0.0063	0.7610	0.0059	0.7569	0.0059	0.7610	0.0059	0.7570	0.0059
[8]	0.5097	0.0037	0.5092	0.0036	0.5061	0.0035	0.5092	0.0036	0.5062	0.0035	0.7625	0.0064	0.7672	0.0071	0.7657	0.0060	0.7673	0.0071	0.7658	0.0061
[9]	0.5046	0.0023	0.5051	0.0022	0.5043	0.0022	0.5051	0.0023	0.5043	0.0022	0.7584	0.0043	0.7570	0.0045	0.7528	0.0042	0.7570	0.0045	0.7528	0.0042
[10]	0.5117	0.0025	0.5076	0.0024	0.5087	0.0024	0.5075	0.0024	0.5088	0.0024	0.7852	0.0058	0.7841	0.0058	0.7814	0.0054	0.7840	0.0059	0.7815	0.0054
[11]	0.5031	0.0022	0.5048	0.0021	0.5050	0.0022	0.5049	0.0021	0.5050	0.0022	0.7580	0.0043	0.7575	0.0041	0.7528	0.0041	0.7573	0.0041	0.7529	0.0041
[12]	0.5074	0.0024	0.5035	0.0022	0.5061	0.0022	0.5036	0.0022	0.5061	0.0022	0.7643	0.0046	0.7643	0.0048	0.7620	0.0043	0.7642	0.0048	0.7620	0.0044
[13]	0.5048	0.0016	0.5026	0.0016	0.5029	0.0015	0.5026	0.0016	0.5029	0.0015	0.7571	0.0030	0.7544	0.0033	0.7544	0.0030	0.7544	0.0033	0.7543	0.0030
[14]	0.5062	0.0016	0.5026	0.0015	0.5037	0.0016	0.5025	0.0015	0.5038	0.0016	0.7638	0.0037	0.7640	0.0038	0.7629	0.0033	0.7641	0.0039	0.7628	0.0033

Table 8.3 The AE and MSE of θ_1 and θ_2 when $(\theta_1, \theta_2) = (0.75, 1.5)$.

	$\hat{\theta}_1$										$\hat{\theta}_2$									
			Bayes										Bayes							
			T-K				M-H						T-K				M-H			
	MLE		Prior A		Prior B		Prior A		Prior B		MLE		Prior A		Prior B		Prior A		Prior B	
CS	AE	MSE	AE	MSE	AE	MSE	AE	MSE	AE	MSE	AE	MSE	AE	MSE	AE	MSE	AE	MSE	AE	MSE
[1]	0.7821	0.0161	0.7724	0.0176	0.7699	0.0147	0.7725	0.0175	0.7700	0.0147	1.5373	0.0441	1.5232	0.0443	1.5212	0.0395	1.5233	0.0442	1.5211	0.0396
[2]	0.7001	0.0173	0.7768	0.0178	0.7739	0.0150	0.7772	0.0178	0.7741	0.0149	1.6153	0.0660	1.5901	0.0563	1.5844	0.0498	1.5903	0.0563	1.5846	0.0498
[3]	0.7870	0.0187	0.7695	0.0178	0.7671	0.0147	0.7699	0.0178	0.7672	0.0147	1.5370	0.0430	1.5127	0.0405	1.5113	0.0363	1.5128	0.0405	1.5113	0.0363
[4]	0.7805	0.0180	0.7651	0.0183	0.7630	0.0153	0.7652	0.0183	0.7633	0.0153	1.5875	0.0544	1.5835	0.0539	1.5783	0.0478	1.5838	0.0541	1.5788	0.0480
[5]	0.7644	0.0083	0.7591	0.0075	0.7605	0.0070	0.7592	0.0075	0.7604	0.0070	1.5233	0.0261	1.5179	0.0270	1.5225	0.0248	1.5182	0.0270	1.5224	0.0248
[6]	0.7675	0.0080	0.7587	0.0078	0.7590	0.0070	0.7588	0.0078	0.7592	0.0070	1.5741	0.0328	1.5563	0.0310	1.5689	0.0327	1.5563	0.0310	1.5689	0.0327
[7]	0.7596	0.0080	0.7653	0.0078	0.7560	0.0073	0.7654	0.0078	0.7561	0.0073	1.5142	0.0243	1.5137	0.0241	1.5148	0.0212	1.5138	0.0240	1.5148	0.0213
[8]	0.7670	0.0086	0.7604	0.0081	0.7599	0.0066	0.7606	0.0081	0.7597	0.0066	1.5485	0.0285	1.5317	0.0285	1.5407	0.0239	1.5319	0.0286	1.5410	0.0240
[9]	0.7628	0.0056	0.7545	0.0049	0.7530	0.0043	0.7546	0.0049	0.7529	0.0043	1.5166	0.0174	1.5119	0.0169	1.5118	0.0168	1.5118	0.0169	1.5118	0.0169
[10]	0.7608	0.0050	0.7566	0.0054	0.7582	0.0048	0.7567	0.0054	0.7582	0.0048	1.5873	0.0269	1.5911	0.0290	1.5870	0.0268	1.5913	0.0290	1.5871	0.0260
[11]	0.7590	0.0051	0.7594	0.0054	0.7564	0.0044	0.7594	0.0054	0.7565	0.0044	1.5154	0.0161	1.5109	0.0158	1.5078	0.0166	1.5109	0.0158	1.5079	0.0166
[12]	0.7647	0.0054	0.7585	0.0050	0.7522	0.0047	0.7586	0.0050	0.7523	0.0047	1.5373	0.0189	1.5290	0.0185	1.5269	0.0175	1.5291	0.0185	1.5270	0.0175
[13]	0.7580	0.0042	0.7527	0.0035	0.7544	0.0034	0.7528	0.0035	0.7546	0.0034	1.5078	0.0117	1.5140	0.0141	1.5123	0.0132	1.5142	0.0141	1.5121	0.0132
[14]	0.7551	0.0034	0.7551	0.0034	0.7554	0.0035	0.7551	0.0034	0.7555	0.0035	1.5375	0.0155	1.5287	0.0149	1.5318	0.0144	1.5287	0.0149	1.5319	0.0144

Table 8.4 The AL and CP of θ_1 and θ_2 when $(\theta_1, \theta_2) = (0.5, 0.75)$.

	$\hat{\theta}_1$						$\hat{\theta}_2$					
			HPD						HPD			
	ACI		Prior A		Prior B		ACI		Prior A		Prior B	
CS	AL	CP	AL	CP	AL	CP	AL	CP	AL	CP	AL	CP
[1]	0.3248	0.969	0.3149	0.951	0.3068	0.956	0.3905	0.944	0.3825	0.935	0.3731	0.941
[2]	0.3224	0.957	0.3206	0.933	0.3126	0.938	0.4073	0.946	0.3996	0.939	0.3877	0.939
[3]	0.3240	0.940	0.3146	0.943	0.3064	0.951	0.3905	0.950	0.3780	0.948	0.3682	0.953
[4]	0.3254	0.956	0.3159	0.934	0.3073	0.955	0.4006	0.952	0.3952	0.927	0.3859	0.951
[5]	0.2244	0.948	0.2188	0.940	0.2138	0.934	0.2985	0.964	0.2935	0.939	0.2871	0.948
[6]	0.2254	0.960	0.2190	0.922	0.2160	0.946	0.3087	0.937	0.3010	0.936	0.2959	0.946
[7]	0.2246	0.955	0.2175	0.943	0.2145	0.949	0.2976	0.941	0.2924	0.955	0.2873	0.941
[8]	0.2234	0.941	0.2191	0.947	0.2147	0.942	0.2993	0.943	0.2956	0.942	0.2905	0.947
[9]	0.1807	0.951	0.1767	0.947	0.1746	0.935	0.2534	0.943	0.2472	0.942	0.2431	0.935
[10]	0.1832	0.944	0.1774	0.939	0.1765	0.940	0.2616	0.940	0.2552	0.920	0.2521	0.930
[11]	0.1801	0.940	0.1760	0.946	0.1750	0.954	0.2516	0.949	0.2456	0.950	0.2415	0.948
[12]	0.1816	0.937	0.1760	0.932	0.1755	0.946	0.2535	0.948	0.2475	0.941	0.2448	0.939
[13]	0.1565	0.953	0.1519	0.942	0.1512	0.943	0.2216	0.955	0.2151	0.942	0.2139	0.954
[14]	0.1569	0.947	0.1514	0.940	0.1506	0.944	0.2235	0.951	0.2185	0.927	0.2164	0.950

Table 8.5 The AL and CP of θ_1 and θ_2 when $(\theta_1, \theta_2) = (0.5, 0.75)$.

	$\hat{\theta}_1$						$\hat{\theta}_2$					
	ACI		HPD				ACI		HPD			
			Prior A		Prior B				Prior A		Prior B	
CS	AL	CP	AL	CP	AL	CP	AL	CP	AL	CP	AL	CP
[1]	0.4850	0.964	0.4731	0.953	0.4555	0.960	0.7887	0.961	0.7716	0.942	0.7519	0.945
[2]	0.4839	0.960	0.4759	0.944	0.4573	0.949	0.8282	0.941	0.8068	0.938	0.7836	0.941
[3]	0.4879	0.957	0.4708	0.945	0.4531	0.952	0.7825	0.953	0.7623	0.949	0.7423	0.952
[4]	0.4839	0.961	0.4679	0.929	0.4507	0.937	0.8092	0.945	0.7983	0.941	0.7771	0.943
[5]	0.3351	0.955	0.3273	0.948	0.3212	0.945	0.6034	0.949	0.5933	0.944	0.5859	0.940
[6]	0.3364	0.954	0.3261	0.944	0.3205	0.952	0.6219	0.938	0.6064	0.934	0.6018	0.931
[7]	0.3330	0.950	0.3293	0.947	0.3192	0.945	0.5948	0.943	0.5854	0.948	0.5774	0.948
[8]	0.3362	0.939	0.3274	0.941	0.3210	0.958	0.6080	0.945	0.5927	0.938	0.5884	0.957
[9]	0.2730	0.947	0.2642	0.945	0.2602	0.957	0.5091	0.950	0.4990	0.943	0.4935	0.942
[10]	0.2723	0.960	0.2649	0.931	0.2622	0.944	0.5307	0.916	0.5231	0.903	0.5171	0.903
[11]	0.2716	0.949	0.2660	0.941	0.2621	0.954	0.5033	0.961	0.4941	0.948	0.4870	0.941
[12]	0.2737	0.951	0.2657	0.942	0.2607	0.944	0.5102	0.951	0.5002	0.944	0.4928	0.945
[13]	0.2349	0.937	0.2271	0.943	0.2263	0.948	0.4418	0.952	0.4363	0.937	0.4322	0.940
[14]	0.2340	0.956	0.2287	0.946	0.2263	0.954	0.4501	0.945	0.4395	0.947	0.4359	0.941

the capital and time involved in the experiment. All the interval estimates attain their nominal level of significance.

8.6 Real-Life Application

For the illustration of the theory proposed in the article here, we analyze a real life problem. Here, we have used the data set of failure times (in minutes) for two types of electrical insulation from Lawless [18]. The two data sets are shown below

Type A: 219.3, 79.4, 86.0, 150.2, 21.7, 18.5, 121.9, 40.5, 147.1, 35.1, 42.3, 48.7.

Type B: 21.8, 70.7, 24.4, 138.6, 151.9, 75.3, 12.3, 95.5, 98.1, 43.2, 28.6, 46.9.

First, we check if the given data set fits inverse Rayleigh distribution or not. To check the fitting of the model, we use Kolmogorov-Smirnov (K-S) test, and the corresponding K-S distance and p-values are obtained as 0.2864 and 0.2301 for data X, and 0.3348 and 0.1061 for the data set Y, respectively. From the obtained p-values, we can say that the given data set fits the model well. For studying the above data set under joint progressive type-II censoring, we discuss some censoring schemes given as:

Scheme 1: (m_1=12, m_2=12, m=14) S=(0*5, 2*4, 2, 0*4).
Scheme 2: (m_1=12, m_2=12, m=18) S=(0*8, 2*3, 0*7).
Scheme 3: (m_1=12, m_2=12, m=20) S=(0*9, 2*2, 0*9).

Table 8.6 The ML and Bayes estimates with their corresponding confidence/credible intervals of θ_1 and θ_2 for the real data set.

				Bayes		
CS	$\hat{\theta}$	MLE	ACI	T-K	M-H	HPD
[1]	$\hat{\theta_1}$	40.6598	(28.8735,52.4461)	40.281	40.3702	(27.6449,52.5761)
[2]		39.0715	(28.0155,50.1275)	38.6583	37.6643	(26.8293,49.6951)
[3]		39.5665	(28.3424,50.7906)	39.1539	38.3336	(28.1267,48.7299)
[1]	$\hat{\theta_2}$	30.6232	(21.9329,39.3136)	30.3053	29.7444	(27.6449,52.5761)
[2]		30.6221	(21.9323,39.3118)	30.3042	30.3088	(26.8293,49.6951)
[3]		30.3776	(21.7752,38.9800)	30.0583	29.189	(28.1267,48.7299)

To study various estimates obtained in the above sections, we use the above censoring schemes to calculate ML and Bayes estimates with their respective ACI and HPD credible intervals. Bayes estimates for the real data are calculated in case of non-informative prior due to unavailability of any prior information. In the case of the M-H algorithm, 10,000 samples were generated, and 4000 were taken as burn-in period. The various estimates calculated for the data set under different censoring schemes are shown in Table 8.6.

8.7 Conclusions

In this chapter, problems of statistical inferences on inverse Rayleigh distribution in the case of joint progressive type II censored was studied. Both classical and Bayesian methods were followed to draw the inferences. The maximum likelihood and asymptotic confidence interval estimates were calculated in case of classical estimation methods. The Bayes estimates with corresponding highest posterior density (HPD) credible intervals were calculated for in case of Bayesian paradigm. The Tierney-Kadane's approximation and M-H algorithm were used to obtain Bayes estimates. For the numerical exploration of the theory developed in the article, a simulation study was performed, and the average estimates with their corresponding mean squared errors for the point estimates, average length, and corresponding coverage probabilities for the interval estimates were calculated. For the illustration of the theory developed, a real-life application was analyzed. On the basis of results obtained in the simulation study, we can conclude that the Bayes estimates when we have some prior information, provide better estimates than the maximum likelihood estimates. Similarly, in the case of prior information, HPD credible intervals provide better coverage probability and shorter length for the interval estimates of the parameters.

References

1. Athirakrishnan, R. and Abdul-Sathar, E., E-Bayesian and hierarchical Bayesian estimation of inverse Rayleigh distribution. *Am. J. Math. Manage. Sci.*, 41, 1, 70–87, 2022.
2. Balakrishnan, N. and Aggarwala, R., *Progressive censoring: Theory, methods, and applications*, Birkhäuser, Boston, 2000.

3. Balakrishnan, N. and Rasouli, A., Exact likelihood inference for two exponential populations under joint type-II censoring. *Comput. Stat. Data Anal.*, 52, 5, 2725–2738, 2008.
4. Balakrishnan, N. and Su, F., Exact likelihood inference for k exponential populations under joint type-II censoring. *Commun. Stat. Simul. Comput.*, 44, 3, 591–613, 2015.
5. Bhattacharyya, G.K. and Mehrotra, K.G., On testing equality of two exponential distributions under combined type II censoring. *J. Am. Stat. Assoc.*, 76, 376, 886–894, 1981.
6. Chaturvedi, A., Kumar, N., Kumar, K., Statistical inference for the reliability functions of a family of lifetime distributions based on progressive type II right censoring. *Statistica*, 78, 1, 81–101, 2018.
7. Chen, M.-H. and Shao, Q.-M., Monte Carlo estimation of Bayesian credible and hpd intervals. *J. Comput. Graph. Stat.*, 8, 1, 69–92, 1999.
8. Cohen, A.C., Progressively censored samples in life testing. *Technometrics*, 5, 3, 327–339, 1963.
9. Dey, S., Bayesian estimation of the parameter and reliability function of an inverse Rayleigh distribution. *Malaysian J. Math. Sci.*, 6, 1, 113–124, 2012.
10. Doostparast, M., Ahmadi, M.V., Ahmadi, J., Bayes estimation based on joint progressive type II censored data under LINEX loss function. *Commun. Stat. Simul. Comput.*, 42, 8, 1865–1886, 2013.
11. El-Helbawy, A., Bayesian estimation and prediction for the inverse Rayleigh lifetime distribution, in: *Proceeding of the 40st annual conference of Statistics, Computer sciences and Operation Research*, Cairo University, pp. 45–59, 2005.
12. Gelman, A., Carlin, J.B., Stern, H.S., Dunson, D.B., Vehtari, A., Rubin, D.B., *Bayesian Data Analysis*, CRC press, 2013.
13. Hastings, W.K., Monte Carlo sampling methods using Markov chains and their applications. *Biometrika*, 57, 97–109, 1970.
14. Krishna, H. and Goel, R., Inferences for two Lindley populations based on joint progressive type-II censored data. *Commun. Stat. Simul. Comput.*, 1–18, 2020, https://doi.org/10.1080/03610918.2020.1751851.
15. Krishna, H. and Kumar, K., Reliability estimation in generalized inverted exponential distribution with progressively type II censored sample. *J. Stat. Comput. Simul.*, 83, 6, 1007–1019, 2013.
16. Kumar, I. and Kumar, K., On estimation of P(V<U) for inverse Pareto distribution under progressively censored data. *Int. J. Syst. Assur. Eng. Manage.*, 13, 1, 189–202, 2022.
17. Kumar, K., Garg, R., Krishna, H., Nakagami distribution as a reliability model under progressive censoring. *Int. J. Syst. Assur. Eng. Manage.*, 8, 1, 109–122, 2017.
18. Lawless, J.F., *Statistical models and methods for lifetime data*, Wiley, New York, 2003.

19. Metropolis, N. and Ulam, S., The Monte Carlo method. *J. Am. Stat. Assoc.*, 44, 247, 335–341, 1949.
20. Mondal, S. and Kundu, D., On the joint type-II progressive censoring scheme. *Commun. Stat. Theory Methods*, 49, 4, 958–976, 2020.
21. Parsi, S. and Bairamov, I., Expected values of the number of failures for two populations under joint type-II progressive censoring. *Comput. Stat. Data Anal.*, 53, 10, 3560–3570, 2009.
22. Rao, U., Savage, I.R., Sobel, M. *et al.*, Contributions to the theory of rank order statistics: the two-sample censored case. *Ann. Math. Stat.*, 31, 2, 415–426, 1960.
23. Rasouli, A. and Balakrishnan, N., Exact likelihood inference for two exponential populations under joint progressive type-II censoring. *Commun. Stat. Theory Methods*, 39, 12, 2172–2191, 2010.
24. Shawky, A. and Badr, M., Estimations and prediction from the inverse Rayleigh model based on lower record statistics. *Life Sci. J.*, 9, 2, 985–990, 2012.
25. Sindhu, T.N., Khan, H.M., Hussain, Z., Lenzmeier, T., Bayesian prediction from the inverse Rayleigh distribution based on type-II trim censoring. *J. Stat. Manage. Syst.*, 20, 5, 995–1008, 2017.
26. Soliman, A., Amin, E.A., Abd-El Aziz, A.A., Estimation and prediction from inverse Rayleigh distribution based on lower record values. *Appl. Math. Sci.*, 4, 62, 3057–3066, 2010.
27. Tierney, L. and Kadane, J.B., Accurate approximations for posterior moments and marginal densities. *J. Am. Stat. Assoc.*, 81, 393, 82–86, 1986.
28. Voda, R.G., On the inverse Rayleigh distributed random variable. *Rep. Statis. App. Res. JUSE*, 19, 4, 15–21, 1972.

9

Component Reliability Estimation Through Competing Risk Analysis of Fuzzy Lifetime Data

Rashmi Bundel[1], M. S. Panwar[2*] and Sanjeev K. Tomer[2]

[1]*Department of Statistics, University of Rajasthan, JLN Marg, Jaipur, Rajasthan, India*
[2]*Department of Statistics, Institute of Science, Banaras Hindu University, Varanasi, U.P., India*

Abstract

The competing risk analysis of lifetime data from multi-component series systems is often carried out to assess the reliability measures of systems' components. In many cases, systems' lifetimes are not observed precisely, or they are reported in a "vague" form. This imprecision or vagueness in data can be dealt with higher accuracy by incorporating fuzzy concepts. In the present chapter, we perform a competing risk analysis of lifetime data by considering lifetimes as fuzzy numbers. Using different membership functions, we provide procedures for maximum likelihood and Bayesian estimation of component reliability. We also evaluate bootstrap confidence intervals and the highest posterior density intervals. To observe the impact of various membership functions on the considered estimators, a comprehensive simulation study has been carried out. Finally, a real data set of small electric appliance has been analyzed.

Keywords: Series system, Bayesian inference, trapezoidal membership function, floored observation, small electric appliance data

**Corresponding author*: vir.panwar@gmail.com

S. C. Malik, Deepak Sinwar, Ashish Kumar, S. R. Gadde, Prasenjit Chatterjee and Bui Thanh Hung (eds.) Computational Intelligence in Sustainable Reliability Engineering, (181–216) © 2023 Scrivener Publishing LLC

9.1 Introduction

Estimation of the reliability function of a particular component in a series system by using systems failure time data is often considered in reliability theory. By definition, a series system fails as soon as one of its components fails. The component responsible for the failure of the system is usually termed as systems' cause of failure. When a system fails, two quantities may be noted: its failure time and an indicator denoting the index of failed component, i.e., its cause of failure. Thus, in such studies, one may observe data in pairs of random variables (X, C), where the random variable X denotes the failure time of the system, and C is an indicator of the cause of failure. In the presence of such data, component reliability measures may be estimated by employing the notion of competing risks. To get more insight into the theory of competing risks, one may refer to Crowder [5], Flehinger *et al.* [8], Deshpande and Purohit [6], Lagakos [14].

In many studies, failure times of systems may not be observed precisely or they are reported in "vague" form. For example, the lifetime of a ball bearing can be reported as "around 20000 cycles" and that of a printer cartridge as "between 400 and 500 pages." Such lifetimes are vague in nature and cannot be analyzed with usual statistical methods. However, such data can be dealt with higher accuracy by incorporating fuzzy concepts. Statistical analysis with this notion has been considered by many authors. Mon and Cheng [15] addressed the system reliability analysis for components by using different fuzzy membership functions. Instead of assuming classical probability distribution for the components, they employed different fuzzy membership function to assess system reliability and solved the problem via non-linear programming techniques. Huang *et al.* [11] discuss Bayesian reliability analysis for fuzzy data and proposed a new method to determine the membership function of the estimates of the parameters and the reliability function of multiparameter lifetime distributions. Pak *et al.* [16] considered the estimation of the stress-strength reliability when strength and stress are statistically independent exponential random variables, and observed data from both the variables are reported in the form of fuzzy numbers. Hryniewicz [10] discussed an application in reliability using Bayes statistical decisions with random fuzzy data. Recently, Bundel and Tomer [2] employed fuzzy concepts in presence of competing risk for the reliability estimation of components when the given data are in the form of fuzzy numbers.

In this chapter, we focus on competing risk analysis of series system lifetime data (X, C) when system lifetime X is observed in "vague" or imprecise form. Considering component lifetimes to be s-independent exponential random variables, we first introduce the likelihood function assuming each observation to be in the form of a trapezoidal fuzzy number. Then we derive maximum likelihood and Bayes estimators of the model parameters as well as component reliabilities. Through simulation study, we determine the effect of different membership functions on considered estimators which may help in selecting an appropriate membership function for any estimation problem. Rest of the chapter is organised as follows. In Section 9.2, we discuss fuzzy membership functions and related concepts in detail. Estimation procedures in classical as well as Bayesian framework has been developed in Section 9.4 and Section 9.5, respectively. To observe the impact of various membership functions on considered estimators, a comprehensive simulation study is provided in Subsection 9.6.1. A real dataset is analyzed in Section 9.7 and finally we conclude the chapter in Section 9.8.

9.2 Fuzzy Lifetime Data

In the literature, most of the inferential procedures for estimating the parameters of various distributions are limited to crisp data ($\in \mathbb{R}$). In real-world situations when the data cannot be measured or recorded precisely due to machine errors, human errors, or unexpected situations, the conventional procedures used for estimating model parameters need to be updated [17].

9.2.1 Fuzzy Set

When a set X is expressed, an element of the universe either belongs to it or not. Accordingly, this element has two possibilities associated with it, either it is an element of set X or it is not. It can be inferred in terms of the membership of an element in a set. In this case, the membership is binary in nature. In real-life situations, it might not always be possible to find the belongingness and non-belongingness of an element in a set. This motivates one to think about associating a degree of membership of an element to a set.

A prevalent instance of imprecision is observed when realizations are reported on the lower bound of a continuous random variable. One may expect such data as an answer when a person's age or income is asked. Sometimes, data repositories also report data in form of floored numbers which might be because storing large datasets along with lots of decimal parts of a real number may occupy ample storage space. Such imprecise real values may result in substantive changes in desired estimates.

A fuzzy set is a collection of elements with associated degrees of membership. In many aspects, the fuzzy set is comparable to the framework employed in crisp sets, but the fuzzy set is more comprehensive and potentially has a much greater range of use, particularly in pattern recognition and information processing.

9.2.2 Fuzzy Numbers and Membership Function

As real numbers are to a crisp set, the fuzzy set has fuzzy numbers associated with it [18] defines fuzzy number as follows:

Definition 1. A fuzzy number $\tilde{x}$ on X is a fuzzy subset of X which is characterized by a Borel-measurable membership function $\mu_{\tilde{x}}(.)$ which associates with each observation in X a real number in the interval [0,1]. The value of $\mu_{\tilde{x}}(.)$ represent the "grade of membership" of X in $\tilde{x}$.

There are various fuzzy numbers available in the literature. Zhao and Bose [20] evaluated sensitivity of fuzzy controller for different membership functions based on the triangular function as the foundation. For a speed-controlled induction motor drive, the performance of fuzzy control is then tested for each type of membership function. Hammed [9] employed different fuzzy membership function to improve the reliability and robustness of students' evaluation systems. Abdullah and Chong [1] proposed fuzzy human development index using various fuzzy number. Some of the frequently used membership are as follows:

1. **Triangular Fuzzy Number:** A fuzzy number $\tilde{x}$ is said to be triangular fuzzy number if its membership function $\mu_{\tilde{x}}(.)$ is;

$$\mu_{\tilde{x}}(x;a,b,c) = \begin{cases} \dfrac{x-a}{b-a} & ,a \le x \le b \\ \dfrac{c-x}{c-b} & ,b \le x \le c \\ 0 & ,otherwise \end{cases} \tag{9.1}$$

or

$$\mu_{\tilde{x}}(x;a,b,c) = \max\left(\min\left(\frac{x-a}{b-a}, \frac{c-x}{c-b}\right), 0\right) \tag{9.2}$$

where the parameters a and c locate the "feet" of the triangle and b locates the "peak" [see Figure 9.1(a)].

2. **Trapezoidal Fuzzy Number:** A fuzzy number $\tilde{x}$ is said to be trapezoidal fuzzy number if its membership function is;

$$\mu_{\tilde{x}}(x;a,b,c,d) = \begin{cases} \dfrac{x-a}{b-a} & ,a \le x \le b \\ 1 & ,b \le x \le c \\ \dfrac{d-x}{d-c} & ,c \le x \le d \\ 0 & ,otherwise \end{cases} \tag{9.3}$$

or

$$\mu_{\tilde{x}}(x;a,b,c,d) = \max\left(\min\left(\frac{x-a}{b-a}, 1, \frac{d-x}{d-c}\right), 0\right) \tag{9.4}$$

where a and d locate the "feet" of the trapezoid and b and c locate the "shoulder" [see Figure 9.1(b)]. Triangular fuzzy number is a special case of trapezoidal membership function when $b = c$.

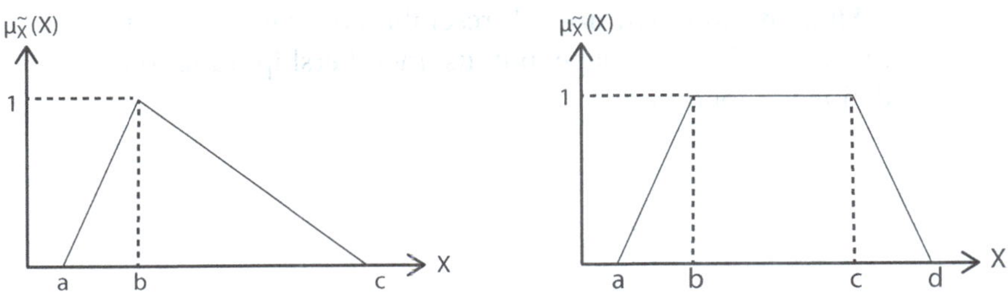

Figure 9.1 Fuzzy membership function.

3. **Gaussian Fuzzy Number:** A gaussian fuzzy number has the following membership function

$$\mu_{\tilde{x}}(x;\sigma,c)=\exp\left(-\frac{(x-c)^2}{2\sigma^2}\right) \tag{9.5}$$

where the parameter c locates the distance from the origin and σ indicates width of the curve.

4. **Sigmoidal Fuzzy Number:** A fuzzy number $\tilde{x}$ having the following membership function $\mu_{\tilde{x}}(.)$;

$$\mu_{\tilde{x}}(x;a,c)=\frac{1}{1+exp(-a(x-c))} \tag{9.6}$$

is defined as sigmoidal fuzzy number. Where c locates distance from the origin and a determines steepness of the function.

5. **Bell-Shaped Fuzzy Number:** A bell-shaped fuzzy number has membership function

$$\mu_{\tilde{x}}(x;a,b,c)=\frac{1}{\left|\frac{x-c}{a}\right|^{2b}+1} \tag{9.7}$$

where the parameter b is usually positive, parameter c locates the center of the curve and a shows width of the curve.

6. **S-Shaped Fuzzy Number:** It resembles to that of sigmoidal fuzzy number in shape but its membership function is defined as follows:

$$\mu_{\tilde{X}}(x;a,b) = \begin{cases} 0 & ,x \leq a \\ 2\left(\dfrac{x-a}{b-a}\right)^2 & ,a \leq x \leq \dfrac{a-b}{2} \\ 1-2\left(\dfrac{x-b}{b-a}\right)^2 & ,\dfrac{a+b}{2} \leq x \leq b \\ 1 & ,x \geq b \end{cases} \tag{9.8}$$

where a and b are parameters.

For this chapter we have considered triangular and trapezoidal membership functions.

9.2.3 Fuzzy Event and its Probability

Lofti A. Zadeh in his seminal paper ([19]) on "Probability measures of fuzzy event" defines fuzzy event and its probability as:

1. **Fuzzy Event:** Let $(\mathbb{R}^n, \mathbb{B}^n, \mathrm{P})$ be a probability space in which $\mathbb{B}^n$ is the Borel σ-field in $\mathbb{R}^n$, and P is a probability measure over $\mathbb{R}^n$. Then, a fuzzy event in $\mathbb{R}^n$, is a fuzzy set $\tilde{X}$, whose membership function $\mu_{\tilde{X}}$ is Borel measurable.
2. **Probability of Fuzzy Event:** The probability of a fuzzy event A is defined by the Lebesgue-Stieltjes integral:

$$\mathrm{P}(\tilde{X}) = \int_{\mathbb{R}^n} \mu_{\tilde{X}}(x)\,d\mathrm{P} \tag{9.9}$$

$$= E(\mu_{\tilde{X}}). \tag{9.10}$$

9.3 Modeling with Fuzzy Lifetime Data in Presence of Competing Risks

Consider a nonrepairable series system consisting of J components. Let the random variable T_j denotes the lifetime of jth component, $j = 1,2,\ldots,J$. Since a series system fails as soon as any of its components fails, the system's lifetime X is equal to the minimum of lifetimes of its components, that is,

$X = \min\{T_1, T_2, \ldots, T_J\}$ and $C \in \{1,2, \ldots, J\}$ denotes the cause of failure of the system. Once a system fails, we obtain realizations on a random vector (X, C) consisting of time to failure as well as cause of failure of the system. As discussed in previous sections, we assume that time to failure of units cannot be measured or recorded precisely.

Consider that T_j has probability density function (pdf) $f_j(.|\theta_j)$ with respect to the Lebesgue measure in R^+, the positive half of the real line. Here, θ_j denotes the parameter which may be vector valued. The probability that the system fails at time x due to j^{th} cause is given by the joint distribution of (X, C) which comes out to be

$$\lim_{\Delta x \to 0} \frac{1}{\Delta x} P[x \le X \le x + \Delta x, C = j] = f_j(x|\theta_j) \prod_{l \ne j}^{J} \bar{F}_l(x|\theta_l), \quad (9.11)$$

where $\bar{F}_j(x|\theta_j) = \int_t^\infty f_j(x|\theta_j)dx$ represents the reliability function of jth component.

Let us consider a life-testing experiment where the system's lifetime X cannot be observed precisely but it may be conceived in form of a fuzzy number. Under this notion, on using Zadeh's [19] probabilistic definition, the probability of system's failure due to j^{th} cause given by (9.11) can be defined by the Lebesgue-Stieltjes integral

$$\int f_j(x|\theta_j) \prod_{l \ne j}^{J} \bar{F}_l(x|\theta_j) \mu_{\tilde{x}}(x) dx.$$

Suppose that n identical series-systems were put to test and the test was terminated as soon as all the systems fail. So that n independent observations (x_i, c_i); $i = 1,2, \ldots, n$, were realized on (X, C). Here, all x_i's; $i = 1,2, \ldots, n$, are in vague form and we consider them independent fuzzy observations $\tilde{x}_i$ with membership functions $\mu_{\tilde{x}_i}(x)$. The likelihood function of $\Theta = (\theta_1, \theta_2, \ldots, \theta_J)$, in the light of observed data $(\tilde{x}, c)$ can be written as follows;

$$L(\Theta|\tilde{x}, c)) = \prod_{j=1}^{J} \left\{ \prod_{i=1}^{n_j} \int f_j(x|\theta_j) \prod_{l \ne j}^{J} \bar{F}_l(x|\theta_l) \mu_{\tilde{x}_i}(x) dx \right\}, \quad (9.12)$$

where n_j denotes the number of systems that failed due to j^{th} component.

9.4 Maximum Likelihood Estimation with Exponential Lifetimes

We assume that the lifetime of jth component follows exponential distribution with parameter θ_j having *pdf*

$$f_j(x|\vartheta_j) = \theta_j \exp(-\theta_j) \tag{9.13}$$

with its reliability function

$$\bar{F}_j(x|\theta_j) = \exp(-\theta_j x). \tag{9.14}$$

The hazard rate of this distribution is constant which is equal to θ_j for j^{th} component. The assumption of exponential lifetime is quite relevant for the useful period of device. For exponential lifetimes and $J = 2$, the likelihood (9.12) becomes

$$L(\Theta|\tilde{\boldsymbol{x}},\boldsymbol{c})) = \prod_{i=1}^{n_1}\left(\int \theta_1 \exp(-\theta_1 x)\exp(-\theta_2 x)\mu_{\tilde{x}_i}(x)dx\right)$$
$$\cdot\prod_{i=1}^{n_2}\left(\int \theta_2 \exp(-\theta_2 x)\exp(-\theta_1 x)\mu_{\tilde{x}_i}(x)dx\right)$$
$$L(\Theta|\tilde{\boldsymbol{x}},\boldsymbol{c})) = \prod_{i=1}^{n_1}\theta_1\left(\int \exp(-(\theta_1+\theta_2)x)\mu_{\tilde{x}_i}(x)dx\right)$$
$$\cdot\prod_{i=1}^{n_2}\theta_2\left(\int \exp(-(\theta_1+\theta_2)x)\mu_{\tilde{x}_i}(x)dx\right) \tag{9.15}$$

where n_1, n_2 are the number of systems failed due to component 1 and component 2 respectively.

On further simplifying the equation (9.15), we get

$$L(\Theta|(\tilde{\boldsymbol{x}},\boldsymbol{c})) = \theta_1^{n_1}\theta_2^{n_2}\prod_{i=1}^{n}\int \exp(-(\theta_1+\theta_2)x)\mu_{\tilde{x}_i}(x)dx. \tag{9.16}$$

A frequently occurred form of 'lack of precision' in lifetime data is seen when they are reported in the form of floored values (integer values by ignoring decimal part). We can obtain estimates of various parametric functions by considering these floored observations as fuzzy numbers. In light of this, we assign following membership function to the floored lifetime of ith system

$$\mu_{\tilde{x}_i}^{(a,b)}(x)=\begin{cases}\dfrac{x-x_i}{a} & ,x_i\le x\le x_i+a\\ 1 & ,x_i+a\le x\le x_i+b\\ \dfrac{x_i+1-x}{1-b} & ,x_i+b\le x\le x_i+1\\ 0 & ,otherwise\end{cases} \tag{9.17}$$

where $0 \le a \le b \le 1$ are arbitrary real numbers. The membership defined in (9.17) is a membership of a trapezoidal fuzzy number $[x_i, x_i + a, x_i + b, x_i + 1]$. If $a = b$ in (9.17), it reduces to a triangular fuzzy number $[x_i, x_i + a, x_i + 1]$.

Using (9.17), the likelihood (9.16) becomes

$$L(\Theta|(\tilde{\boldsymbol{x}},\boldsymbol{c}))=\theta_1^{n_1}\theta_2^{n_2}\prod_{i=1}^{n}\left(\int_{x_i}^{x_i+a}\exp(-(\theta_+)x)\frac{x-x_i}{a}dx+\int_{x_i+a}^{x_i+b}\exp(-(\theta_+)x)\,dx\right.$$
$$\left.+\int_{x_i+b}^{x_i+1}\exp(-(\theta_+)x)\frac{x_i+1-x}{1-b}dx\right)$$

$$L(\Theta|(\tilde{\boldsymbol{x}},\boldsymbol{c}))=\theta_1^{n_1}\theta_2^{n_2}\prod_{i=1}^{n}\left[\frac{1}{a}\int_{x_i}^{x_i+a}x\exp\left(-(\theta_+)x\right)dx-\frac{x_i}{a}\int_{x_i}^{x_i+a}\exp(-(\theta_+)x)\,dx\right.$$
$$+\int_{x_i+a}^{x_i+b}\exp(-(\theta_+)x)dx+\frac{x_i+1}{1-b}\int_{x_i+b}^{x_i+1}\exp(-(\theta_+)x)\,dx$$
$$\left.-\frac{1}{1-b}\int_{x_i+b}^{x_i+1}x\exp\left(-(\theta_+)x\right)dx\right] \tag{9.18}$$

$$L(\Theta|(\tilde{\boldsymbol{x}},\boldsymbol{c})) = \theta_1^{n_1}\theta_2^{n_2}\prod_{i=1}^{n}\left[\frac{\exp(-\theta_+ x_i)}{\theta_+^2}\left\{\frac{1}{a}[1-\exp(-\theta_+ a)] + \frac{1}{1-b}[\exp(-\theta_+) - \exp(-\theta_+ b)]\right\}\right] \tag{9.19}$$

where $(\theta_+) = (\theta_1 + \theta_2)$. For finding estimates we take logarithm of (9.19)

$$l(\Theta|(\tilde{\boldsymbol{x}},\boldsymbol{c})) = n_1 \log\theta_1 + n_2 \log\theta_2 + \sum_{i=1}^{n}\log\left[\frac{\exp(-\theta_+ x_i)}{\theta_+^2}\left\{\frac{1}{a}[1-\exp(-\theta_+ a)] + \frac{1}{1-b}[\exp(-\theta_+) - \exp(-\theta_+ b)]\right\}\right]$$

$$l(\Theta|(\tilde{\boldsymbol{x}},\boldsymbol{c})) = n_1 \log\theta_1 + n_2 \log\theta_2 - \sum_{i=1}^{n}\theta_+ x_i - 2n\log(\theta_+) + n\log\left\{\frac{1}{a}[1-\exp(-\theta_+ a)] + \frac{1}{1-b}[\exp(-\theta_+) - \exp(-\theta_+ b)]\right\} \tag{9.20}$$

Finally, in order to obtain maximum likelihood estimator (MLEs) of $\hat{\theta}_1$ and $\hat{\theta}_2$, through iterative numerical procedure, we derive following expression by differentiating (9.20) partially wrt θ_1 and θ_2 and solving the corresponding likelihood equations.

$$\hat{\theta}_1 = n_1\left[\sum_{i=1}^{n} x_i + \frac{2n}{\hat{\theta}_+} - na\frac{\left[(1-b)\exp(-\hat{\theta}_+ a) - \exp(-\hat{\theta}_+) + b\exp(-\hat{\theta}_+ b)\right]}{\left[(1-b)(1-\exp(-\hat{\theta}_+ a)) + a(\exp(-\hat{\theta}_+) - \exp(-\hat{\theta}_+ b))\right]}\right]^{-1} \tag{9.21a}$$

$$\hat{\theta}_2 = n_2\left[\sum_{i=1}^{n} x_i + \frac{2n}{\hat{\theta}_+} - na\frac{\left[(1-b)\exp(-\hat{\theta}_+ a) - \exp(-\hat{\theta}_+) + b\exp(-\hat{\theta}_+ b)\right]}{\left[(1-b)(1-\exp(-\hat{\theta}_+ a)) + a(\exp(-\hat{\theta}_+) - \exp(-\hat{\theta}_+ b))\right]}\right]^{-1} \tag{9.21b}$$

9.4.1 Bootstrap Confidence Interval

We obtain confidence intervals for the parameter based on bootstrap percentiles, which are termed as boot-p confidence intervals [7]. Wen-Liang [12, 13] has obtained bootstrap and weighted bootstrap estimates for fuzzy data. Here, using the method of Efron and Tibshirani [7], we evaluate boot-p confidence intervals for the parameters using fuzzy competing risk data. The main steps of the algorithm are as follows.

1. From the original sample $\tilde{x}_1,\tilde{x}_2,\ldots,\tilde{x}_n$, draw a sample $\tilde{y}_1,\tilde{y}_2,\ldots,\tilde{y}_n$, with replacement.
2. Compute the MLE of $\hat{\theta}$ based on $\tilde{y}_1,\tilde{y}_2,\ldots,\tilde{y}_n$, say $\hat{\theta}_j^*$.
3. Repeat step 1 and step 2, B times and arrange the values of $\hat{\theta}_j^*$ in ascending order. Let $\hat{\theta}_{j(1)}^*,\hat{\theta}_{j(2)}^*,\ldots,\hat{\theta}_{j(B)}^*$, be the ordered values.

Let $\hat{\theta}_{j([pB])}^*$ denote the pth empirical percentile, that is, $[pB]^{th}$ value in the ordered list $\hat{\theta}_{j(1)}^*,\hat{\theta}_{j(2)}^*,\ldots,\hat{\theta}_{j(B)}^*$. A two-sided 100(1 – 2α)% percentile bootstrap confidence interval of θ_j, is then given by

$$\left(\hat{\theta}_{j([\alpha B])}^*,\hat{\theta}_{j([1-\alpha)B])}^*\right). \tag{9.22}$$

Here $[x]$ denotes the integer part of x.

9.5 Bayes Estimation

In Bayesian paradigm the parameter of interest is treated as a random variable having its own distribution termed as prior distribution. It is well known that Bayesian methods perform good when sample size is small. We consider the prior distribution of θ_j to be gamma with hyper-parameter (α_j, β_j) given by

$$\pi_j(\theta_j)=\frac{\beta_j^{\alpha_j}\theta_j^{\alpha_{j-1}}exp(-\beta_j\theta_j)}{\Gamma(\alpha_j)},$$
$$\theta_j>0,\alpha_j>0,\beta_j>0\,(j=1,2;k=1,2) \tag{9.23}$$

Assuming θ_j's to be independent, the joint prior for Θ is $\pi(\Theta) = \prod_j \pi_j(\theta_j)$. Merging the joint prior with the likelihood function (9.19) we get the expression for joint posterior of Θ given $(\tilde{\mathbf{x}},\mathbf{c})$ as follows,

$$\pi(\Theta|(\tilde{x},c)) = \frac{\pi(\Theta)L(\Theta|(\tilde{\mathbf{x}},\mathbf{c}))}{\int_0^\infty \pi(\Theta)L(\Theta|(\tilde{\mathbf{x}},\mathbf{c}))d\Theta}$$

$$\propto \left(\frac{\beta_1^{\alpha_1}\theta_1^{\alpha_1-1}\exp(-\beta_1\theta_1)}{\Gamma(\alpha_1)}\right)\left(\frac{\beta_2^{\alpha_2}\theta_2^{\alpha_2-1}\exp(-\beta_2\theta_2)}{\Gamma(\alpha_2)}\right)$$

$$\cdot\theta_1^{n_1}\theta_2^{n_2}p_1^{n_1}p_2^{n_2}\prod_{i=1}^{n}\left[\frac{\exp(-\theta_+x_i)}{\theta_+^2}\left\{\frac{1}{a}[1-\exp(-\theta_+a)]\right.\right.$$

$$\left.\left.+\frac{1}{1-b}[\exp(-\theta_+)-\exp(-\theta_+b)]\right\}\right] \tag{9.24}$$

From (9.24), we see that, it is not possible to derive the expressions for the marginal posterior of any of the θ_j's. We therefore use Gibbs sampler. The full conditionals for θ_1 and θ_2 obtained from (9.24), are as follows;

$$\pi_1(\theta_1|\theta_2,(\tilde{\mathbf{x}},\mathbf{c})) \propto \theta_1^{\alpha_1+n_1-1}\exp(-\beta_1\theta_1)$$

$$\cdot\prod_{i=1}^{n}\left[\frac{\exp(-\theta_+x_i)}{\theta_+^2}\left\{\frac{1}{a}[1-\exp(-\theta_+a)]\right.\right.$$

$$\left.\left.+\frac{1}{1-b}[\exp(-\theta_+)-\exp(-\theta_+b)]\right\}\right] \tag{9.25}$$

$$\pi_2(\theta_2|\theta_1,(\tilde{x},c)) \propto \theta_1^{\alpha_2+n_2-1}\exp(-\beta_2\theta_2)$$

$$\cdot\prod_{i=1}^{n}\left[\frac{\exp(-\theta_+x_i)}{\theta_+^2}\left\{\frac{1}{a}[1-\exp(-\theta_+a)]\right.\right.$$

$$\left.\left.+\frac{1}{1-b}[\exp(-\theta_+)-\exp(-\theta_+b)]\right\}\right]. \tag{9.26}$$

We carryout Gibbs sampling using the following Multistage Gibbs Sampler algorithm [4].

Algorithm 1: At t^{th} iteration, $t = 1,2,\ldots$, using $\Theta^{(t)} = (\theta_1^{(t)},\theta_2^{(t)})$, we generate

1. $\theta_1^{(t+1)} \sim \pi_1(\theta_1|\theta_2^{(t)},(\tilde{\mathbf{x}},\mathbf{c}))$;
2. $\theta_2^{(t+1)} \sim \pi_2(\theta_2|\theta_1^{(t+1)},(\tilde{\mathbf{x}},\mathbf{c}))$.

This algorithm provides us two Markov chains. We find the points of convergence for these chains by evaluating their respective cumulative means. As soon as a chain converges to its stationary distribution, we start sampling. Let $\theta_j^{(1)}, \theta_j^{(2)}, \ldots, \theta_j^{(N)}$, where N is sufficiently large, be the observations drawn from the stationary distribution of jth Markov chain to estimate θ_j. Under squared error loss function, we can obtain Bayes estimate of θ_j by calculating the mean of these observations.

Further, from (9.25) and (9.26) we observe that none of the above full conditionals is in the form of well-known density. Therefore, to generate sample observations from the marginal densities of θ_j, $j = 1,2$, we use Metropolis-Hastings (M-H) algorithm which includes the following steps.

1. Generate a new value using this candidate distribution, say $\theta_j^{new} \sim N(\hat{\hat{\theta}}_j, \widehat{var}(\hat{\theta}_j))$.
2. Calculate the ratio: $\rho = \dfrac{\pi_j(\theta_j^{new} | \Theta_{(-\theta_j^{new})}, (\tilde{x}, c))}{\pi_j(\theta_j^{old} | \Theta_{(-\theta_j^{old})}, (\tilde{x}, c))}$.
3. Draw a random number from $U(0,1)$, say u.
4. Accept the new value θ_j^{new} if $u < min(1, \rho)$.

9.5.1 Highest Posterior Density Confidence Estimates

Here we evaluate highest posterior density (HPD) intervals for the parameters under study based on sample obtained through Gibbs Sampler and M-H algorithm. We first arrange these sample values into ascending order. Then using method of Chen & Shao [3] we obtain the (*HPD*) intervals which has the property that the minimum density of any point within this interval is equal to or larger than the density of any point outside it. To evaluate an HPD interval for θ_j(say) using sample values $\theta_{j(1)}, \theta_{j(2)}, \ldots, \theta_{j(N)}$, we first obtain all the possible $100(1 - 2\alpha)\%$ credible intervals, that is

$$(\theta_{(j)}, \theta_{(j+[(1-2\alpha)N])}); \ j = 1,2, \ldots, [2\alpha N]$$

and evaluate their corresponding length given by

$$l_j = \theta_{(j+[(1-2\alpha)N])} - \theta_{(j)}$$

then pick up the interval having minimum length.

9.6 Numerical Illustration

This chapter mainly concerns with the estimation of component reliability when lifetimes are observed in the form of fuzzy numbers. To show the applicability of the proposed methods various numerical illustrations have been provided. The developed method concerns with the lifetimes which are stored in the form of floored observations. We assume these floored values as fuzzy numbers with trapezoidal membership function given in (9.8). This assumption of membership function left us with infinite possible values of a and $b(0 < a < b < 1)$. Since numerical illustration of all the combinations in not possible, we consider three forms of fuzzy number presented as follows.

$$\mu_{\tilde{x}_i}^{(1)}(x)=\begin{cases}\dfrac{x-x_i}{0.1} & ,x_i \le x \le x_i+0.1\\ 1 & ,x_i+0.1 \le x \le x_i+0.3\\ \dfrac{x_i+1-x}{0.7} & ,x_i+0.3 \le x \le x_i+1\\ 0 & ,otherwise\end{cases}$$

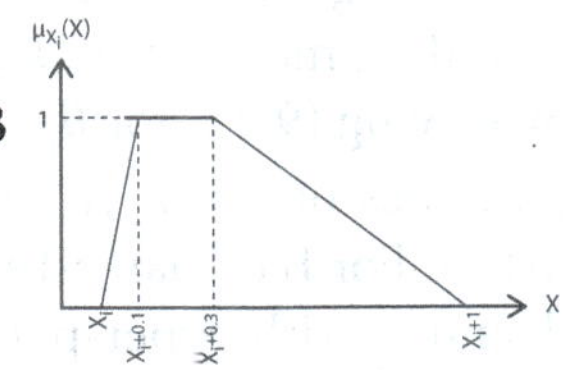

$$\mu_{\tilde{x}_i}^{(2)}(x)=\begin{cases}\dfrac{x-x_i}{0.4} & ,x_i \le x \le x_i+0.4\\ 1 & ,x_i+0.4 \le x \le x_i+0.6\\ \dfrac{x_i+1-x}{0.4} & ,x_i+0.6 \le x \le x_i+1\\ 0 & ,otherwise\end{cases}$$

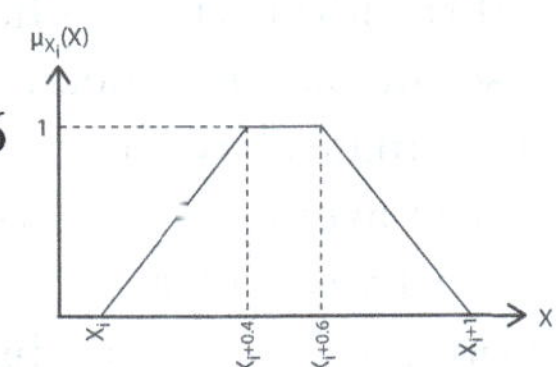

$$\mu_{\tilde{x}_i}^{(3)}(x)=\begin{cases}\dfrac{x-x_i}{0.7} & ,x_i \le x \le x_i+0.7\\ 1 & ,x_i+0.7 \le x \le x_i+0.9\\ \dfrac{x_i+1-x}{0.1} & ,x_i+0.9 \le x \le x_i+1\\ 0 & ,otherwise\end{cases}$$

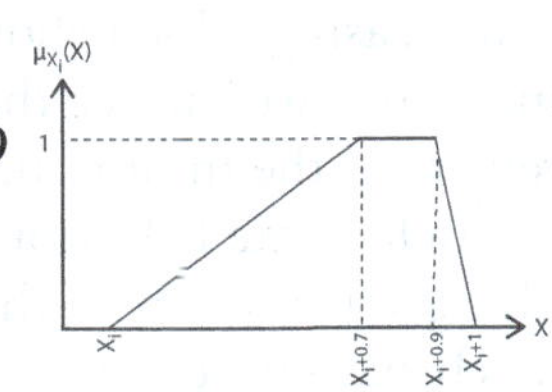

9.6.1 Simulation Study

Simulation study has the capacity to comprehend the behaviour of statistical methods by unravelling the truth about the parameter of interest in due process of generating the data. This enables one to investigate various properties of estimators of various parametric functions like mean-life and reliability which cannot be obtained in explicit form. This simulation study is designed for two-component series systems to study the impact of different membership functions on the concerned estimators. For systems with more than two components, similar methodology can be adopted. We start with generating a competing risk data set, for which lifetimes of component 1 (T_1) and component 2 (T_2) are generated from exponential distributions with rate parameters $\theta_1 = 0.2$ and $\theta_2 = 0.25$, respectively. Rest of the procedure can be followed as given in Bundel and Tomer [2].

Using generated fuzzy data, we obtain ML estimates of rate parameters θ_1 and θ_2, mean-lives ($m_1 = 1/\theta_1$ and $m_2 = 1/\theta_2$) and reliabilities $\overline{F}_1$ and $\overline{F}_2$ by solving (9.21) for θ_1 and θ_2, through iterative numerical procedure. The estimates of $m_1, m_2, \overline{F}_1$ and $\overline{F}_2$ are obtained using invariance property of MLEs. For Bayesian estimation, we employ $Gamma(\alpha_j, \beta_j); j = 1,2$ prior for θ_1 and θ_2 with hyper-parameters $\alpha_1 = 0.8$, $\beta_1 = 4$ and $\alpha_2 = 1$, $\beta_2 = 4$, respectively and evaluate Bayes estimate of rate parameter, mean-life and reliability using Gibbs sampler. For varying samples *viz.* $n = 15, 35, 55$, and 75 the performance of estimators has been checked. For each value of n, $N = 4000$ random samples have been generated and using them, average values, absolute bias and mean squared error (MSE)/posterior expected loss (PEL) of estimates have been evaluated.

The average values of ML estimates of the rate parameters θ_1 and θ_2, mean-lives and reliabilities are presented in Table 9.1 with their corresponding MSEs and absolute bias. Similarly average values of Bayes estimates are shown in Table 9.2 with their respective posterior expected losses and absolute bias. From these two tables one can observe that as sample size increases, MSEs/PELs decreases. Further from same tables, we observe that absolute bias of estimators is decreasing as sample size is increasing which shows that the estimators are consistent. On observing MSEs, we can see that for the rate parameter and reliability, $\mu_{\tilde{x}_i}^{(3)}(x)$ is returning the minimum MSE whereas for mean-life minimum occurs for $\mu_{\tilde{x}_i}^{(1)}(x)$. Same behavior can be observed in respect of PELs in Table 9.2. This gives us a way to think that different parameters/function of parameters behave differently under the same membership function.

For interval estimates the average length of boot-p CIs of the rate parameter, mean-lives and reliability are presented in Table 9.3 with their

Table 9.1 Average values (AV) of ML estimates of parameter, meanlife and reliability of component $1(j = 1)$ and component $2(j = 2)$ with their MSE, bias and absolute bias(|Bias|).

			$\hat{\theta}_j$			$\hat{m}_j$			$\hat{\bar{F}}_j$		
n	j	**Estimates**	$\mu_{\tilde{x}_i}^{(1)}(x)$	$\mu_{\tilde{x}_i}^{(2)}(x)$	$\mu_{\tilde{x}_i}^{(3)}(x)$	$\mu_{\tilde{x}_i}^{(1)}(x)$	$\mu_{\tilde{x}_i}^{(2)}(x)$	$\mu_{\tilde{x}_i}^{(3)}(x)$	$\mu_{\tilde{x}_i}^{(1)}(x)$	$\mu_{\tilde{x}_i}^{(2)}(x)$	$\mu_{\tilde{x}_i}^{(3)}(x)$
15	1	AV	0.2301	0.2157	0.2039	5.0084	5.2938	5.5632	0.7142	0.7286	0.7408
		MSE	0.0089	0.0067	0.0054	3.7568	4.0545	4.5025	0.0087	0.0070	0.0060
		Bias	0.0301	0.0157	0.0039	0.0084	0.2938	0.5632	-0.0266	-0.0122	0.0000
		|Bias|	0.0690	0.0616	0.0570	1.5047	1.5204	1.5686	0.0720	0.0659	0.0622
	2	AV	0.2775	0.2600	0.2457	4.1042	4.3360	4.5561	0.6673	0.6833	0.6971
		MSE	0.0119	0.0088	0.0072	2.3783	2.5913	2.9058	0.0096	0.0079	0.0070
		Bias	0.0275	0.0100	0.0043	0.1042	0.3360	0.5561	-0.0200	-0.0040	0.0098
		|Bias|	0.0772	0.0692	0.0647	1.1380	1.1583	1.2079	0.0745	0.0689	0.0661

(Continued)

Table 9.1 Average values (AV) of ML estimates of parameter, meanlife and reliability of component $1(j = 1)$ and component $2(j = 2)$ with their MSE, bias and absolute bias(|Bias|). (*Continued*)

			$\hat{\theta}_j$			$\hat{m}_j$			$\hat{F}_j$		
n	**j**	**Estimates**	$\mu^{(1)}_{\tilde{x}_i}(x)$	$\mu^{(2)}_{\tilde{x}_i}(x)$	$\mu^{(3)}_{\tilde{x}_i}(x)$	$\mu^{(1)}_{\tilde{x}_i}(x)$	$\mu^{(2)}_{\tilde{x}_i}(x)$	$\mu^{(3)}_{\tilde{x}_i}(x)$	$\mu^{(1)}_{\tilde{x}_i}(x)$	$\mu^{(2)}_{\tilde{x}_i}(x)$	$\mu^{(3)}_{\tilde{x}_i}(x)$
35	1	AV	0.2159	0.2035	0.1931	4.9714	5.2534	5.5216	0.7260	0.7391	0.7505
		MSE	0.0035	0.0028	0.0024	1.8970	2.0855	2.4230	0.0039	0.0032	0.0030
		Bias	0.0159	0.0035	-0.0069	-0.0286	0.2534	0.5216	-0.0148	-0.0017	0.0096
		|Bias|	0.0453	0.0409	0.0390	1.0593	1.0741	1.1340	0.0488	0.0450	0.0436
	2	AV	0.2710	0.2555	0.2424	3.9002	4.1215	4.3319	0.6691	0.6842	0.6974
		MSE	0.0046	0.0034	0.0029	0.8820	0.9297	1.0707	0.0043	0.0034	0.0031
		Bias	0.0210	0.0055	-0.0076	-0.0998	0.1215	0.3319	-0.0182	-0.0031	0.0101
		|Bias|	0.0513	0.0457	0.0434	0.7419	0.7399	0.7782	0.0509	0.0464	0.0450

(*Continued*)

Table 9.1 Average values (AV) of ML estimates of parameter, meanlife and reliability of component $1(j = 1)$ and component $2(j = 2)$ with their MSE, bias and absolute bias(|Bias|). (*Continued*)

n	j	Estimates	$\hat{\theta}_j$			$\hat{m}_j$			$\hat{\bar{F}}_j$		
			$\mu^{(1)}_{\tilde{x}_i}(x)$	$\mu^{(2)}_{\tilde{x}_i}(x)$	$\mu^{(3)}_{\tilde{x}_i}(x)$	$\mu^{(1)}_{\tilde{x}_i}(x)$	$\mu^{(2)}_{\tilde{x}_i}(x)$	$\mu^{(3)}_{\tilde{x}_i}(x)$	$\mu^{(1)}_{\tilde{x}_i}(x)$	$\mu^{(2)}_{\tilde{x}_i}(x)$	$\mu^{(3)}_{\tilde{x}_i}(x)$
55	1	AV	0.2152	0.2031	0.1929	4.8586	5.1348	5.3976	0.7258	0.7388	0.7500
		MSE	0.0023	0.0017	0.0015	1.1359	1.2047	1.4182	0.0026	0.0020	0.0019
		Bias	0.0152	0.0031	-0.0071	-0.1414	0.1348	0.3976	-0.0150	-0.0020	0.0092
		|Bias|	0.0367	0.0325	0.0312	0.8384	0.8316	0.8816	0.0397	0.0358	0.0349
	2	AV	0.2673	0.2524	0.2397	3.8771	4.0975	4.3070	0.6716	0.6865	0.6995
		MSE	0.0029	0.0022	0.0019	0.5649	0.5858	0.6989	0.0028	0.0022	0.0021
		Bias	0.0173	0.0024	-0.0103	-0.1229	0.0975	0.3070	-0.0157	-0.0008	0.0122
		|Bias|	0.0413	0.0365	0.0354	0.6011	0.5934	0.6359	0.0413	0.0373	0.0369

(*Continued*)

Table 9.1 Average values (AV) of ML estimates of parameter, meanlife and reliability of component $1(j = 1)$ and component $2(j = 2)$ with their MSE, bias and absolute bias(|Bias|). (*Continued*)

			$\hat{\theta}_j$			$\hat{m}_j$			$\hat{\bar{F}}_j$		
n	**j**	**Estimates**	$\mu^{(1)}_{\tilde{x}_i}(x)$	$\mu^{(2)}_{\tilde{x}_i}(x)$	$\mu^{(3)}_{\tilde{x}_i}(x)$	$\mu^{(1)}_{\tilde{x}_i}(x)$	$\mu^{(2)}_{\tilde{x}_i}(x)$	$\mu^{(3)}_{\tilde{x}_i}(x)$	$\mu^{(1)}_{\tilde{x}_i}(x)$	$\mu^{(2)}_{\tilde{x}_i}(x)$	$\mu^{(3)}_{\tilde{x}_i}(x)$
75	1	AV	0.2119	0.2002	0.1903	4.8791	5.1552	5.4179	0.7289	0.7416	0.7526
		MSE	0.0016	0.0012	0.0012	0.8444	0.9063	1.1112	0.0019	0.0015	0.0015
		Bias	0.0119	0.0002	-0.0097	-0.1209	0.1552	0.4179	-0.0119	0.0008	0.0118
		|Bias|	0.0314	0.0282	0.0278	0.7304	0.7325	0.7957	0.0342	0.0312	0.0313
	2	AV	0.2659	0.2512	0.2387	3.8568	4.0752	4.2830	0.6725	0.6872	0.7000
		MSE	0.0020	0.0015	0.0014	0.3989	0.4025	0.4966	0.0020	0.0016	0.0015
		Bias	0.0159	0.0012	-0.0113	-0.1432	0.0752	0.2830	-0.0148	-0.0001	0.0127
		|Bias|	0.0350	0.0306	0.0304	0.5090	0.4976	0.5436	0.0351	0.0314	0.0318

Table 9.2 Average values (AV) of Bayes estimates of parameter, meanlife and reliability of component 1($j = 1$) and component 2($j = 2$) with their PEL, bias and absolute bias(|Bias|).

n	j	Estimates	$\hat{\theta}_j$			$\hat{m}_j$			$\hat{\bar{F}}_j$		
			$\mu^{(1)}_{\tilde{x}_i}(x)$	$\mu^{(2)}_{\tilde{x}_i}(x)$	$\mu^{(3)}_{\tilde{x}_i}(x)$	$\mu^{(1)}_{\tilde{x}_i}(x)$	$\mu^{(2)}_{\tilde{x}_i}(x)$	$\mu^{(3)}_{\tilde{x}_i}(x)$	$\mu^{(1)}_{\tilde{x}_i}(x)$	$\mu^{(2)}_{\tilde{x}_i}(x)$	$\mu^{(3)}_{\tilde{x}_i}(x)$
15	1	AV	0.2286	0.2138	0.2016	5.6822	5.9930	6.2916	0.7195	0.7338	0.7460
		PEL	0.0091	0.0068	0.0055	6.3865	7.2869	8.3620	0.0084	0.0069	0.0061
		Bias	0.0286	0.0138	0.0016	0.6822	0.9930	1.2916	-0.0213	-0.0070	0.0052
		|Bias|	0.0696	0.0623	0.0577	1.8268	1.9178	2.0362	0.0712	0.0658	0.0627
	2	AV	0.2775	0.2594	0.2446	4.4954	4.7414	4.9760	0.6714	0.6874	0.7011
		PEL	0.0122	0.0090	0.0073	3.5897	4.0718	4.6542	0.0093	0.0078	0.0070
		Bias	0.0275	0.0094	0.0051	0.4954	0.7414	0.9760	-0.0159	0.0001	0.0138
		|Bias|	0.0783	0.0700	0.0655	1.3173	1.3862	1.4788	0.0740	0.0690	0.0667

(Continued)

Table 9.2 Average values (AV) of Bayes estimates of parameter, meanlife and reliability of component 1($j = 1$) and component 2($j = 2$) with their PEL, bias and absolute bias(|Bias|). (*Continued*)

			$\hat{\theta}_j$			$\hat{m}_j$			$\hat{\bar{F}}_j$		
n	**j**	**Estimates**	$\mu_{\tilde{x}_i}^{(1)}(x)$	$\mu_{\tilde{x}_i}^{(2)}(x)$	$\mu_{\tilde{x}_i}^{(3)}(x)$	$\mu_{\tilde{x}_i}^{(1)}(x)$	$\mu_{\tilde{x}_i}^{(2)}(x)$	$\mu_{\tilde{x}_i}^{(3)}(x)$	$\mu_{\tilde{x}_i}^{(1)}(x)$	$\mu_{\tilde{x}_i}^{(2)}(x)$	$\mu_{\tilde{x}_i}^{(3)}(x)$
35	1	AV	0.2156	0.2030	0.1924	5.1865	5.4821	5.7620	0.7277	0.7409	0.7523
		PEL	0.0036	0.0028	0.0024	2.2824	2.6387	3.1488	0.0039	0.0032	0.0030
		Bias	0.0156	0.0030	-0.0076	0.1865	0.4821	0.7620	-0.0131	0.0001	0.0115
		|Bias|	0.0456	0.0412	0.0394	1.1238	1.1761	1.2710	0.0487	0.0451	0.0440
	2	AV	0.2714	0.2556	0.2423	4.0144	4.2426	4.4595	0.6702	0.6855	0.6987
		PEL	0.0047	0.0035	0.0030	0.9667	1.0762	1.2819	0.0042	0.0034	0.0032
		Bias	0.0214	0.0056	-0.0077	0.0144	0.2426	0.4595	-0.0171	-0.0018	0.0114
		|Bias|	0.0518	0.0459	0.0436	0.7630	0.7837	0.8432	0.0508	0.0464	0.0452

(*Continued*)

Table 9.2 Average values (AV) of Bayes estimates of parameter, meanlife and reliability of component $1(j=1)$ and component $2(j=2)$ with their PEL, bias and absolute bias(|Bias|). (*Continued*)

n	j	Estimates	$\hat{\theta}_j$			$\hat{m}_j$			$\hat{\bar{F}}_j$		
			$\mu^{(1)}_{\tilde{x}_i}(x)$	$\mu^{(2)}_{\tilde{x}_i}(x)$	$\mu^{(3)}_{\tilde{x}_i}(x)$	$\mu^{(1)}_{\tilde{x}_i}(x)$	$\mu^{(2)}_{\tilde{x}_i}(x)$	$\mu^{(3)}_{\tilde{x}_i}(x)$	$\mu^{(1)}_{\tilde{x}_i}(x)$	$\mu^{(2)}_{\tilde{x}_i}(x)$	$\mu^{(3)}_{\tilde{x}_i}(x)$
55	1	AV	0.2150	0.2028	0.1925	4.9785	5.2628	5.5330	0.7268	0.7398	0.7511
		PEL	0.0023	0.0017	0.0015	1.2267	1.3760	1.6757	0.0025	0.0020	0.0019
		Bias	0.0150	0.0028	-0.0075	-0.0215	0.2628	0.5330	-0.0140	-0.0010	0.0103
		|Bias|	0.0368	0.0326	0.0313	0.8543	0.8739	0.9495	0.0395	0.0358	0.0351
	2	AV	0.2677	0.2525	0.2396	3.9446	4.1694	4.3830	0.6722	0.6872	0.7002
		PEL	0.0030	0.0022	0.0020	0.5871	0.6419	0.7911	0.0028	0.0022	0.0021
		Bias	0.0177	0.0025	-0.0104	-0.0554	0.1694	0.3830	-0.0151	-0.0001	0.0129
		|Bias|	0.0416	0.0366	0.0355	0.6064	0.6153	0.6736	0.0412	0.0373	0.0371

(*Continued*)

Table 9.2 Average values (AV) of Bayes estimates of parameter, meanlife and reliability of component 1($j = 1$) and component 2($j = 2$) with their PEL, bias and absolute bias(|Bias|). *(Continued)*

n	j	Estimates	$\hat{\theta}_j$			$\hat{m}_j$			$\hat{\bar{F}}_j$		
			$\mu^{(1)}_{\tilde{x}_i}(x)$	$\mu^{(2)}_{\tilde{x}_i}(x)$	$\mu^{(3)}_{\tilde{x}_i}(x)$	$\mu^{(1)}_{\tilde{x}_i}(x)$	$\mu^{(2)}_{\tilde{x}_i}(x)$	$\mu^{(3)}_{\tilde{x}_i}(x)$	$\mu^{(1)}_{\tilde{x}_i}(x)$	$\mu^{(2)}_{\tilde{x}_i}(x)$	$\mu^{(3)}_{\tilde{x}_i}(x)$
75	1	AV	0.2118	0.2000	0.1900	4.9645	5.2462	5.5146	0.7296	0.7424	0.7534
		PEL	0.0016	0.0013	0.0012	0.8861	1.0038	1.2678	0.0019	0.0015	0.0015
		Bias	0.0118	0.0000	-0.0100	-0.0355	0.2462	0.5146	-0.0112	0.0015	0.0126
		|Bias|	0.0315	0.0283	0.0280	0.7394	0.7631	0.8470	0.0341	0.0313	0.0315
	2	AV	0.2662	0.2514	0.2388	3.9038	4.1254	4.3362	0.6729	0.6876	0.7005
		PEL	0.0021	0.0015	0.0014	0.4038	0.4301	0.5489	0.0020	0.0016	0.0016
		Bias	0.0162	0.0014	-0.0112	-0.0962	0.1254	0.3362	-0.0144	0.0004	0.0132
		|Bias|	0.0352	0.0307	0.0305	0.5093	0.5111	0.5704	0.0351	0.0315	0.0319

Table 9.3 Average lengths (AL) of boot-p interval estimates of rate parameter, meanlife and reliability with their respective shapes and coverage probabilities (CP).

			$\hat{\theta}_j$			$\hat{m}_j$			$\hat{\bar{F}}_j$		
n	j	**Estimates**	$\mu_{\tilde{x}_i}^{(1)}(x)$	$\mu_{\tilde{x}_i}^{(2)}(x)$	$\mu_{\tilde{x}_i}^{(3)}(x)$	$\mu_{\tilde{x}_i}^{(1)}(x)$	$\mu_{\tilde{x}_i}^{(2)}(x)$	$\mu_{\tilde{x}_i}^{(3)}(x)$	$\mu_{\tilde{x}_i}^{(1)}(x)$	$\mu_{\tilde{x}_i}^{(2)}(x)$	$\mu_{\tilde{x}_i}^{(3)}(x)$
15	1	AL	0.3145	0.2734	0.245	10.2508	10.6738	11.1079	0.2173	0.2043	0.1938
		Shape	4.874	4.5854	4.3917	6.8655	6.9768	7.0427	2.3158	2.3673	2.4009
		CP	0.6665	0.7148	0.7372	0.7665	0.7312	0.7105	0.7665	0.7312	0.7105
	2	AL	0.3491	0.3005	0.2675	6.252	6.4913	6.7142	0.2208	0.2079	0.1971
		Shape	5.0598	4.7439	4.5499	5.4205	5.5052	5.5624	2.2671	2.3148	2.3515
		CP	0.6835	0.7385	0.7585	0.746	0.7062	0.6728	0.746	0.7062	0.6728
35	1	AL	0.1727	0.1555	0.1425	4.7039	4.8736	5.045	0.1453	0.1367	0.1296
		Shape	4.0165	3.8888	3.8012	4.6046	4.6492	4.6811	2.432	2.4717	2.4998
		CP	0.652	0.7108	0.7635	0.7588	0.714	0.6607	0.7588	0.714	0.6607

(Continued)

Table 9.3 Average lengths (AL) of boot-p interval estimates of rate parameter, meanlife and reliability with their respective shapes and coverage probabilities (CP). (*Continued*)

n	j	Estimates	$\hat{\theta}_j$			$\hat{m}_j$			$\hat{\bar{F}}_j$		
			$\mu^{(1)}_{\tilde{x}_i}(x)$	$\mu^{(2)}_{\tilde{x}_i}(x)$	$\mu^{(3)}_{\tilde{x}_i}(x)$	$\mu^{(1)}_{\tilde{x}_i}(x)$	$\mu^{(2)}_{\tilde{x}_i}(x)$	$\mu^{(3)}_{\tilde{x}_i}(x)$	$\mu^{(1)}_{\tilde{x}_i}(x)$	$\mu^{(2)}_{\tilde{x}_i}(x)$	$\mu^{(3)}_{\tilde{x}_i}(x)$
	2	AL	0.1953	0.1744	0.1585	2.9583	3.0454	3.1332	0.1501	0.141	0.1336
		Shape	4.1466	4.0059	3.9059	4.0031	4.0427	4.0659	2.3854	2.4246	2.4514
		CP	0.6382	0.7092	0.7625	0.7648	0.7173	0.6617	0.7648	0.7173	0.6617
55	1	AL	0.1315	0.1192	0.1097	3.2835	3.3977	3.5096	0.1174	0.1105	0.1048
		Shape	3.7494	3.6603	3.5944	4.0756	4.1123	4.1328	2.5022	2.5381	2.5617
		CP	0.634	0.7155	0.7725	0.7725	0.718	0.6482	0.7725	0.718	0.6482
	2	AL	0.148	0.1332	0.1219	2.2176	2.2815	2.3435	0.1216	0.1144	0.1083
		Shape	3.8372	3.7426	3.6736	3.7114	3.744	3.7552	2.4685	2.5037	2.5226
		CP	0.631	0.7245	0.7808	0.768	0.7042	0.6302	0.768	0.7042	0.6302

(*Continued*)

Table 9.3 Average lengths (AL) of boot-p interval estimates of rate parameter, meanlife and reliability with their respective shapes and coverage probabilities (CP). (*Continued*)

			$\hat{\theta}_j$			$\hat{m}_j$			$\hat{\bar{F}}_j$		
n	j	**Estimates**	$\mu^{(1)}_{\tilde{x}_i}(x)$	$\mu^{(2)}_{\tilde{x}_i}(x)$	$\mu^{(3)}_{\tilde{x}_i}(x)$	$\mu^{(1)}_{\tilde{x}_i}(x)$	$\mu^{(2)}_{\tilde{x}_i}(x)$	$\mu^{(3)}_{\tilde{x}_i}(x)$	$\mu^{(1)}_{\tilde{x}_i}(x)$	$\mu^{(2)}_{\tilde{x}_i}(x)$	$\mu^{(3)}_{\tilde{x}_i}(x)$
75	1	AL	0.1095	0.0998	0.092	2.7259	2.8155	2.9057	0.1013	0.0953	0.0903
		Shape	3.6161	3.5485	3.4808	3.8569	3.8732	3.9034	2.5451	2.5688	2.5993
		CP	0.6408	0.7208	0.7742	0.7618	0.6985	0.6095	0.7618	0.6985	0.6095
	2	AL	0.1236	0.1117	0.1025	1.8348	1.8818	1.9346	0.105	0.0986	0.0934
		Shape	3.6922	3.6208	3.5583	3.5595	3.5719	3.5959	2.5167	2.5395	2.566
		CP	0.628	0.732	0.793	0.7882	0.7133	0.6058	0.7882	0.7133	0.6058

Table 9.4 Average lengths (AL) of HPD interval estimates of rate parameter, meanlife and reliability with their respective shapes and coverage probabilities (CP).

			$\hat{\theta}_j$			$\hat{m}_j$			$\hat{\bar{F}}_j$		
n	j	**Estimates**	$\mu^{(1)}_{\tilde{x}_i}(x)$	$\mu^{(2)}_{\tilde{x}_i}(x)$	$\mu^{(3)}_{\tilde{x}_i}(x)$	$\mu^{(1)}_{ex_i}(x)$	$\mu^{(2)}_{\tilde{x}_i}(x)$	$\mu^{(3)}_{\tilde{x}_i}(x)$	$\mu^{(1)}_{\tilde{x}_i}(x)$	$\mu^{(2)}_{\tilde{x}_i}(x)$	$\mu^{(3)}_{\tilde{x}_i}(x)$
15	1	AL	0.2505	0.2285	0.2119	7.1488	7.398	7.6571	0.2573	0.2416	0.2291
		Shape	1.0514	1.0475	1.0467	1.3589	1.3491	1.344	1.0001	0.9987	0.9972
		CP	0.8632	0.8528	0.8398	0.8722	0.8858	0.895	0.8505	0.846	0.8338
	2	AL	0.2775	0.2522	0.233	4.9852	5.1428	5.3067	0.2637	0.2482	0.2357
		Shape	1.047	1.0439	1.0398	1.3014	1.2911	1.287	1.0062	1.0041	1.0052
		CP	0.854	0.8433	0.8222	0.8662	0.8735	0.8798	0.841	0.8378	0.823
35	1	AL	0.1539	0.1425	0.1333	3.9465	4.1077	4.2697	0.1646	0.1557	0.1482
		Shape	1.0303	1.0296	1.0287	1.1779	1.1752	1.1754	0.9983	0.9994	0.999
		CP	0.8258	0.8258	0.8065	0.8345	0.8482	0.848	0.819	0.825	0.8095
	2	AL	0.1734	0.1599	0.1492	2.6689	2.7696	2.869	0.1705	0.1615	0.154
		Shape	1.0267	1.0248	1.025	1.1513	1.1464	1.1463	1.004	1.0061	1.0038
		CP	0.8348	0.8383	0.8145	0.8235	0.8475	0.852	0.8242	0.8362	0.8162

(Continued)

Table 9.4 Average lengths (AL) of HPD interval estimates of rate parameter, meanlife and reliability with their respective shapes and coverage probabilities (CP). (*Continued*)

			$\hat{\theta}_j$			$\hat{m}_j$			$\hat{\bar{F}}_j$		
n	j	**Estimates**	$\mu^{(1)}_{\tilde{x}_i}(x)$	$\mu^{(2)}_{\tilde{x}_i}(x)$	$\mu^{(3)}_{\tilde{x}_i}(x)$	$\mu^{(1)}_{ex_i}(x)$	$\mu^{(2)}_{\tilde{x}_i}(x)$	$\mu^{(3)}_{\tilde{x}_i}(x)$	$\mu^{(1)}_{\tilde{x}_i}(x)$	$\mu^{(2)}_{\tilde{x}_i}(x)$	$\mu^{(3)}_{\tilde{x}_i}(x)$
55	1	AL	0.1215	0.1129	0.1059	2.922	3.0454	3.1671	0.1309	0.124	0.1182
		Shape	1.0254	1.0228	1.0217	1.1288	1.1303	1.1277	0.9996	0.9997	1.0007
		CP	0.8285	0.83	0.8075	0.8178	0.8413	0.8383	0.8235	0.8272	0.8102
	2	AL	0.1363	0.1261	0.1179	2.0577	2.1364	2.2148	0.1355	0.1285	0.1226
		Shape	1.0211	1.0217	1.0189	1.1122	1.1119	1.1096	1.0056	1.0033	1.0034
		CP	0.821	0.8285	0.7992	0.814	0.836	0.8262	0.814	0.8232	0.8018
75	1	AL	0.1029	0.0958	0.0899	2.4792	2.5844	2.6864	0.1116	0.1058	0.1009
		Shape	1.0212	1.0184	1.0177	1.1069	1.1063	1.1043	1.0004	1.0001	1.0011
		CP	0.8122	0.8242	0.7852	0.8085	0.8302	0.8192	0.8072	0.823	0.7905
	2	AL	0.1156	0.1073	0.1005	1.7265	1.7936	1.8608	0.1156	0.1098	0.1049
		Shape	1.0172	1.019	1.0168	1.0929	1.0901	1.0902	1.005	1.0028	1.0027
		CP	0.8235	0.8308	0.7995	0.808	0.8378	0.8282	0.8168	0.8298	0.8025

corresponding shape and coverage probabilities. Similarly, in Table 9.4 average length of highest posterior density intervals for the three parameters with their respective shape and coverage probability has been presented. Here the shape $=(\hat{\theta}_U-\hat{\theta})/(\hat{\theta}-\hat{\theta}_L)$ is evaluated as a measure of asymmetry for the interval estimates $[\hat{\theta}_L,\hat{\theta}_U]$ of the parameter θ. From these two tables one can observe that as sample size increases, average length of intervals decreases. For rate parameter and reliability, minimum average length is obtained in $\mu_{\tilde{x}_i}^{(3)}(x)$ and for mean life it is observed for $\mu_{\tilde{x}_i}^{(3)}(x)$. From same table, we also observe that shape of boot-p estimators is heavily positively skewed whereas that of HPD is fairly normal. On observing coverage probability HPD intervals are giving a better coverage than that of boot-p intervals.

9.6.2 Reliability Analysis Using Simulated Data

In this chapter we are focussing mainly on the reliability measures of the unit under consideration. In simulation study, we have provided various estimates of mean lives and reliabilities of the two component series system. Here we present a study based on simulated sample to check the effect of different membership on estimates of reliability.

A sample of size 50 is generated from exponential distributions with rates 0.02 and 0.03 individually. Finally minimum of the two has been found and floored to get the desired data set (Table 9.5). Table 9.6 is constructed for ML and Bayes estimators of reliability function under different membership function at the time point $T = 25$.

Table 9.5 One sample study: Floored failure times with their respective cause of failure.

(0,1)	(0,1)	(0,1)	(0,1)	(0,2)	(0,2)	(1,1)	(1,1)	(1,2)	(1,2)
(1,2)	(2,1)	(2,1)	(4,1)	(4,2)	(4,2)	(5,1)	(5,1)	(8,1)	(8,2)
(9,1)	(10,2)	(12,2)	(12,2)	(15,1)	(15,1)	(18,1)	(20,2)	(20,2)	(20,2)
(21,1)	(22,2)	(22,2)	(24,2)	(25,1)	(25,1)	(27,1)	(27,2)	(28,1)	(31,1)
(32,2)	(35,2)	(36,2)	(39,2)	(40,2)	(43,2)	(48,2)	(48,2)	(56,1)	(58,1)

Table 9.6 Reliability estimates for different membership function.

Estimates	Triangular membership			Trapezoidal membership		
		$\hat{R}_1$	$\hat{R}_2$		$\hat{R}_1$	$\hat{R}_2$
MLE	$\mu_{\tilde{x}_i}^{(0.05,0.05)}(x)$	0.5143	0.4866	$\mu_{\tilde{x}_i}^{(0.25,0.75)}(x)$	0.5171	0.4895
Boot-p		(0.4648, 0.6502)	(0.4437, 0.6029)		(0.4679, 0.6522)	(0.4468, 0.6053)
Bayes		0.5181	0.4888		0.521	0.4917
HPD		(0.4226, 0.6206)	(0.3986, 0.5801)		(0.426, 0.6228)	(0.401, 0.5821)
MLE	$\mu_{\tilde{x}_i}^{(0.5,0.5)}(x)$	0.5171	0.4895	$\mu_{\tilde{x}_i}^{(0.25,0.5)}(x)$	0.5162	0.4885
Boot-p		(0.4679, 0.6522)	(0.4468, 0.6053)		(0.4669, 0.6516)	(0.4458, 0.6046)
Bayes		0.521	0.4917		0.5201	0.4908
HPD		(0.426, 0.6228)	(0.401, 0.5821)		(0.4248, 0.622)	(0.4001, 0.5813)
MLE	$\mu_{\tilde{x}_i}^{(0.95,0.95)}(x)$	0.5199	0.4924	$\mu_{\tilde{x}_i}^{(0.5,0.75)}(x)$	0.5181	0.4905
Boot-p		(0.4709, 0.6542)	(0.45, 0.6074)		(0.4689, 0.6529)	(0.448, 0.6061)
Bayes		0.5238	0.4946		0.5219	0.4927
HPD		(0.4266, 0.6228)	(0.4046, 0.5847)		(0.4271, 0.6235)	(0.4024, 0.5831)

9.7 Real Data Study

As a real life example, we consider a competing risk dataset of 36 small electric appliances as discussed in [2]. The data set is shown in Table 9.7. Here, we have found estimates of mean-life and reliability using the proposed methodology for the membership function discussed in Subsection 9.6.2. Table 9.8 and Table 9.9 provide the estimates of mean-life and reliability for the small electric appliances.

9.8 Conclusion

In this chapter, we have considered competing risk analysis of series-system lifetime data when lifetimes are not observed precisely. Considering such lifetimes to be fuzzy numbers, we have provided procedures to obtain maximum likelihood and Bayesian point, as well as interval estimates of parameters, mean-lives and reliability functions of components of system. We carried out an extensive simulation study and observed that different membership function affect the considered estimators differently in terms of bias and mean squared error. Finally, a real data set of small electric appliances is analyzed, and various estimates of mean-lives and reliabilities of components have been obtained.

Table 9.7 Floored data of small electric appliances.

T	0	0	0	1	3	3	7	9	11	12	16	19	20	22	23	24	25	25
C	1	1	1	1	1	1	1	1	1	2	1	2	2	2	1	2	1	2
T	26	26	26	27	27	28	28	30	31	31	32	35	35	43	64	70	78	134
C	2	2	2	2	1	1	1	2	1	2	2	2	2	2	2	2	2	2

Table 9.8 ML and Bayes estimates of mean life for real data.

Estimates	**Triangular membership**			**Trapezoidal membership**		
		$\hat{m}_1$	$\hat{m}_2$		$\hat{m}_1$	$\hat{m}_2$
MLE	$\mu_{\tilde{x}_i}^{(0.05,0.05)}(x)$	61.7832	49.4265	$\mu_{\tilde{x}_i}^{(0.25,0.75)}(x)$	62.1208	49.6966
Boot-p		(50.84, 118.41)	(50.62, 92.97)		(50.87, 115.63)	(50.42, 94.99)
Bayes		64.3614	55.7303		64.2549	56.0903
HPD		(41.64, 91.39)	(40.7, 73.27)		(40.54, 90.08)	(40.66, 73.73)
MLE	$\mu_{\tilde{x}_i}^{(0.5,0.5)}(x)$	62.1216	49.6973	$\mu_{\tilde{x}_i}^{(0.25,0.5)}(x)$	62.0088	49.607
Boot-p		(50.35, 120.45)	(50.66, 92.64)		(51.48, 117.26)	(50.44, 94.37)
Bayes		64.8359	55.879		64.7407	55.8264
HPD		(42.22, 92.5)	(40.51, 73.17)		(41.03, 91.14)	(39.84, 73.11)
MLE	$\mu_{\tilde{x}_i}^{(0.95,0.95)}(x)$	62.4582	49.9666	$\mu_{\tilde{x}_i}^{(0.5,0.75)}(x)$	62.2338	49.787
Boot-p		(51.4, 116.61)	(50.82, 93.8)		(50.46, 117.64)	(49.99, 92.98)
Bayes		64.9924	56.322		64.5252	55.9075
HPD		(41.33, 92.12)	(40.79, 74.1)		(41.56, 91.77)	(40.2, 73.13)

Table 9.9 ML and Bayes estimates of reliability at T=25 for real data.

Estimates	Triangular membership			Trapezoidal membership		
		$\hat{\tilde{F}}_1$	$\hat{\tilde{F}}_2$		$\hat{\tilde{F}}_1$	$\hat{\tilde{F}}_2$
MLE	$\mu_{\tilde{x}_i}^{(0.05,0.05)}(x)$	0.6672	0.603	$\mu_{\tilde{x}_i}^{(0.25,0.75)}(x)$	0.6687	0.6047
Boot-p		(0.6088, 0.8026)	(0.6067, 0.7646)		(0.6119, 0.8085)	(0.6116, 0.7624)
Bayes		0.6691	0.6319		0.6701	0.6343
HPD		(0.5673, 0.7693)	(0.548, 0.7167)		(0.5716, 0.7742)	(0.5507, 0.7197)
MLE	$\mu_{\tilde{x}_i}^{(0.5,0.5)}(x)$	0.6687	0.6047	$\mu_{\tilde{x}_i}^{(0.25,0.5)}(x)$	0.6682	0.6041
Boot-p		(0.6108, 0.8068)	(0.6113, 0.7674)		(0.6146, 0.8061)	(0.6086, 0.7624)
Bayes		0.6706	0.6342		0.6703	0.6331
HPD		(0.57, 0.7729)	(0.5486, 0.7181)		(0.5678, 0.7712)	(0.5499, 0.7169)
MLE	$\mu_{\tilde{x}_i}^{(0.95,0.95)}(x)$	0.6701	0.6063	$\mu_{\tilde{x}_i}^{(0.5,0.75)}(x)$	0.6692	0.6052
Boot-p		(0.6129, 0.8067)	(0.612, 0.7678)		(0.6121, 0.8035)	(0.6127, 0.761)
Bayes		0.6707	0.6362		0.6717	0.6355
HPD		(0.5685, 0.7699)	(0.5561, 0.7185)		(0.5683, 0.7699)	(0.5556, 0.7222)

References

1. Abdullah, L. and Chong, Y.H., Human development index using membership functions of fuzzy sets. *Int. J. Ecol. Econ. Stat.*, 37, 105–118, 2016.
2. Bundel, R. and Tomer, S.K., Competing risk analysis of fuzzy lifetime data. *J. Sci. Res.*, 65, 299–312, July 2021.
3. Chen, M.-H. and Shao, Q.-M., Monte Carlo estimation of Bayesian credible and HPD intervals. *J. Comput. Graph. Stat.*, 8, 69–92, 1999.
4. Casella, G. and Robert, C., *Introducing monte carlo methods with R*, Springer-Verlag, 2010.
5. Crowder, M.J., *Classical competing risks*, Hall, Boca Raton, Florida, 2001.
6. Deshpande, J.V. and Purohit, S.G., Survival, hazard and competing risks. *Curr. Sci.*, 80, 1991–1202, May 2001.
7. Efron, B. and Tibshirani, R.J., *An introduction to the bootstrap*, CRC press, 1993.
8. Flehinger, B.J., Reiser, B., Yashchin, E., Parametric modeling for survival with competing risks and masked failure causes. *Lifetime Data Anal.*, 8, 177–203, 2002.
9. Hameed, I.A., Using gaussian membership functions for improving the reliability and robustness of students' evaluation systems. *Exp. Syst. Appl.*, 38, 7135–7142, 2011.
10. Hryniewicz, O., Bayes statistical decisions with random fuzzy data—an application in reliability. *Reliab. Eng. Syst. Saf.*, 151, 20–33, 2016.
11. Huang, H., Zuo, M., Sun, Z., Bayesian reliability analysis for fuzzy lifetime data. *Fuzzy Sets Syst.*, 157, 1674–1686, 2006.
12. Hung, W.-L., Bootstrap method for some estimators based on fuzzy data. *Fuzzy Sets Syst.*, 119, 337–341, April 2001.
13. Hung, W.-L., Weighted bootstrap method for fuzzy data. *Soft Comput.*, 10, 140–143, Jan 2006.
14. Lagakos, S.W., A Covariate Model for Partially Censored Data Subject to Competing Causes of Failure. *J. R. Stat. Soc Ser. C (Appl. Stat.)*, 27, 235–241, 1978.
15. Mon, D.-L. and Cheng, C.-H., Fuzzy system reliability analysis for components with different membership functions. *Fuzzy Sets Syst.*, 64, 145–157, 1994.
16. Pak, A., Parham, G.A., Saraj, M., Inferences on the competing risk reliability problem for exponential distribution based on fuzzy data. *IEEE Trans. on Reliab.*, 63, 2–12, March 2014.
17. Ross, T.J., *Fuzzy logic with engineering applications*, John Wieley & Sons Ltd, 2005.
18. Viertl, R., *Fuzzy numbers and fuzzy vectors*, in: *Statistical Methods for Fuzzy Data*, pp. 7–19, John Wiley & Sons, Ltd, 2011.
19. Zadeh, L.A., Probability measures of fuzzy events. *J. Math. Anal. Appl.*, 23, 421 – 427, 1968.

20. Zhao, J. and Bose, B.K., Evaluation of membership functions for fuzzy logic controlled induction motor drive, in: *IEEE 2002 28th Annual Conference of the Industrial Electronics Society*, IECON 02, vol. 1, pp. 229–234, 2002.

10

Cost-Benefit Analysis of a Redundant System with Refreshment

M.S. Barak[1]* and Dhiraj Yadav[2]

[1]*Department of Mathematics, Indira Gandhi University, Meerpur, Rewari(Hr.), India*
[2]*Department of Mathematics, Govt. College for Women, Rewari(Hr.), India*

Abstract

We are ushering in a digitally empowered society and knowledge economy under the umbrella of the digital India program initiated by the Government of India and marching ahead in an era of 5G technology. But one cannot guarantee an error-free and complete perfection of a system or a unit. Keeping in view the efficiency of the system, the concept of inspection/replacement/repair is introduced in this system model. Sometimes due to some system-oriented snags and glitches, system performance may be hindered but can be overcome by repair. At the same time, the failure of the server can also happen depending upon the situation faced by the server. The goal of this article is to look at the survey of cost-benefit of the system having two-unit, a single unit can operate the system, and another unit hold as a spare with server failure and refreshment provide to the server on his demand. Before being repaired or replaced, the failing unit is inspected by the server, who also takes a break (whenever feels tired). It is assumed that the operative unit can break down after entering the standard mode, but the second spare unit can fail after a period of inactivity or due to a technical fault. A single-server facility is available at any time for the system, which serves the purposes of repair, and inspection of the unit. In the steady-state, the formulation of variegated stochastic measures is investigated using a "semi-Markov process" and the regeneration point approach. The way of acting on certain critical performance metrics is displayed using a visual representation of MTSF, availability, and profit with repair rate to verify the capacity of the system.

Keywords: Server failure, cost-benefit analysis, cold-standby, inspection, regenerative point technique

**Corresponding author*: ms_barak@igu.ac.in

S. C. Malik, Deepak Sinwar, Ashish Kumar, S. R. Gadde, Prasenjit Chatterjee and Bui Thanh Hung (eds.) Computational Intelligence in Sustainable Reliability Engineering, (217–234) © 2023 Scrivener Publishing LLC

10.1 Introduction

In this modern era of science and technology, the probability of failure of a system or unit always exists. The success and failure of every system are an integral part of the system. One fails, the other succeeds. So, some alternative i.e., redundancy is required to enhance the availability of the system. Due to its multiple advantages in improving a system's dependability and availability, redundancy is widely used in practically all sorts of systems. Depending on the practicality, many types of redundancy may be implemented in a system. Many elements influence the installation of technology, including the working of components or systems.

Many scholars have argued many systems in the literature, such as Subramanian *et al.* [1], who looked at the dependability of a rectifiable system. Gopalan & Naidu [2, 3] inspected the systems having a single server for cost-benefit. Taneja [4] looked at the theoretical and economic examination of several reliability system models with the various facility. A system model with maintenance for a two-unit was established by Lam [5]. Kumar [6] discussed reliability models with various failure kinds and repair procedures. The dependability system model for its cost was emphasized by Malik [7] for a case study. Dhankhar & Malik [8] looked at the system models in the course of the job made by the server. Dhiraj & Barak [9] looked at the economic analysis of the system model with the assumption that the server can fail during his job. The stochastic study of a system model with conditional server failure was emphasized by Barak *et al.* [10]. Barak and his co-researchers [11–13] performed the study of the same system under inspection taking precedence over the repair activity. Barak *et al.* [14] analyzed reliability measures of a milk plant using RPGT. Kumar *et al.* [15] evaluated some reliability measures of a cold standby system subject to refreshment. Some of them have the impression that the server is always in good health and never gets sick while doing his duties. But, given that he must deal with a variety of difficulties in a variety of environments, this image appears to be impractical. We have seen a lot of instances where the server loses his cool while working.

When the operational unit breaks down, then the spare unit automatically grasps over and the process of repair of the breakdown unit has started. However, it is conceivable that the backup unit has already failed due to inactivity, erosion, or another technical issue. So, based on the aforementioned study, a system model is developed having two identical units with a refreshment facility whenever the server is required during his job. The model has two identical units, one can operate the system. When the operational unit gets broken down, the immediately spare unit starts in to help. The single server inspects the failing unit immediately before repairing or replacing it. The server also receives refreshment/treatment as needed to keep the system running smoothly. The server has access to refreshments both during maintenance and during the inspection. After being refreshed, the server resumes full functionality. The duration utilized for repair activities and refreshment act under the only distribution of negative exponential whenever other activities act under distributions of probability density functions of arbitrary nature. The process of semi-Markov and regenerating points approach are used to evaluate the various measures as indicated in tables that show the conduct of certain effecting indicators to test a system model's efficiency for arbitrary parameter values (θ, φ, λ, μ).

10.2 Notations

E:	S_0, S_1, S_2, S_3, S_5, and S_{14} are a set of regeneration states
O/Cs:	The unit is up and running/on cold standby.
FUi/*FUI*:	From its former status, the failing unit has been under constant examination.
FWi/*FWI*:	The failed unit has been waiting for inspection/ waiting for inspection from the previous status.
FUr/*FUR*:	The failing device is being repaired/continuously being repaired from its former state.
SUt/SUT:	Server undergoes refreshment/treatment in order to meet his needs/is undergoing refreshment/treatment in order to recover from a prior state.
FWr/*FWR*:	Server is engaged in taking refreshment as the result failed unit is waiting for repair/waiting for repair continuously from that state.
λ/μ:	Constant unit failure rate/rate at which the server requires refreshment/treatment.
a/b:	The probability of spare unit which can use/not able to use.
p/q:	Probability that repairing of the spare unit is not viable so it must be replaced/that it is doable, so it must be sent for repair.
f(t)/F(t):	The refreshing rate at which the server regains its freshness in pdf or cdf format.
g(t)/G(t):	pdf/cdf of repair rate of failed unit.

h(t)/H(t):	pdf/cdf of replacement rate of defective unit.
$q_{ij}(t)/Q_{ij}(t)$:	Direct transition time from a regenerative state Si to a regenerative state Sj or a failed state Sj in pdf/cdf format in (0,t), Sj does not visit any other regenerative state.
$q_{i,j;k}(t)/Q_{i,j;k}(t)$:	pdf/cdf of first passage time from a regenerative state S_i to a regenerative state S_j or to a failed state S_j visiting state S_k once in (0,t].
$q_{i,j;k(r,s)}(t)/\ Q_{i,j;k(r,s)}(t)$:	pdf/cdf of first passage time from a regenerative state S_i to a regenerative state S_j or to a failed state S_j visiting state S_k, S_r and S_s once or more than one times in (0,t].
$M_i(t)$:	Probability that the system is up at time t without reaching any other regenerative state or returning to the same state via one or more non-regenerative stages.
$W_i(t)$:	Probability that the server will remain active in state Si until time t without changing to another regenerative state or returning to the same state via one or more non-regenerative stages.
m_{ij}:	Contribution to mean sojourn time μ_i in state S_i when the system transits directly to state S_j so that $\mu_i = \sum_j m_{ij} = \int t dQ_{ij}(t) = -q_{ij}^{*}{'}(0)$
Ⓢ/Ⓒ:	Symbol for Stieltjes convolution/Laplace convolution
*/**:	Laplace transformation (L.T.)/Laplace stieltjes transformation (L.S.T.)

10.3 Average Sojourn Times and Probabilities of Transition States

The probability considerations cause the non-zero elements to acquire the following expressions when considering the potential states of the system model as shown in Figure 10.1.

$$P_{ij} = Q_{ij}(\infty) = \int_0^{\infty} q_{ij}(t)dt \tag{10.1}$$

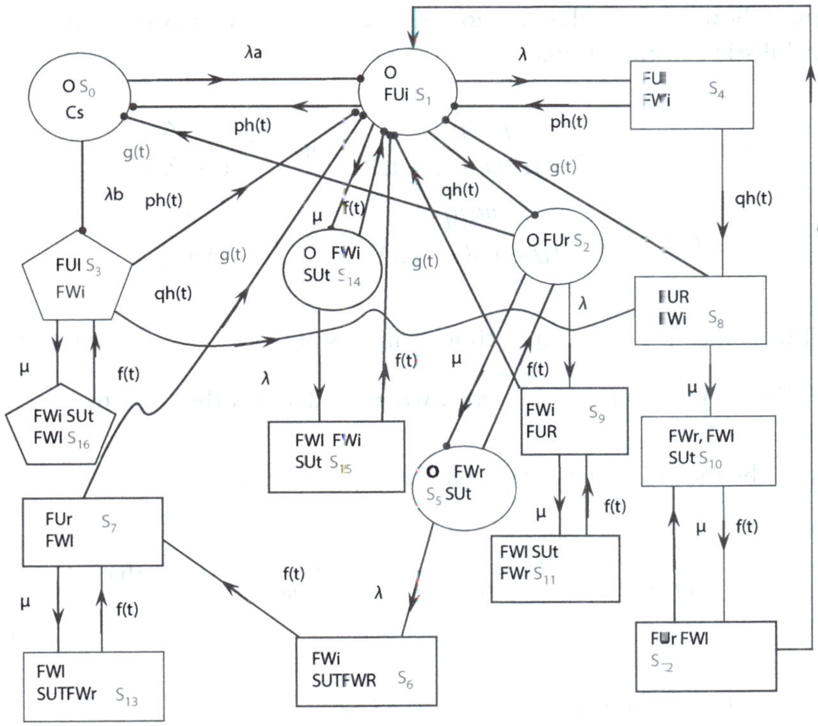

Figure 10.1 State transition diagram.

$$p_{10} = \frac{\psi}{\psi+\lambda+\mu}, p_{14} = \frac{\psi}{\psi+\lambda+\mu}, p_{1,14} = \frac{\mu}{\psi+\lambda+\mu}, p_{12} = \frac{\psi}{\psi+\lambda+\mu}q, p_{29} = \frac{\lambda}{\theta+\lambda+\mu}$$

$$p_{20} = \frac{\theta}{\theta+\lambda+\mu}, p_{25} = \frac{\mu}{\theta+\lambda+\mu}, p_{31} = \frac{\psi}{\psi+\mu}p, p_{38} = \frac{\psi}{\psi+\mu}q, p_{41} = p, p_{52} = \frac{\phi}{\phi+\lambda} = p_{14,1}$$

$$p_{56} = \frac{\lambda}{\phi+\lambda} = p_{14,15}\text{ , } p_{01} = a\text{ , } p_{03} = b\text{ , } p_{48} = q$$

$$p_{71} = \frac{\theta}{\theta+\mu}, p_{7,13} = \frac{\mu}{\theta+\mu}\text{ , } p_{67} = p_{10,12} = p_{11,9} = p_{13,7} = p_{15,1} = p_{16,3} = 1$$

$$p_{81} = p_{91} = p_{12,1} = \frac{\theta}{\theta+\mu}, p_{8,10} = p_{9,11} = p_{12,10} = \frac{\mu}{\mu+\theta}, p_{14,15} = \frac{\lambda}{\phi+\lambda}, p_{3,15} = \frac{\mu}{\psi+\mu} \tag{10.2}$$

It is possible to verify that these transition probabilities are correct.

$$p_{03} + p_{01} = 1, p_{10} + p_{12} + p_{14} + p_{14} + p_{1,14} = 1, p_{29} + p_{25} + p_{20} = 1, p_{31} + p_{3,16} + p_{38} = 1$$
$$p_{41} + p_{48} + p_{52} + p_{56} = 1, p_{71} + p_{7,13} = 1, p_{8,10} + p_{81} = 1, p_{9,11} + p_{91} = 1, p_{14,1} + p_{14,15} = 1,$$
$$p_{20} + p_{2,1(9,11)} + p_{2,1;9}, p_{10} + p_{12} + p_{1,1:4} + p_{1,14} + p_{1,1;4,8} + p_{1,1;4,8(10,12)} = 1$$

The following formulas for non-zero elements are also derived from probabilistic considerations.

$$p_{1,1;4,8(10,12)} = \frac{\mu\lambda q}{(\theta+\mu)(\psi+\lambda+\mu)},\ p_{2,1(9,11)} = \frac{\lambda}{\theta+\lambda+\mu},$$
$$p_{3,1;8(10,12)} = \frac{\psi\mu q}{(\theta+\mu)(\psi+\mu)}\quad p_{5,1;6(7,13)} = \frac{\lambda}{\phi+\lambda} \tag{10.3}$$

The mean Sojourn time in state S_i is given by $\mu_i = E(t) = \int_0^\infty P(T>t)dt = \sum_j m_{ij}$, where T denotes the time to system failure. The mean sojourn times $(\mu_i\ and\ \mu_i^{'})$ in the state S_i are:

$$\mu_o = \frac{1}{\lambda},\mu_1 = \frac{1}{\psi+\lambda+\mu},\mu_2 = \frac{1}{\theta+\lambda+\mu},\mu_3 = \frac{1}{\psi+\mu},\mu_5 = \mu_{14} = \frac{1}{\phi+\lambda} \tag{10.4}$$

$$\mu_1' = \frac{\begin{array}{l}[\theta\phi\{(\theta+\mu)\lambda\psi^2(\theta+p\mu)+\mu\psi(\theta+\mu)^2(\psi+\lambda+\mu)\\ +\lambda\mu(\theta+\mu)(\lambda+\mu)(\theta+p\mu)(\lambda+\mu+2\psi)-q\lambda\theta\psi(\mu+1)(\psi+\lambda+\mu)(\lambda+\mu)\}\\ +q\lambda\mu(\lambda+\mu)(\theta+\mu)(\psi+\lambda+\mu)\{\theta\phi\psi\\ +(\lambda+\mu)(\phi\psi(\theta+\mu)+\mu^2\psi+\theta(\phi+\psi))\}]\end{array}}{\theta\phi\psi(\theta+\mu)^2(\lambda+\mu)(\psi+\lambda+\mu)^2}$$

$$\mu_3' = \frac{\begin{array}{c}q\phi\psi[\theta\phi(\theta\psi-\mu^2)+\mu\{\mu(\theta+\mu)(\psi+\mu)(\theta+\theta+\mu)+\mu^2-\theta\psi\}]\\ +(\theta+\mu)^2(\psi+\mu)(\phi+\mu)\end{array}}{\phi(\theta+\mu)^2(\psi+\mu)^2}$$

$$\mu_2' \frac{\theta\phi+\lambda(\phi+\mu)}{\theta\phi+(\theta+\lambda+\mu)},\ \mu_5' \frac{\theta\phi+\lambda(\theta+\phi+\mu)}{\theta\phi(\phi+\lambda)},\ \mu_{14}' \frac{\lambda}{\phi(\phi+\lambda)} \tag{10.5}$$

10.4 Mean Time to Failure of the System

Let $\phi_i(t)$ be the c.d.f. of the time it takes to get from the regenerative state S_i to a failing state. We have the following recursive relations $\phi_i(t)$ for the failed state as an absorbing state.

$$\phi_0(t)=Q_{03}(t)+Q_{01}(t)Ⓢ\phi_1(t)$$

$$\phi_1(t)=Q_{12}(t)Ⓢ\phi_2(t)=Q_{10}(t)Ⓢ\phi_0(t)=Q_{1.14}(t)Ⓢ\phi_{14}(t)=Q_{14}(t)$$

$$\phi_2(t)=Q_{20}(t)Ⓢ\phi_0(t)=Q_{25}(t)Ⓢ\phi_5(t)=Q_{29}(t)$$

$$\phi_5(t)=Q_{52}(t)Ⓢ\phi_2(t)=Q_{56}(t)$$

$$\phi_{14}(t)=Q_{14,1}(t)Ⓢ\phi_1(t)=Q_{14,15}(t) \quad (10.6)$$

For the simplification applying L.S.T. on equation (10.6) and obtaining for $\phi_0^{**}(s)$:

$$R^*(s)=\frac{1-\phi_0^{**}}{s} \quad (10.7)$$

Laplace inverse transformation may be used to determine its reliability (10.7). MTSF (mean time to system failure) is calculated as follows:

$$\text{MTSF} = \lim_{s\to 0} R^*(s)=\lim_{s\to 0}\frac{1-\phi_0^{**}(s)}{s}$$

$$=\frac{(1-p_{52}p_{25})(p_{01}\mu_1+\mu_0)-\mu_0 p_{1,14}p_{14,1}+p_{01}p_{1,14}+p_{12}p_{10}(p_{25}\mu_5+\mu_2)}{(1-p_{52}p_{25})(1-p_{10}p_{01}-p_{1,14}p_{14,1})-p_{12}p_{20}p_{01}} \quad (10.8)$$

10.5 Steady-State Availability

Let $A_i(t)$ be the probability that at an instant the system is available for use or in up-state, given that the system entered the regenerative state S_i at t = 0. The recursive relations $A_i(t)$ are as follows:

$$A_0(t)=M_0(t)+q_{03}(t)©A_3(t)+q_{01}(t)©A_1(t)$$

$$A_1(t)=M_1(t)+\{q_{1,1;4}(t)+q_{1,1;4,8}(t)+q_{1,1;4,8}(t)+q_{1,1;4;8(10,12)}\}©A_1(t)+q_{12}(t)© A_2(t)+q_{10}(t)©A_0(t)+q_{1,14}(t)©A_{14}(t)$$

$$A_2(t)=M_2(t)+\{q_{2,1(9,11)}(t)+q_{2,1;9}(t)\}©A_1(t)+q_{25}(t)©A_5(t)+q_{20}(t)©A_0(t)$$

$$A_3(t)= q_{3,3;16}(t)©A_3(t)+\{q_{31}(t)+q_{3,1;8,(10,12)}(t)+q_{3,1;8}(t)\}©A_1(t)$$

$$A_5(t)= M_5(t)+\{q_{5,1;6(7,13)}(t)+q_{5,1;6,7}(t)\}©A_1(t)+q_{52}(t)©A_2(t)$$

$$A_{14}(t)= M_{14}(t)+q_{14,1}(t)©A_1(t)+q_{14,1;15}(t)©A_1(t) \tag{10.9}$$

$M_i(t)$ is the probability that the system is available for use initially in the state $S_i \in E$ (Set of regenerative state) is available for use at time t without visiting any other regenerative state, where

$$M_{14}(t)=M_5(t)=\overline{F}(t)e^{-\lambda t},\ M_2(t)=\overline{G}(t)e^{-(\lambda+\mu)t},M_1(t)=\overline{H}(t)e^{-(\lambda+\mu)t}, M_0(t)=e^{-\lambda t} \tag{10.10}$$

Applying L.T. on equations (10.9 and 10.10) and obtaining $A^*_0(s)$:

$$A_0=\lim_{s\to 0} sA_0^*(s)=\frac{\begin{array}{c}(1-p_{33,16})[\mu_0\{p_{10}(1-p_{25}p_{52})+p_{12}p_{20}\}\\ +(\mu_1+p_{1,14}\mu_{14})(1-p_{25}p_{52})+p_{12}(\mu_2+\mu_2+p_{25})\end{array}}{\begin{array}{c}\left[\left\{\mu_0(1-p_{33,16})+p_{03}\mu_3'\right\}\{1-p_{25}p_{52})p_{10}+p_{12}p_{20}\right\}\\ +(1-p_{33,16})\left\{(1-p_{25}p_{52}(\mu_1'+\mu_{14}'p_{1,14})+p_{12}(\mu_2'+\mu_3'p_{25})\right\}\Big]\end{array}} \tag{10.11}$$

10.6 The Period in Which the Server is Busy With Inspection

(a) Busy period due to inspection

"Let $B_i^I(t)$ be the probability that the server is engaged regarding the inspection of the failed unit at instant 't' given that the system entered the regenerative state S_i at t = 0. The recursive relations $B_i(t)$ are as follows":

$$B_0^I(t)=q_{03}(t)©B_3^I(t)+q_{01}(t)©B_1^I(t)$$

$$B_2^I(t)=W_1(t)+q_{10}(t) © B_0^I(t)+q_{12}(t) © B_2^{I1}(t)+q_{1,14}(t) © B_{14}^I(t)$$

$$+\{q_{1,1;4}(t)+q_{1,1,4;8}(t)+q_{1,1;4,8,10,12)}(t)\} © B_1^I(t)$$

$$B_2^I(t)=q_{20}(t) © B_0^I(t)+\{q_{2,1;9}(t)+q_{2,1(9,11)}(t)\} © B_1^I(t)+q_{25}(t) © B_5^I(t)$$

$$B_3^I(t)=W_3(t)+\{q_{31}(t)+q_{3,1;8,(10,12))}(t) © B_1^I(t)+q_{3,3,16}(t) © B_3^I(t)$$

$$B_5^I(t)=q_{52}(t) © B_2^I(t)+\{q_{5,1;6,7}(t)+q_{5,1;6(7,13)}(t)\} © B_1^I(t)$$

$$B_{14}^I(t)=q_{14,1}(t) © B_1^I(t)+q_{14,1;15}(t) © B_1^I(t) \tag{10.12}$$

where $W_i(t)$ is the representation of the probability that the server is engaged in the state S_i due to repairing of the unit up to time 't' without making any transition to any other regenerative state or before returning to the same state via one or more non-regenerative states so

$$W_1(t)=\overline{H}(t)e^{-\lambda t}e^{-\mu t}+\overline{H}(t)\lambda e^{-\lambda t} © \overline{H}(t), \quad W_3(t)=\overline{H}(t)e^{-\mu t} \tag{10.13}$$

applying L.T. of the relations (6.1 and 6.2) and evaluating $B_0^{*I}(s)$:

$$B_0^I=\lim_{s\to 0} sB_0^{*I}(s)=\frac{W_1^*(0)(1-p_{52}p_{25})(1-p_{33,16})+W_3^*(0)\{(1-p_{52}p_{25})p_{10}+p_{20}p_{12}\}p_{03}}{[\{\mu_0(1-p_{33,16})+p_{03}\mu_3'\}\{1-p_{25}p_{52})p_{10}+p_{12}p_{20}\} \\ +(1-p_{33,16})\{(1-p_{25}p_{52})(\mu_1'+\mu_{14}'p_{1,14})+p_{12}(\mu_{12}'+\mu_5'p_{25})\}]} \tag{10.14}$$

(b) Busy due to repair.

Let $B_0^R(t)$ be the probability that the server is busy due to repair of the failed unit at instant 't' given that the system entered the regenerative state S_i at t = 0. The recursive relations $B_i(t)$ are as follows:

$$B_0^R(t)=q_{03}(t) © B_3^R(t)+q_{01}(t) © B_1^R(t)$$

$$B_1^R(t) = q_{10}(t) © B_0^R(t)+q_{12}(t) ©$$
$$B_2^R(t)+\{q_{1,1;4}(t)+q_{1,1;4,8}(t))+q_{1,1;4,8(10,12)}(t)\} © B_1^R(t) +q_{1,14}(t) © B_{14}^R(t)$$

$$B_2^R(t)+W_2(t)+q_{20}(t) © B_0^R(t)+q_{25}(t) © B_5^R(t)+\{q_{2,1;9}(t)+q_{2,1(9,11)}(t)\}$$
$$© B_1^R(t)$$

$$B_3^R(t)=\{q_{31}(t)+q_{3,1;8}(t)+q_{3,1;8(10,12)}(t)\} © B_1^R(t)+q_{3,3;16}(t) © B_3^R(t)$$

$$B_5^R(t)=W_5+q_{52}(t) © B_2^R(t)+\{q_{5,1;67}(t)q_{5,1;6(7,11)}(t)\} © B_1^R(t)$$

$$B_{14}^*(t)=\{q_{14,1}(t)+q_{14,1;15}(t)\} © B_1^R(t) \tag{10.15}$$

where $W_i(t)$ is the probability that the server is busy in the state S_i due to repairing the unit up to time 't' without making any transition to any other regenerative state or before returning to the same via one or more non-regenerative states so

$$W_2(t)=e^{-(\lambda+\mu)t}\overline{G(t)}+\lambda e^{-(\lambda+\mu)t}\overline{G(t)} © e^{-\mu t}\overline{G(t)}+\lambda e^{-(\lambda+\mu)t}\overline{G}(t) ©$$
$$\mu e^{-\mu t}\overline{G(t)} © f(t)© e^{-\mu t}\overline{G(t)}+\lambda e^{-(\lambda+\mu)t}\overline{G(t)} © \mu e^{-\mu t}\overline{G(t)} © f(t)©$$
$$\mu e^{-\mu t}\overline{G(t)} © f(t)© e^{-\mu t}\overline{G(t)}+\text{-------------}$$

$$W_5(t)=\lambda e^{-\lambda t}\overline{F(t)} © f(t) © e^{-\mu t}\overline{G(t)}+\lambda e^{-\lambda t}\overline{F}(t) © \mathrm{f(t)} © f(t) ©$$
$$e^{-\mu t}\overline{G(t)} © \mathrm{f(t)} © e^{-\mu t}\overline{G(t)}+\lambda e^{-\lambda t}\overline{F}(t) © \mathrm{f(t)} © \mu e^{-\mu t}\overline{G(t)} © \mathrm{f(t)} ©$$
$$\mu e^{-\mu t}\overline{G(t)} © \mathrm{f(t)} © e^{-\mu t}\overline{G(t)}+\text{----------} \tag{10.16}$$

applying L.T.on equations (10.15 and 10.16) and obtaining for $B_0^{*R}(s)$:

$$B_0^R = \lim_{s\to 0} sB_0^{*R}(\mathrm{s}) = \frac{p_{12}(1-p_{33,16})\{W_2^*(0)+W_5^*(0)p_{25}\}}{\begin{array}{l}[\mu_0(1-p_{33,16})+p_{03}\mu_3\}\{1-p_{25}p_{52})p_{10}+p_{12}p_{20}\} \\ +(1-p_{33,16})\{(1-p_{25}p_{52})(\mu_1+\mu_{14}p_{1,14})+p_{12}(\mu_2+\mu_5 p_{25})\}]\end{array}} \tag{10.17}$$

10.7 Expected Number of Visits for Repair

"Let $R_i(t)$ be the anticipated number of visits by the server during (0,t], given that the system entered the regenerative state S_i at t=0. The recursive relations for $R_i(t)$ are as follows":

$$R_0(t)=Q_{01}(t)\circledS R_1(t)+1)+Q_{03}(t)\circledS (R_3(t)+1)$$

$$R_1(t)=Q_{10}(t)\circledS R_0(t)=Q_{12}(t)\circledS R_2(t)$$

$$+\{Q_{1,1;4}(t)+Q_{1,1;48}(t)+Q_{1,1;4(8,10)}(t)\}\circledS R_1(t)+Q_{1,14}(t)\circledS R_{14}(t)$$

$$R_2(t)=Q_{20}(t)\circledS R_0(t)+Q_{25}(t\circledS R_5(t)+Q_{2,1;9}(t)\circledS R_1(t)+Q_{2,1(9,11)}(t)\circledS R_1(t)$$

$$R_3(t)=Q_{31(t)}\circledS R_1(t)+Q_{3,1;8}(t)\circledS R_1(t)+Q_{3,3;16}(t)\circledS R_3(t)+Q_{3,1;8(10,12)}\circledS R_1(t)$$

$$R_5(t)=Q_{52}(t)\circledS R_2(t)+Q_{5,1;67}(t)\circledS R_1(t)+Q_{5,1;6(7,13)}(t)\circledS R_1(t)$$

$$R_1(t)+Q_{14,1;15}(t)\circledS R_1(t)\circledS R_0^{**}(s) \tag{10.18}$$

applying L.S.T. on the equations (10.18) and obtaining for $R_0^{**}(s)$

$$R_0 \lim_{s\to 0} sR_0^{**}(s)=\frac{(1-p_{33,16})\{p_{10}(1-p_{25}p_{52})+p_{12}p_{20}\}}{\begin{array}{l}[\{\mu_0(1-p_{33,16})+p_{03}\mu_3'\}\{(1-p_{25}p_{52})p_{10}+p_{12}p_{20}\}\\ +(1-p_{33,16})\{(1-p_{25}p_{52})(\mu_1'+\mu_{14}p_{1,14})+p_{12}(\mu_2'+\mu_5'p_{25})\}]\end{array}} \tag{10.19}$$

10.8 Expected Number of Refreshments

Let $T_i(t)$ be the expected number of treatments given to server in (0,t] such that the system entered the regenerative state at t = 0. The recursive relations $T_i(t)$ are as follows:

$$T_0(t) = Q_{01}(t) \circledS T_1(t) + Q_{03}(t) \circledS T_3(t)$$

$$T_1(t) = Q_{10}(t) \circledS T_0(t) = Q_{12}(t) \circledS T_2(t) = Q_{1,1;4}(t) \circledS T_1(t) = Q_{1,14}(t) \circledS$$

$$T_{14}(t) + 1)Q_{1,1;4,8}(t) \circledS T_1(t) + Q_{1,1;4,8(10,12)}(t)] \circledS [1 + T_1(t)]$$

$$T_2(t) = Q_{20}(t) \circledS T_0(t) + Q_{25}(t) \circledS [T_5(t) + 1] + Q_{2,1;9}(t) \circledS$$

$$T_1(t) = Q_{2,1(9,11)}(t) \circledS [T_1(t)1]$$

$$T_3(t) = Q_{31}(t) \circledS T_1(t) + Q_{3,1;18}(t) \circledS T_1(t) + Q_{3,3;16}(t) \circledS$$

$$(T_3(t) + 1) + Q_{3,1;8(10,12)}(t) \circledS [T_1(t) + 1]$$

$$T_5(t) = Q_{52}(t) \circledS T_2(t) = Q_{5,1;67}(t) \circledS T_1(t) = Q_{5,1;6(7,13)}(t) \circledS T_1(t)$$

$$T_{14}(t) = Q_{14,1}(t) \circledS T_1(t) + Q_{14,1;15}(t) \circledS T_1(t) \tag{10.20}$$

Applying L.S.T. on equation (10.20) and obtaining $T_0^{**}(s)$

$$T_0 \lim_{s\to 0} T_0^{**}(s) = \frac{\begin{matrix}[(1-p_{33,16})\{(1-p_{25}p_{52})(1-p_{12}-p_{11,4}-p_{10}-p_{11,4,8})+p_{12}(1-p_{21,9}-p_{25})\} \\ +p_{03}(1-p_{31}-p_{31,8})\{(1-p_{25}p_{52})(p_{10}+p_{11,4}-p_{1,14})+p_{12}p_{20}\}]\end{matrix}}{\begin{matrix}[\{\mu_0(1-p_{33,16})+p_{03}\mu_3'\}\{(1-p_{25}p_{52})p_{10}+p_{12}-p_{20}\} \\ +(1-p_{33,16})\{(1-p_{25}p_{52})(\mu_1'+\mu_{14}'p_{1,14})+p_{12}(\mu_2'+\mu_5'p_{25})\}\end{matrix}} \tag{10.21}$$

10.9 Particular Case

Taking into account $h(t) = \psi e^{-\phi 1}$, $f(t) = \phi e^{-\phi 1}$, and $g(t) = \theta e^{-\theta 1}$ for particular case

$$\text{“}\mathbf{MTSF} = \lim_{s\to 0} R^*(s) = \lim_{s\to 0} \frac{1-\phi_0^{**}(s)}{s}\text{”}$$

$$= \frac{\begin{matrix}(\phi+\lambda)[\{(\theta+\lambda)(\phi+\lambda)+\lambda\mu\}\{\psi+\mu+(1+a)\lambda\} \\ +\mu(a\lambda-\phi)(\theta+\lambda+\mu)+aq\psi\lambda(\phi+\lambda+\mu)]\end{matrix}}{\lambda[\{(\theta+\lambda)(\phi+\lambda)+\lambda\mu\}\{(\phi+\lambda-ap\psi)+\lambda\mu\}-aq\theta\psi(\phi+\lambda)^2]} \tag{10.22}$$

$$\textbf{Availability } A_0 = \frac{\begin{array}{l}\mu\theta\phi[\psi(\phi+\lambda)\{\theta\phi+\theta\lambda+p\lambda(\phi+\lambda+\mu)\}+\\ \lambda\{(\phi+\lambda+\mu)(\theta\phi+\lambda(\theta+\phi+\mu))+q\psi(\phi-\lambda)(\phi+\lambda+\mu)\}]\end{array}}{\begin{array}{l}\psi[\theta\phi(\phi+\lambda)\{(\theta\phi+\theta\lambda+p\lambda(\phi+\lambda+\mu)\}(\psi+b\lambda\mu_3')\\ +\lambda\{\theta((\theta+\lambda)(\phi+\lambda)+\lambda\mu)(\phi(\phi+\lambda)((\psi+\lambda+\mu)\mu_1'+\lambda\mu)\\ +q\psi(\phi+\lambda)((\phi+\lambda)(\theta\phi+\lambda(\phi+\mu))+\mu(\theta\phi+\lambda(\phi+\mu)))\}]\end{array}} \tag{10.23}$$

$$B_0^I = \frac{W_1^*(0)(1-p_{25}p_{52})(1-p_{33,16})+W_3^*(0))p_{03}\{p_{10}(1-p_{25}p_{52})+p_{12}p_{20}\}}{\begin{array}{l}[\{\mu_0(1-p_{33,16})+p_{03}\mu_3'\}\{(1-p_{25}p_{52})p_{10}+p_{12}p_{20}\}\\ +(1-p_{33,16})(\mu_1'+\mu_{14}'p_{1,14})+p_{12}(\mu_2'\mu_5'p_{25}))\}]\end{array}} \tag{10.24}$$

$$B_0^R = \frac{q\lambda\phi(\phi+\lambda)\{\psi(\lambda+\phi)+\lambda\mu\theta\}}{\begin{array}{l}[\theta\phi(\phi+\lambda)\{(\theta\phi+\lambda+p\lambda)(\phi+\lambda+\mu)\}(\psi+b\lambda\mu_3')\\ +\lambda\{\theta((\theta+\lambda)(\phi+\lambda)+\lambda\mu)\phi(\phi+\lambda)(\psi+\lambda+\mu)\mu_1'+\lambda\mu)\\ +q\psi(\phi+\lambda)((\phi+\lambda)(\theta\phi+\lambda(\phi+\mu))+\mu(\theta\phi+\lambda(\theta+\phi+\mu)))\}]\end{array}} \tag{10.25}$$

$$R_0 = \frac{\lambda\theta\phi\mu(\phi+\lambda)\{\theta\phi+p\lambda(\phi+\lambda+\mu)\}}{\begin{array}{l}[\theta\phi(\phi+\lambda)\{\theta\phi+\theta\lambda+p\lambda(\phi+\lambda+\mu)\}(\psi+b\lambda\mu_3')\\ +\lambda\{\theta((\theta+\lambda)(\phi+\lambda)+\lambda\mu)(\phi(\phi+\lambda)(\psi+\lambda+\mu)(\mu_1'+\lambda\mu)\\ +q\psi(\phi+\lambda)((\theta+\lambda)(\theta\phi+\lambda(\phi+\mu))+\mu)(\theta\phi+\lambda(\theta+\phi+\mu)))\}]\end{array}} \tag{10.26}$$

$$T_0 = \frac{\begin{array}{l}\lambda\theta\phi(\phi+\lambda)[(\theta\phi+\lambda(\theta+\phi+\lambda+\mu))\{\psi\mu(q\lambda+\theta+\mu)\\ +b\mu(q\psi+\theta+\mu)(p\psi+p\lambda-\mu)\}\\ +q\psi(\phi+\lambda)\{\psi(\theta^2+\theta\mu+\lambda\mu)+b\mu\theta(q\psi+\theta+\mu)\}]\end{array}}{\begin{array}{l}[\theta\phi(\phi+\lambda)\{(\theta\phi+\theta\lambda+p\lambda(\phi+\lambda+\mu)\}(\psi+b\lambda\mu_3')\\ +\lambda\{\theta((\theta+\lambda(\phi+\lambda)+\lambda\mu)(\phi(\phi+\lambda)(\psi+\lambda+\mu)(\mu_1'+\lambda\mu)\\ +q\psi(\phi+\lambda)((\phi+\lambda)(\theta\phi+\lambda(\phi+\mu))+\mu(\theta\phi-\lambda(\theta+\phi+\mu)))\}]\end{array}} \tag{10.27}$$

10.10 Cost-Benefit Examination

The profit function of the system model

$$P_0 = K_0A_0 - K_1B_0^I - K_2B_0^R - K_3R_0 - K_4T_0 \tag{10.28}$$

where K_0 = (10000) Revenue per unit up-time of the system.

K_1 = (300) Cost per unit time for which server is engaged causing inspection.

K_2 = (300) Cost per unit time for which server is busy due to repair.

K_3 = (500) Cost per unit visits by the server.

K_4 = (300) Cost per unit time refreshment arranged for server.

10.11 Discussion

The present model is based on a visual evaluation of the failed unit before repair. Tables 10.1 to 10.3 analyze the numerical behavior of the mean-time to system break down, availability, and profit function. The idea of refreshment/treatment is extended to inspection activities in this model, in addition to assessing the system's availability and efficacy. The influence of various parameters on the performance measure of the system model is envisioned in this research. Table 10.1 demonstrates the fact that the higher the repair rate, the higher the MTSF value. On the other hand, as the rate of unit failure falls, the value of MTSF rises. This table also shows that when the rate of server failure falls from .45 to .35 (second column), the numerical values of MTSF rise. In the third column, it appears that the value of MTSF drops when the refreshment rate lowers from .65 to .35 increase with fixed values of other parameters (however pattern is increasing w.r.t. repair rate). It happens due to the increasing frequency of refreshments/treatments during the performance of the system model.

Table 10.2 shows that the system's availability rises in proportion to the unit's repair rate. The third column of the table shows that when the refreshment rate/treatment rate lowers from .65 to .35, the system's availability declines. The availability of the system dramatically increases anytime the unit failure rate reduces from .55 to .40 with fixed values of other parameters, as seen in the fourth column of this table. It is worth noting that when the server failure rate drops from .45 to .35, availability

Table 10.1 MTSF (mean time to system failure vs repair rate (θ).

θ↓	λ=55,μ=.45 ø=.65,ψ=.30	λ=55,μ=.35, ø=.65,ψ=.30	λ=.65,μ=.45 ø=.35,ψ=.30	λ=.40,μ=.45, ø=.65,ψ=.30	λ=.55,μ=.45, ø=.65,ψ=.45
.1	2.051752	2.153653	2.638638	3.238898	2.087812
.2	2.05768	2.159041	2.643387	3.270946	2.093981
.3	2.062529	2.163407	2.647331	3.296018	2.099042
.4	2.066569	2.167016	2.650658	3.316169	2.10327
.5	2.069986	2.170049	2.653503	3.332718	2.106855
.6	2.072914	2.172634	2.655963	3.346551	2.109933
.7	2.075452	2.174863	2.658112	3.358287	2.112604
.8	2.077671	2.176805	2.660004	3.368369	2.114944
.9	2.07963	2.178513	2.661684	3.377123	2.117012
1.	2.08137	2.180025	2.663185	3.384795	2.118851

Table 10.2 Availability vs repair rate (θ).

θ↓	λ=.55,μ=.45, ø=.65ψ=.30	λ=.55,μ=.35, ø=.65,ψ=.30	λ=.55,μ=.45, ø=.35,ψ=.30	λ=.40,μ=.45, ø=.65,ψ=.30	λ=,.55,μ=.45, ø=.65,ψ=.45
.1	0.415285	0.370861	0.339779	0.542887	0.292818
.2	0.574478	0.504801	0.475948	0.728888	0.432416
.3	0.659875	0.573884	0.550481	0.820197	0.514543
.4	0.712285	0.614402	0.596997	0.872342	0.567726
.5	0.746967	0.639944	0.628288	0.904818	0.604238
.6	0.7711	0.656851	0.650419	0.926247	0.630364
.7	0.788523	0.668452	0.666663	0.940991	0.649662
.8	0.801467	0.676636	0.678932	0.951463	0.664287
.9	0.811309	0.682536	0.688419	0.95909	0.675607
1	0.818937	0.686863	0.695896	0.964756	0.684529

drops as well (however, availability is an increasing pattern with respect to repair rate). It occurs as a result of an increase in the frequency of failures/treatments; the system may experience more downtime, and the system's availability decreases, i.e. the system becomes less available. As a result of the interruption, the system is down for a longer period. The system is also impacted if the rate of replacement rises from .30 to .45 as seen in the fifth column. Availability of system deteriorates by the recurrence of inspection/replacement thus reducing operating time.

Table 10.3 depicts the numerical trend of profit in relation to repair rate, which is similar to availability. The table may be used to examine the behavior of parameters and their separate effects on the system's profit. This table shows that the system's profit rises in proportion to the unit's repair rate. The third column of the table shows that when the refreshment rate/treatment rate lowers from .65 to .35, the system's profit reduces as well. The fourth column of this table also shows that when the unit failure rate lowers from .55 to .40 with fixed values of other parameters, the profit of the system increases dramatically. The second column is catching attention since the server failure rate has decreased from .45 to .35, and profit has decreased as well (however, profit is increasing pattern with respect to repair rate).

Table 10.3 Profit vs repair rate (θ).

θ↓	λ=.55,μ=.45, ø=.65ψ=.30	λ=.55,μ=.35, ø=.65,ψ=.30	λ=.55,μ=.45, ø=.35,ψ=.30	λ=.40,μ=.45, ø=.65,ψ=.30	λ=,.55,μ=.45, ø=.65,ψ=.45
.1	4005.474	3540.633	3286.399	5269.059	2765.782
.2	5560.428	4845.112	4617.207	7094.814	4116.347
.3	6389.972	5513.213	5342.169	7986.463	4906.014
.4	6895.746	5901.695	5792.065	8492.303	5413.717
.5	7227.778	6143.851	6092.627	8804.644	5759.269
.6	7456.55	6301.777	6303.463	9008.445	6003.937
.7	7619.725	6408.033	6456.668	9146.64	6182.369
.8	7739.172	6481.06	6571.022	9242.966	6315.516
.9	7828.369	6531.91	6658.19	9311.429	6416.688
1	7896.003	6567.508	6725.758	9360.723	6494.673

In the fifth column, it is stated that an increase in replacement rate results in a drop in profit. It is concluded that when the frequency of failures/treatments increases, the system's profit decreases, implying that the system becomes less productive. The system is down for a longer period, which reduces its efficacy. As a result, we can make the system more lucrative by keeping an eye on unit failure recurrence and keeping a close eye on server failure/refreshment rate.

10.12 Conclusion

By replacing different values for various parameters with fixed values for other parameters, the present model, as well as refreshment/treatment during inspection activities, is numerically appraised. It turns out that by intelligently offering refreshment/treatment facilities to the server, the model may become more available and productive. If there are some rigorous checks on server failure/unit failure to save downtime and extra investment, the system may be made more lucrative. To assess server failure/refreshment/treatment rate, we will need an ordinary and experienced repairman; however, the latter may not always be accessible. The study has applications in a variety of domains, including water pipe pumps in a boosting station, remote sensing systems with backup power, computer network operation, and communication satellites, among others.

References

1. Subramanian, R., Venkatakrishan, K.S., Kistner, K.P., Reliability of a repairable system with standby failure. *Oper. Res.*, 24, 1, 169–176, 1976.
2. Gopalan M, N. and Naidu, R.S., Cost-benefit analysis of a one-server system subject to inspection. *Microelectron. Reliab.*, 22, 699–705, 1982.
3. Naidu R, S. and Gopalan, M.N., Cost-benefit analysis of one-server two-unit system subject to arbitrary failure, inspection, and repair. *J. Reliabil. Eng.*, 22, 8–11, 1984.
4. Taneja, G.L., *Stochastic and profit analysis of some reliability models with different types of failure and repair*, Ph.D. Thesis, M.D. University, Rohtak, 1992.
5. Lam, Y.F., A maintenance model for the two-unit redundant system. *Microelectron. Reliab.*, 37, 3, 497–504, 1997.
6. Kumar, R., *Analysis of reliability models with different types of failure and repair policies*, Ph.D. Thesis, M.D. University, Rohtak, 2005.
7. Malik, S.C., Reliability modeling and cost-benefit analysis of a system—A case study. *J. Math. Syst. Sci.*, 5, 2, 26–38, 2009.

8. Dhankhar A, K. and Malik, S.C., Cost-benefit analysis of system reliability models with server failure during inspection and repair. *Int. J. Stat. Anal.*, 1, 3, 265–278, 2011.
9. Dhiraj, Y. and Barak, M.S., Stochastic analysis of a cold standby system with server failure. *Int. J. Math. Stat. Invent.*, 4, 6, 18–22, 2016.
10. Barak, M.S., Dhiraj, Y., Barak, S.K., Cost-benefit analysis of a redundant system with the server having refreshment facility subject to inspection. *Int. J. Adv. Eng. Res. Sci. (IJAERS)*, 4, 6, 24–31, 2017.
11. Barak, M.S., Dhiraj, Y., Barak Sudesh, K., Stochastic analysis of a cold standby system with conditional failure of the server. *Int. J. Stat. Reliab. Eng.*, 4, 1, 65–69, 2017.
12. Barak, M.S., Yadav, D., Kumari, S., Stochastic analysis of a two-unit system with standby and server failure subject to inspection, in: *Life Cycle Reliability and Safety Engineering*, vol. 7, 1, pp. 23–32, Springer, 2018.
13. Barak, M.S., Yadav, D., Kumari, S., Stochastic analysis of a two-unit redundant system with priority to inspection over repair, in: *Life Cycle Reliability and Safety Engineering*, vol. 7, 2, pp. 71–79, Springer, 2018.
14. Barak, M.S., Garg, R., Kumar, A., Reliability measures analysis of a milk plant using RPGT, in: *Life Cycle Reliability and Safety Engineering*, vol. 10, 3, pp. 295–302, Springer, 2021.
15. Kumar, A., Garg, R., Barak, M.S., Reliability measures of a cold standby system subject to refreshment. *Int. J. Syst. Assur. Eng. Manage.*, 12, 1-9, 2021.

11

Fuzzy Information Inequalities, Triangular Discrimination and Applications in Multicriteria Decision Making

Ram Naresh Saraswat and Sapna Gahlot*

Department of Mathematics and Statistics Manipal University Jaipur, Jaipur (Rajasthan), India

Abstract
This chapter introduced few novel inequalities of fuzzy measures and established the bounds in term of triangular discrimination. We obtained some new relations between new and existing fuzzy divergence measures with the help of the properties of convex function, new f-divergence measure. Also, we showed utility of new fuzzy divergence measures in multicriteria decision-making problem.

Keywords: Fuzzy set, fuzzy new f-divergence measure, Jensen's inequality, fuzzy triangular divergence measure, MCDM

2000 Mathematics Subject Classification: 40H05, 46A45 62B86, 03E72, 68T10

11.1 Introduction

In information theory, Shannon [4] had defined the expected value of information which is called entropy. Entropy is defined on the pair of probability distribution. Afterward, Zadeh [21] developed the nature of entropy in fuzzy set form, which is called fuzzy entropy. There are many scientists [33–36] and researchers discussed or cited fuzzy divergence measures and existing in the literatures of measures. After that De Luca and Termini [24] introduced entropy on the fuzzy set (FS), which is called Shannon's

*Corresponding author: sapna1994gahlot@gmail.com

S. C. Malik, Deepak Sinwar, Ashish Kumar, S. R. Gadde, Prasenjit Chatterjee and Bui Thanh Hung (eds.) Computational Intelligence in Sustainable Reliability Engineering, (235–250) © 2023 Scrivener Publishing LLC

function. Afterward, numerous researchers made additional effort during this area. Among its membership function and nearby classical sets. Kaufmann [2] proposed fully distinct fuzziness measure of FS. Entropy is a critical concept for identifying ambiguous data that has received a great deal of interest in the past. It has a broad variety of applications, including higher cognitive processes, diagnosis, bioinformatics, and so on.

Following that, a lot of scholars investigated fuzzy divergence metrics in a variety of ways and proposed applications in a variety of fields [3, 5, 6, 11, 12, 23, 29–31]. Fan and Xie [15] presented the fuzzy divergence measure, which is established on the exponential process and its relationship with the fuzzy divergence measure [2]. Prakash *et al.* [26] proposed two fuzzy measures established on Ferreri's probabilistic directed divergence measure [10]. According to Taneja [13], Bhatia and Singh [27] suggested the fuzzy measures.

Tomar and Ohlan [37, 38] used inequalities between fuzzy mean difference measures to generate a series of fuzzy mean difference divergence measures, which they applied to linguistic variables and pattern recognition. Among the applications to strategic decision making, Tomar and Ohlan [38] developed a novel parametric generalised exponential fuzzy divergence measure equivalent to Verma and Sharma [32].

Although the literature on the applications of information and divergence measures [10, 14, 15, 20, 25, 28] among FS has grown significantly in recent years, there is still the possibility of generating superior divergence measures that can be applied to a variety of sectors. The essential idea of Information Theory is entropy. Shannon [4] defined entropy as an expected value of information or the degree of randomness uncertainty in a probability distribution (p.d). Let us consider $p_1, p_2, \ldots, p_n$ are the set of probability distribution and sum of all probability distribution should be 1, then the entropy is as follows:

$$H(P)=\sum_{i=1}^{n}-p_i log p_i$$

$$D(P|Q)=\sum_{i=1}^{n}p_i log\left(\frac{p_i}{q_i}\right)$$

where D(P, Q) is directed divergence or cross entropy measure. The new f-divergence measure incorporates divergences that are utilized to determine the affinity between two probability distribution. Jain and Saraswat [16–18] and Csiszar [8, 9] introduced for the convex function f in $(0, \infty)$, which is given by

$$S_f(P,Q)=\sum_{i=1}^{n} q_i f\left(\frac{p_i+q_i}{2q_i}\right)$$

Fuzziness is a characteristic of uncertainty that arises from the lack of a clear distinction between being a member of the set and not being a member of the set. Set of fuzziness investigated by Zadeh [21].

$$A = \{x, \mu_A(x) / x \in X\},$$

where X is universe of discourse.

Here, the membership function is defined on set A and as follows: $\mu_A(x)$: $X \rightarrow [0, 1]$ is the membership function of A and Shannon's entropy defined as

$$H(N)=\sum_{i=1}^{n} -(\mu_A(x_i) log \mu_A(x_i)+(1-\mu_A(x_i) log(1-\mu_A(x_i)))$$

The above equation is the expected value of information. It is for discriminating a fuzzy set B from another fuzzy set A. Let A and B be two fuzzy sets with membership values A (xi) and B (xi), i = 1 to m, respectively, defined in $X = x_1, x_2, \ldots, x_m$. Here $\mu_A(x)$ and $\mu_B(x)$ is denoted by μ_{Ai} and μ_{Bi} throughout the paper.

11.2 New f-Divergence Measure on Fuzzy Sets

Here, discuss about the new f-divergence functional. Jain and Saraswat [16, 17] investigated the divergence functional, which is defined on convex and normalized functionals.

$$S_f(A,B)=\sum_{i=1}^{n}\left[\mu_{B_i} f\left(\frac{\mu_{A_i}+\mu_{B_i}}{2\mu_{B_i}}\right)+(1-\mu_{B_i}) f\left(\frac{2-\mu_{A_i}-\mu_{B_i}}{2(1-\mu_{B_i})}\right)\right] \tag{11.1}$$

where f: $(0.5, \infty) \rightarrow R_+$ is convex function and A, B $\in$ X.

Proposition 2.1 Let us consider for the convex and normalized function f: $(0.5, \infty) \rightarrow R$ be and A, B $\in$ X then we have the relation

$$S_f(A,B) \geq f(1) \tag{11.2}$$

Equality hold iff $\mu_{Ai} = \mu_{Bi} \forall i = 1, 2,\ldots,m$

Corollary 2.1 Let us consider for the convex and normalized function f: $(0.5, \infty) \rightarrow R$ be and A, B $\in$ X then we have the relation

$$S_f(A,B) \geq 0 \tag{11.3}$$

If f is strictly convex, equality holds *iff* $\mu_{Ai} = \mu_{Bi}$, $i = 1$ *to* m

$$S_f(A,B) \geq 0 \text{ and } S_f(A,B) = 0 \text{ iff } A = B \tag{11.4}$$

Proposition 2.2 If $g = af_1 + bf_2$ and f_1 and f_2 are two convex functions, then $aS_{f1}(A, B) + bS_{f2}(A, B)$ is a convex function, where A and B are the set of X and a and b are constants. We now have a few examples of well-known information measures that have been discovered using the fuzzy new f-divergence measure.

- **Fuzzy K-L Divergence** [33]-
  If $f(y)=(2y-1)\log(2y-1), \forall y > \frac{1}{2}$ then,

$$S_f(A,B)=\sum_{i=1}^{n}\left[\mu_{A_i}log\left(\frac{\mu_{A_i}}{\mu_{B_i}}\right)+(1-\mu_{A_i})log\left(\frac{1-\mu_{A_i}}{1-\mu_{B_i}}\right)\right]=K(A,B) \tag{11.5}$$

- **Fuzzy Chi-Square Divergence** [19]:
  If $f(y) = (y-1)^2$ then,

$$S_f(A,B)=\sum_{i=1}^{n}\left[\frac{(\mu_{A_i})^2}{\mu_{B_i}}+\frac{(1-\mu_{A_i})^2}{(1-\mu_{B_i})}\right]-1=\chi^2(A,B) \tag{11.6}$$

- **Fuzzy Relative JS Divergence** [7]:
  If $f(y) = -\log y$ then,

$$S_f(A,B)=\sum_{i=1}^{n}\left[\mu_{B_i}log\left(\frac{2\mu_{B_i}}{\mu_{A_i}+\mu_{B_i}}\right)+(1-\mu_{B_i})log\left(\frac{2(1-\mu_{B_i})}{2-\mu_{A_i}-\mu_{B_i}}\right)\right] \tag{11.7}$$

$$=S_t(A,B)$$

- **Fuzzy Relative AG Divergence** [14]:
  If f (y) = y log y then,

$$S_f(A,B)=\sum_{i=1}^{n}\left[\left(\frac{\mu_{A_i}+\mu_{B_i}}{2}\right)\log\left(\frac{\mu_{A_i}+\mu_{B_i}}{2\mu_{A_i}}\right)+\left(\frac{2-\mu_{A_i}-\mu_{B_i}}{2}\right)log\left(\frac{2-\mu_{A_i}-\mu_{B_i}}{2(1-\mu_{A_i})}\right)\right]=H(A,B) \tag{11.8}$$

$$=S_t(A,B)$$

- **Fuzzy Hellinger Divergence** [6]:
  If f (y) = $(1-\sqrt{y})$ then,

$$S_f(A,B)=\frac{1}{2}\sum_{i=1}^{n}\left[\left(\sqrt{\mu_{A_i}}-\sqrt{\mu_{B_i}}\right)^2+\left(\sqrt{(1-\mu_{A_i})}-\sqrt{(1-\mu_{B_i})}\right)^2\right] \tag{11.9}$$

$$=M(A,B)$$

- **Fuzzy Triangular Divergence** [16]:

If $f(y)=\frac{(y-1)^2}{y},\forall\ y>0$ then,

$$S_f(A,B)=\sum_{i=1}^{n}\left[\frac{(\mu_{A_i}-\mu_{B_i})^2}{(\mu_{A_i}+\mu_{B_i})}+\frac{(\mu_{B_i}-\mu_{A_i})^2}{(2-\mu_{A_i}-\mu_{B_i})}\right]=\Delta(A,B) \tag{11.10}$$

11.3 New Fuzzy Information Inequalities Using Fuzzy New f-Divergence Measure and Fuzzy Triangular Divergence Measure

Theorem 3.1 Let us consider for the convex and normalized function *f*: (0.5, ∞) → *R* and A, B ϵ X and satisfied the below properties

(i) *f* is double derivable and defined on (*r*, *R*), where $0 \le r \le 1 \le R < \infty$
(ii) ∃ constants n, N such that

$$n \le y^3 f''(y) \le N \tag{11.11}$$

If fuzzy set A and B in the discourse world, then X = $x_1, x_2, \ldots, x_m$ meets the assumptions.

$$r < \frac{1}{2} \le r_i = \sum_{i=1}^{n} \left[\frac{\mu_{A_i} + \mu_{B_i}}{2\mu_{B_i}} + \frac{2 - \mu_{A_i} - \mu_{B_i}}{2(1 - \mu_{B_i})} \right] \le R, \forall_i \{1, 2, \ldots m\} \quad (11.12)$$

Then, we have the following relations

$$\frac{n}{2}\Delta(A,B) \le S_f(A,B) \le \frac{N}{2}\Delta(A,B) \quad (11.13)$$

Proof. Let a function $F_n:(0,\infty) \to R, F_n(y) = f(y) - n\frac{(y-1)^2}{y}, \forall y > 0$. Then, F_n (.) is
f (1) = 0 and double derivable, then

$$F_n''(y) = f''(y) - \frac{n}{y^3} = \frac{1}{y^3}[y^3 f''(y) - n] \ge 0 \quad (11.14)$$

The function F is convex function and defined on (r, R), then using the property of non-negativity and proposition 2.2, then we get

$$0 < S_{F_n}(A,B) = S_f(A,B) - nS\frac{(y-1)^2}{y}(A,B) = S_f(A,B) \quad (11.15)$$

$$-\frac{n}{2}\Delta(A,B) = 0 \le S_f(A,B) - \frac{n}{2}\Delta(A,B)$$

Again,
Let a mapping $F_N:(0,\infty) \to R, F_N(y) = N\frac{(y-1)^2}{y} - f(y), \forall y > 0$. The function F is convex on (r, R) and defined on (r, R), then using the property of non-negativity and proposition 2.2, then we get the II part of (11.13) i.e.

$$= 0 \le \frac{N}{2}\Delta(A,B) - S_f(A,B) \quad (11.16)$$

From (11.15) and (11.16) give the result (11.13).

11.4 Applications for Some Fuzzy f-Divergence Measures

Using inequality (11.13) of Theorem 3.1, we determined bounds for some existing fuzzy divergence measures in terms of fuzzy triangular discrimination.

Proposition 4.1 Let us consider the two fuzzy A and B and are the elements of X and satisfying (11.11), then we find the relation

$$r<\frac{1}{2}\leq r_i=\sum_{i=1}^{n}\left[\frac{\mu_{A_i}+\mu_{B_i}}{2\mu_{B_i}}+\frac{2-\mu_{A_i}-\mu_{B_i}}{2(1-\mu_{B_i})}\right]\leq R,\forall i\varepsilon\{1,2,\ldots,n\} \quad (11.17)$$

Then we can find the following relations

$$\frac{r}{2}\Delta(A,B)\leq S_t(A,B)\leq\frac{R}{2}\Delta(A,B) \quad (11.18)$$

Proof. Let the function *f*: (0, ∞) – → R

$$\text{f(y)}=-\log \text{y}, \text{f}''(\text{y})=\frac{1}{y^2}>0, \forall\ \text{y}>0$$

$$\text{f}''\ (\text{y})\geq 0,\ \text{f}\ (1)=0,$$

Define $g(y)=y^3 f''(y)=y^{3\left(\frac{1}{y^2}\right)}=y$

Then,

$$N=\sup_{y\in[r,R]} g(y)=R, n=\inf_{y\in[r,R]} g(y)=r \quad (11.19)$$

Also S_f (A, B) = S_t (B, A) from eq. (11.7), (11.3) and (11.19)

$$\frac{r}{2}\Delta(A,B)\leq S_t(\text{B},\text{A})\leq\frac{R}{2}\Delta(\text{A},\text{B})$$

The interchange of A and B result in (11.18).

Proposition 4.2 Let us consider the two fuzzy A and B and are the elements of X and satisfying (11.11) then we find the relation

$$\frac{r^2}{2}\Delta(A,B) \leq H(A,B) \leq \frac{R^2}{2}\Delta(A,B) \tag{11.20}$$

Proof. Let the convex and normalized function f: (0.5, ∞) – → R

$$f(y) = y \log y, f''(y) = \frac{1}{y} > 0, \forall\, y > 0$$

$$f''(y) \geq 0, f(1) = 0,$$

Define $g(y) = y^3 f''(y) = y^2$
Then,

$$N = \sup_{y \in [r,R]} g(y) = R^2, n = \inf_{y \in [r,R]} g(y) = r^2 \tag{11.21}$$

Also $S_f(A, B) = H(B, A)$
From eq. (11.8), (11.13) and (11.25)

$$\frac{r^2}{2}\Delta(A,B) \leq \mathrm{H(B,A)} \leq \frac{R^2}{2}\Delta(\mathrm{A,B})$$

The interchange of A and B result in (11.20).

Proposition 4.3 Let us consider the two fuzzy A and B and are the elements of X and satisfying (11.11) then we find the relation

$$r^3\Delta(A,B) \leq \frac{1}{4}\chi^2(A,B) \leq R^3\Delta(A,B) \tag{11.22}$$

Proof. Let the function *f*: (0.5, ∞) – → R

$$f(y) = y^2 - 1, f''(y) = 2 > 0, \forall\, y > 0$$

Define $g(y) = y^3 f''(y) = 2\,y^3 = 2\,y^3$

Then,

$$N=\sup_{y\in[r,R]} g(y)=2R^3, n=\inf_{y\in[r,R]} g(y)=2r^3 \tag{11.23}$$

Also $S_f(A,B)=\frac{1}{4}\chi^2(A,B)$ from eq. (11.6), (11.13) and (11.23) give the result in (11.22).

Proposition 4.4. Let us consider the two fuzzy A and B and are the elements of X and satisfying (11.11) then we find the relation

$$r(r+1)\,\Delta\,(A,B) \le J_m(A,B) \le R\,(R+1)\,\Delta\,(A,B) \tag{11.24}$$

Proof. Let the function f: $(0.5,\infty) \dashrightarrow R$

$$f(y)=(y-1)\log y, f''(y)=\frac{1+y}{y^2}>0, \forall\ y>0$$

Define $g(y)=y^3 f''(y)=y^3\frac{1+y}{y^2}=y(y+1)$
Then,

$$N=\sup_{y\in[r,R]} g(y)=R(R+1), n=\inf_{y\in[r,R]} g(y)=r(r+1) \tag{11.25}$$

Also, $S_f(A,B)=\frac{1}{2}J_m(A,B)$ from [10]
From [10], (11.13) and (11.25) proved (11.24).

Proposition 4.5 Let us consider the two fuzzy A and B and are the elements of X and satisfying (3.1) then we find the relation

$$\frac{1}{2}r^{\beta+1}\Delta(A,B)\le MG_\beta(A,B)\le\frac{1}{2}R^{\beta+1}\Delta(A,B) \tag{11.26}$$

Proof. Let the function f: $R_+ \dashrightarrow R$

$$f(y)=[\beta(\beta-1)]^{-1}[y^\beta-1], f''(y)=y^{\beta-2}>0, \forall\ y>0$$

$$f''(y)\ge 0, f(1)=0$$

Define $g(y)=y^3 f''(y)=y^{\beta+1}$

Then,

$$N = \sup_{y\in[r,R]} g(y) = R^{\beta+1}, n = \inf_{y\in[r,R]} g(y) = r^{\beta+1} \tag{11.27}$$

Also S_f (A, B) = MG_β (B, A) from (11.9)
From eq. (11.9), (11.13) and (11.27)

$$\frac{1}{2}\mathrm{r}^{\beta+1}\Delta(A,B) \le \mathrm{MG}_\beta(\mathrm{B},\mathrm{A}) \le \frac{1}{2}\mathrm{R}^{\beta+1}\Delta(\mathrm{A},\mathrm{B})$$

The interchange of A and B result in (11.26).

Corollary 4.1 For $\beta = \frac{1}{2}$ and Consider $A, B \in X_M$ be two fuzzy sets satisfying (11.11) then we have the following relations

$$\frac{1}{2}f^{\frac{3}{2}}\Delta(A,B) \le 4\left[1 - W\left(\frac{A+B}{2}, B\right)\right] \le \frac{1}{2}R^{\frac{3}{2}}\Delta(A,B)$$

Where $W\left(\frac{A+B}{2}, B\right)$ is Bhattacharya [1] measure

Proof. Let the function *f*: (0, ∞) − → *R*

If $\beta = \frac{1}{2}$ Then Function,

$$N = \sup_{y\in[r,R]} g(y) = R^{\frac{3}{2}}, n = \inf_{y\in[r,R]} g(y) = r^{\frac{3}{2}}$$

Also $S_f(\mathrm{A},\mathrm{B}) = 4\left[1 - \mathrm{W}\left(\frac{A+B}{2}, B\right)\right] = 4\mathrm{h}\left(\frac{A+B}{2}, \mathrm{B}\right)$

From eq. (11.13) and (11.27) give the result (11.26).

11.5 Applications in MCDM

Decision-making process and problem are concerned of selecting the best option based on the greatest degree of fulfilment of the whole relevant criteria. The most well-known branch of DM is MCDM, which allows the rank for decision-makers and choose options based on a distinct criterion with the help of the Triangular discrimination, we can solve MCDM problems using the way below.

Let R = (R_1, R_2, R_3... R_m) be a set of options, D = (D_1, D_2, D_3, …, D_m) are the sets of criteria and the R_i are the characteristics of criteria D and characterized by the following FSs:

$R_i = \{\langle D_j, \mu_{ij} \rangle, D_j \in D\}$, {i = 1 to m and j = 1 to n}. where μ_{ij} denotes the degree that the R_i fulfils the criteria D_j. Using the equation (11.10) and obtained the solution of discussed and problems by using Eq. (11.10).

Step 1: Find out the PIS R^+ and NIS R–

$$R^+ = \{\langle \mu_1^+ \rangle, \langle \mu_2^+ \rangle, \ldots, \langle \mu_n^+ \rangle\}$$

$$R^- = \{\langle \mu_1^- \rangle, \langle \mu_2^- \rangle, \ldots, \langle \mu_n^- \rangle\}$$

Here, ? j = 1, 2, 3,…, n.

$$\mu_{j+} = <\max_i \mu_{ij}\}$$

$$\mu_{j-} = \langle \min_i \mu_{ij}\}$$

where R+ is the positive-ideal solution and R– is the negative-ideal solution.

Step 2: Compute $\Delta(R^+, R_i)$ and $\Delta(R^-, R_i)$

Step 3: Compute the measure $\Delta(R_i)$ of alternative R_i w.r.t. R^+ and R–, where

$$\Delta(R_i) = \frac{\Delta(R^+, R_i)}{\Delta(R^+, R_i) + \Delta(R^-, R_i)}, i = 1 \text{ to } m.$$

Step 4: Order of the rank preference for all alternatives as per the $\Delta(R_i)$, i = 1 to m in ascent order.

Step 5: Select the greatest alternative R_i with smallest $\Delta(R)$.

To solve MCDM problems, established applications of novel fuzzy measure.

For this we are going to consider costumer decision-making problem given below.

Allow a customer who needs to buy an air cooler to do so. Let us say there are four options (air cooler systems) available on the market Ri, i = 1, 2, 3, 4. Assume that the collecting procedure takes into account three criteria: I (D1) cost-effective, (ii) (D2) serviceable, and (iii) (D3) operational.

The consumer must value the four possible options based on three criteria in the form of sets:

11.5.1 Case Study

$$R_1 = \{\langle D_1, .2\rangle, \langle D_2, .4\rangle, \langle D_3, .6\rangle\}$$

$$R_2 = \{\langle D_1, .5\rangle, \langle D_2, .8\rangle, \langle D_3, .3\rangle\}$$

$$R_3 = \{\langle D_1, .7\rangle, \langle D_2, .3\rangle, \langle D_3, .6\rangle\}$$

$$R_4 = \{\langle D_1, .8\rangle, \langle D_2, .2\rangle, \langle D_3, .4\rangle\}$$

The one by one process and evaluate the solution of discussed MCDM problems.

Step 1: PIS (R^+) and NIS (R^-) respectively are

$$R^+ = \{\langle D_1, .8\rangle, \langle D_2, .8\rangle, \langle D_3, .6\rangle\}$$

$$R_- = \{\langle D_1, .2\rangle, \langle D_2, .2\rangle, \langle D_3, .3\rangle\}$$

Step 2: The numerical values $\Delta(R^+, R_i)$ and $\Delta(R^-, R_i)$ have been calculated and represent in Tables 11.1 and 11.2.

Step 3: Calculated value $\Delta(R_i)$ where i = 1 to 4.

Table 11.1 $\Delta(R^+, R_i)$.

$\Delta(R^+, R_1)$	.7200	.3333	0.0
$\Delta(R^+, R_2)$	.1976	0.0	.1818
$\Delta(R^+, R_3)$	.0267	.1818	0.0
$\Delta(R^+, R_4)$	0.0	.7200	.0800

Table 11.2 $\Delta(R-, R_i)$, $t \geq 0$.

$\Delta(R-, R_1)$	0.0	0.0953	.1818
$\Delta(R-, R_2)$	.1978	.7200	0.0
$\Delta(R-, R_3)$	.5051	.0267	.1818
$\Delta(R-, R_4)$	.7200	0.0	.022

$$\Delta(R_1) = .7917$$

$$\Delta(R_2) = .2925$$

$$\Delta(R_3) = .2261$$

$$\Delta(R_4) = .5188$$

Step 4: Rank the alternatives.

$$R1 > R4 > R2 > R3$$

Step 5: Hence, the optimum alternative is R_1.

11.6 Conclusion

Established various bounds on new fuzzy measures by employing fuzzy new f-divergence measures properties with demonstration of validity. Derived other bounds of new and existing fuzzy measures in terms of triangular discriminations. Inequalities between new and well-known fuzzy divergence measures have also been derived. The direct applications of proposed fuzzy in MCDM have been solved by using real-world examples and case study. These results have been compared with existing data.

References

1. Bhattacharyya, A., Some analogues to the amount of information and their uses in statistical estimation. *Sankhya*, 8, 1–14, 1946.
2. Kaufmann, A., *Introduction to the Theory of Fuzzy Sets-Vol. 1, Fundamental Theoretical Elements*, Academic Press, New York, 1975.
3. Bhandari, D. and Pal, N.R., Some new information measures for fuzzy sets. *Inf. Sci.*, 67, 3, 209–228, 1993.
4. Shannon, C.E., A mathematical theory of communication. *Bell Syst. Tech. J.*, 27, 3, 379–423, 1948.
5. Ferreri, C., Hyperentropy and related heterogeneity divergence and information measures. *Statistica*, 40, 2, 155–168, 1980.
6. Hellinger, E., Neue begrundung der, theorie der quadratischen formen von unendlichen vielen ver anderlichen. *J. Reine Angew. Math.*, 136, 210–271, 1909.
7. Topse, F., Some inequalities for information divergence and related measures of discrimination. *Res. Coll. RGMIA*, 2, 85–98, 1999.
8. Csisza, I., On Topological Properties of f-Divergences. *Studia Math. Hungarica*, 2, 329-339, 1967.
9. Csiszar, I., Information-type measures of difference of probability functions and indirect observations. *Studia Sci. Math. Hungarica*, 2, 299–318, 1961.
10. Taneja, I.J. and Kumar, P., Relative information of type s, Csiszar's f-divergence, and information inequalities. *Inf. Sci.*, 166, 105–125, 2004.
11. Taneja, I.J., New developments in generalized information measures, in: *Advances in imaging and Electron Physics*, vol. 91, Hawkes, P.W. (Ed.), pp. 37–135, 1995.
12. Taneja, I.J. and Kumar, P., Generalized non-symmetric divergence measures and inequalities, Journal of Interdisciplinary Mathematics, 2006, 9, 3, 581-599, DOI: 10.1080/09720502.2006.10700466.
13. Taneja, I.J., On mean divergence measures. *Eprint Arxiv Math*, 1–18, 2005, arXiv:math/0501298v2 [math.ST].
14. Burbea, J. and Rao, C.R., On convexity of some divergence measures based on entropy functions. *IEEE Trans. Inf. Theory*, 28, 489–495, 1982.
15. Fan, J. and Xie, W., Distance measures and induced fuzzy entropy. *Fuzzy Set. Syst.*, 104, 2, 305–31, 1999.
16. Jain, K.C. and Saraswat, R.N., Inequalities among well known divergence measures and triangular discrimination. *Int. J. Comput. Sci. Math.*, 4, 3, 215–224, 2012.
17. Jain, K.C. and Saraswat, R.N., Some well-known inequalities and its applications in information theory. *Jordan J. Math. Stat.*, 6, 2, 157–167, 2013.
18. Jain, K.C. and Saraswat, R.N., Some bounds of information divergence measures in terms of relative-arithmetic divergence measure. *Int. J. Appl. Math. Stat.*, 32, 2, 48–58, 2013.

19. Pearson, K., On the criterion that a give system of deviations from the probable in the case of correlated system of variables in such that it can be reasonable supposed to have arisen from random sampling. *Philos. Mag.*, 50, 157–172, 1900.
20. Loo, S.G., Measures of fuzziness. *Cybernetica*, 20, 201–210, 1977.
21. Zadeh, L.A., Probability measures of fuzzy events. *J. Math. Anal. Appl.*, 23, 421–427, 1968.
22. Zadeh, L.A., Toward a generalized theory of uncertainty (GTU): An outline. *Inf. Sci.*, 172, 1-2, 1–40, June 9, 2005.
23. Zadeh, L.A., Fuzzy sets. *Inf. Control*, 8, 3, 338–353, 1965.
24. De Luca, A. and Termini, S., A definition of non-probabilistic entropy in the setting of fuzzy set theory. *Inf. Control*, 20, 4, 301–312, 1972.
25. Ghosh, M., Das, D., Chakraborty, C., Roy, A.K., Automated leukocyte recognition using fuzzy divergence. *Micron*, 41, 840–846, 2010.
26. Prakash, O. and Kumar, R., Modified fuzzy divergence measure and its application to medical diagnosis and MCDM. *Risk Decis. Anal.*, 6, 3, 231–237, 2017.
27. Bhatia, P.K. and Singh, S., Three families of generalized fuzzy directed divergence. *Adv. Model. Optim.*, 14, 3, 599–614, 2012.
28. Beran, R., Minimum hellinger distance estimates for parametric models. *Ann. Stat.*, 5, 445–463, 1977.
29. Saraswat, R.N. and Khatod, N., New fuzzy divergence measures, series, its bounds and applications in strategic decision-making. *Lecture Notes Electrical Engineering,* Springer, Singapore, 607, 641–653, 2020.
30. Saraswat, R.N. and Tak, A., New f-divergence and jensen-ostrowski's type inequalities. *Tamkang J. Math.*, 50, 1, 111–118, 2019.
31. Sibson, R., Information radius. *Z. Wahrscheinlichkeit*, 14, 149–160, 1969.
32. Verma, R. and Sharma, B.D., On generalized exponential fuzzy entropy. *World Acad. Sci. Eng. Technol.*, 60, 1402–1405, 2011.
33. Dragomir, S.S., Sunde, J., Buse, C., New inequalities for jeffreys divergence measure. *Tamusi Oxf. J. Math. Sci.*, 16, 2, 295–309, 2000.
34. Dragomir, S.S., Bounds of f-divergences under likelihood ratio constraints, Applications of Mathematics, 48, 3, 205–223, 2003.
35. Kullback, S. and Leibler, R.A., On information and sufficiency. *Ann. Math. Stat*, 22, 1, 79–86, 1951.
36. Montes, S., Couso, I., Gil, P., Bertoluzza, C., Divergence measures between fuzzy sets. *Int. J. Approx. Reason.*, 30, 91–105, 2002.
37. Tomar, V.P. and Ohlan, A., Sequence of inequalities among fuzzy mean difference divergence measures and their applications. *Springer Plus*, 3, 623, 1–20, 2014.
38. Tomar, V.P. and Ohlan, A., New parametric generalized exponential fuzzy div ergence measure. *J. Uncertainty Anal. Appl.*, 2, 24, 1–14, 2014.

12

Contribution of Refreshment Provided to the Server During His Job in the Repairable Cold Standby System

M.S. Barak[1]*, Ajay Kumar[2] and Reena Garg[2]

[1]Department of Mathematics, Indira Gandhi University, Meerpur, Rewari, India
[2]Department of Mathematics, J. C. Bose UST, YMCA, Faridabad, India

Abstract
Due to a human being, the server is not always working like a robot or machine; sometimes the server feels tiredness or some emotional things that affect the performance of the server during his job like repair or replacement of the unit. In this manuscript, refreshment is offered to the server to enhance his working capacity and to determine the reliability measures of a cold standby system of identical units in which one is operative and another is a cold standby unit as shown in Figure 12.1. Also, after getting refreshments, the repairman may ultimately provide better service effectively and after getting the renovation, the unit works like a new unit. Let us suppose that renovation and refreshment actions have an exponential distribution whereas unit failure, and repairman failure have a random distribution with a distinct probability density function. The random variables are statistically independent. Figures are used to show the nature of some essential reliability measures, such as mean time to system failure (MTSF), availability, and profit of the system.

Keywords: Refreshment, performance, repairable system, cold-standby, reliability measures

**Corresponding author*: ms_barak@igu.ac.in

S. C. Malik, Deepak Sinwar, Ashish Kumar, S. R. Gadde, Prasenjit Chatterjee and Bui Thanh Hung (eds.) Computational Intelligence in Sustainable Reliability Engineering, (251–268) © 2023 Scrivener Publishing LLC

12.1 Introduction

Reliability is the probability by which the system performs the specified task in a given period without failure, or reliability is the availability of the system to provide a quality product at a low cost. Any system or unit made by the manufacturer works with full capacity to perform a particular task when it is functioning under desired limits. Redundancy provides the parallel path to perform the desired task. For this purpose, we have to use the cold standby redundant system. A large amount of research has been done with a variety of repair strategies using cold standby systems of identical units and stochastic modeling, regardless of server fatigue or his feelings affecting the performance. In 1971, a redundant cold standby system having two identical units with the assumption of standby failure have been analyzed by Osaki and Nakagawa [1]. They explored [2] some more characteristics of a redundant system with the repair facility. Subramanian and his coworkers [3] threw light on the analytic behavior of the standby failure of a repairable system. In continuation of the recent study, Gopalan and Nagarwalla [4] highlighted the profit of the standby system having two-unit with the repair and age replacement facility done by a single server.

The common causes of warm standby system failures and measures of system reliability have been scrutinized by Dhillon [5]. Dhankhar and Malik [6] analyzed the reliability system models subjected to inspection and repair with a single server facility. Kumar and his co-researchers [7] described some measures of 2 nonidentical units systems with replacements of the unit and repair of the unit having correlated failures and repairs to enhance the reliability of the system. The reliability of a cold standby system with repair and preventive maintenance of the unit under different weather conditions has been discussed by Malik and Barak [8]. Dhall *et al.* [9] inspected the failed unit for possible maintenance or the replacement of the failed unit by cold standby identical unit before executing his job. Bhardwaj and Singh [10] explained the performance asymptotically of a standby system having the semi-Markov approach and server failure facility. Bhardwaj *et al.* [11] threw light on stochastic system performance having provision of maintenance and standby failure. Bhardwaj and Singh [12] scrutinized the stochastic behavior of cold-standby redundant systems subjected to server failure and delayed treatment. Kumar and Barak [13] threw light on availability and reliability measures sensitivity of harvesting systems towards different weather conditions. Kumar *et al.* [14] highlighted a cold standby system under the provision of maximum repair time. At the same time, Malik [15] evaluated the behavior of a redundant system stochastically having provision of service failure. Barak and Barak

[16] analyzed the repairable redundant system subjected to inspection under abnormal weather conditions. Barak and his team [17] discussed the profit of the standby redundant system having provision of the inspection before repair or preventive maintenance of the unit and the cold standby parallel system with the provision of preventive maintenance has been studied by Garg and Kadyan [18].

In 2016, Kumar and Goel [19] calculated some reliability measures of a cold standby redundant system having two units for general distribution. Yadav and Barak [20] examined a cold standby system behavior with the assumption of server failure. Barak *et al.* [21] explained the idea to provide refreshments to the server during his job to enhance his efficiency to perform his job. Barak and Barak [22] discussed a two-unit cold standby stochastic system working under distinct weather conditions. Barak *et al.* [23] highlighted the provision of inspection with the repair in a cold standby system. Kumar and Saini [24] explored some reliability measures of the computer system under the provision of preventive maintenance and using the concept of repair times. Yadav and Malik [25] analyzed the computer having unit wise cold standby system subjected to the failure of the server. Malik and Yadav [26] described some essential reliability measures of a computer system having unit wise cold standby system with priority to hardware repair with assumption failure of the server. Malik *et al.* [27] examined the software up-gradation of a computer system with server failure during a job under the provision of priority to hardware repair. Saini *et al.* [28–31] scrutinized the sugar manufacturing plant having reliability parameters estimation, and evaluated a redundant system of non-identical units having priority to repair, and under the provision of different repair facilities.

Further, according to the above study of different kinds of systems, the server is always available for his service without any tiredness which seems to be unrealistic. The server does not always act like a robot or a machine because of humans; sometimes the server feels fatigued or some emotional things that affect his performance during the job like repair, maintenance, and replacement of the particular unit. The contribution of refreshment provided to the server during his work has not been studied so far. In this chapter, to override the server's tiredness during the job refreshment is offered to him that increases his performance and attention toward the job as shown in Figure 12.1. Also, after receiving refreshments, the server can effectively provide better service and after getting the repair, the unit operates like a new unit. Suppose that the repair and refreshment actions have a negative exponential distribution while the unit failure and server refreshment request rate have a random distribution with a different probability density function. The graphs are used to show the nature of some of the required reliability measures such as the meantime to system failure (MTSF), availability, and profit of the system.

12.2 The Assumptions and Notations Used to Solve the System

- Initially, suppose the whole system is in good condition and has two identical units out of which -one unit is sufficient to operate the system and the other unit keeping as cold standby. Model in Figure 12.1.
- When the operative unit gets fails, then the cold standby identical unit starts functioning immediately.
- A single server is always available to repair the failed unit.
- Server is allowed to take the refreshments whenever necessary.
- The unit is functioning as a new unit after undergoing repair.
- All parameters are statistically independent.
- The repair and refreshment actions have a negative exponential distribution, while the server refreshment is required to rate the unit failure and have a random distribution with a different probability density function.

R	Collection of regenerative states S_i ($i = 0, 1, 2, 3$).
O/C_s	The system's unit is operative/cold standby.
a/b	Probability that an additional redundant unit is functioning/not functioning.
Fur/FUR	Direct failed unit of the system under repair/under regular repair from the precedent state.
Fwr/FWR	The failed unit waiting for repair/continuous waiting for repair from the precedent state.
Sut/SUT	The server is busy getting refreshment/server is continuously busy getting refreshment from the precedent state.
λ/μ	The constant failure rate of the unit/rate by which the server needs refreshment.
$g_1(t)/G_1(t)$	Probability density function/cumulative density function of the repair rate of the unit.

$f_1(t)/F_1(t)$	Probability density function/cumulative density function of refreshments that restore freshness to the server.
$q_{rs}(t)/Q_{rs}(t)$	Probability density function/cumulative density function of first passage time from r[th] to s[th] regenerative state or s[th] failed state without halting in any other $S_i \in R$ in (0, t].
$q_{rs.t}(t)/Q_{rs.t}(t)$	Probability density function/cumulative density function of first passage time from $S_r \in R$ to $S_s \in R$ or to a failed state S_s with temporarily staying at S_t once in (0, t].
$q_{rs.t(u,v)}(t)/Q_{rs.t(u,v)}(t)$	Probability density function/cumulative density function of first passage time from $S_r \in R$ to $S_s \in R$ or to a failed state S_s with temporarily staying at S_t, S_u, S_v, once in (0, t].
μ_i	Suppose "T" notation for failure time of the system and the mean sojourn time in the state S_i becomes $$\mu_i = \sum_j m_{ij} = \int_0^\infty P(T > t)dt.$$
$M_r(t)$	Represents the probability of the system that it originally works $S_r \in R$ at a time (t) without moving through another state $S_i \in R$.
$W_r(t)$	The probability that up to time (t) the server is involved in the state S_r, without transit to another state $S_i \in R$ or before return to the same state through one or more non- regenerative states.
$\oplus/\otimes$	Symbols of Laplace convolution/Laplace Stieltjes Convolution.
LT/LST	Laplace Transformation/Laplace Stieltjes Transformation.
$*/**/'$	Laplace Transform symbol/Laplace Stieltjes Transform symbol/Function's derivative symbol for the model.
○/●/⬠/▭	Notation for Up state/Regenerative state/Reduced state/Failed state.

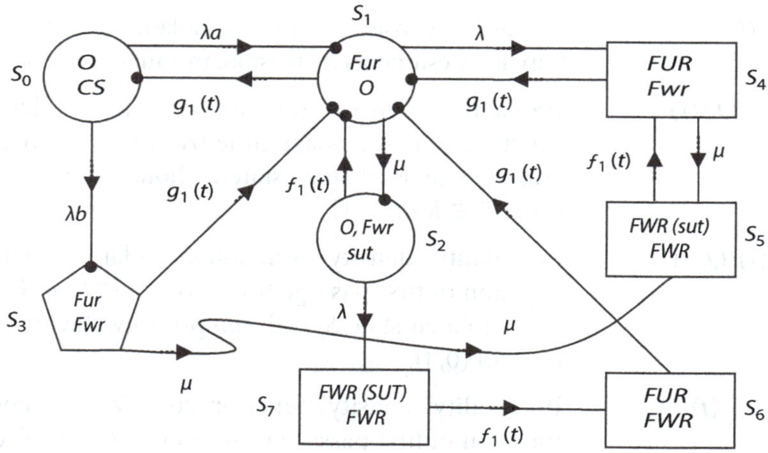

Figure 12.1 State transition diagram.

12.3 The Probabilities of States Transitions

For the non-zero elements, there is smooth probability considerations yield the following expressions;

$$p_{ij} = \int_{0}^{\infty} q_{ij}(t)dt \tag{12.1}$$

$$p_{01} = a, p_{03} = b, p_{54} = p_{61} = p_{76} = 1$$

$$p_{10} = \frac{\phi}{\phi+\mu+\lambda}, p_{12} = \frac{\mu}{\phi+\mu+\lambda}, p_{14} = \frac{\lambda}{\phi+\mu+\lambda}, p_{21} = \frac{\theta}{\theta+\lambda}, p_{27} = \frac{\lambda}{\theta+\lambda}$$

$$p_{31} = \frac{\phi}{\phi+\mu}, p_{35} = \frac{\mu}{\phi+\mu}, p_{41} = \frac{\phi}{\phi+\mu}, p_{45} = \frac{\mu}{\phi+\mu}$$

(12.2)

It has been conclusively established that:

$$p_{01}+p_{03}=1, p_{10}+p_{12}+p_{14}=1, p_{21}+p_{27}=1, p_{31}+p_{35}=1, p_{41}+p_{45}=1,$$
$$p_{10}+p_{12}+p_{11.4}+p_{11.(45)^n}=1, p_{21}+p_{21.(76)}=1, p_{31}+p_{31.(54)^n}=1 \qquad (12.3)$$

12.4 Mean Sojourn Time

Let T be the time at which the system fails at the state S_i, and μ_i be the mean sojourn time, mathematically calculated as $\mu_i=\sum_j m_{ij}=\int_0^\infty P(T>t)dt$. The system has an unconditional mean time to go for anyone $S_i \in R$ while it is calculated from the age of entry into the state S_j's. Then we have

$$\mu_0=m_{01}+m_{03}=\frac{1}{\lambda}, \mu_1=m_{10}+m_{12}+m_{14}=\frac{1}{\phi+\mu+\lambda}, \mu_2=m_{21}+m_{27}=\frac{1}{\phi+\lambda}$$
$$\mu_3=m_{31}+m_{35}=\frac{1}{\phi+\mu}, \mu_4=m_{41}+m_{45}=\frac{1}{\phi+\mu}, \mu_5=\mu_7=\frac{1}{\theta}, \mu_6=\frac{1}{\phi}$$
$$\mu_1'=m_{10}+m_{12}+m_{14}+m_{11.(45)^n}=\frac{\theta\phi+\lambda(\theta+\mu)}{\theta\phi(\phi+\mu+\lambda)}$$
$$\mu_2'=m_{21}+m_{21.(76)}=\frac{\theta\phi+\lambda(\theta+\phi)}{\theta\phi(\theta+\lambda)}, \mu_3'=m_{31}+m_{31.(54)^n}=\frac{\theta+\mu}{\theta\phi} \qquad (12.4)$$

12.5 Mean Time to Failure of the System

Let the cumulative density function of the first elapsed time be $\varphi_i(t)$ from state $S_i \in R$ to the failed state of the system. Treating the failed states as trapping states, then the repetitive interface for $\varphi_i(t)$ are:

$$\varphi_0(t)=Q_{03}(t)+Q_{01}(t)\otimes\varphi_1(t)$$
$$\varphi_1(t)=Q_{14}(t)+Q_{12}(t)\otimes\varphi_2(t)+Q_{10}(t)\otimes\varphi_0(t)$$
$$\varphi_2(t)=Q_{27}(t)+Q_{21}(t)\otimes\varphi_1(t) \qquad (12.5)$$

Taking LST of the relation (12.5), then we get

$$R^*(s) = \frac{1-\varphi_0^{**}(s)}{s} \tag{12.6}$$

Now, the reliability of the system is accessed by using Laplace inverse transformation of equation (12.6), we obtained

$$MTSF = \lim_{s\to 0}\frac{1-\varphi_0^{**}(s)}{s} = \frac{[1-p_{12}p_{21}]\mu_0 + p_{01}\mu_1 + p_{01} + p_{12}\mu_2}{1-p_{01}p_{10} - p_{12}p_{21}} \tag{12.7}$$

12.6 Steady-State Availability

Availability refers to the probability that the system is available for use at the specified time as defined. Suppose $A_i\,(t)$ indicated the system's probability that it is in up-state at a time (t) while the defined system turned $S_i \in R$ at a time (t = 0). Then the repetitive interface for $A_i\,(t)$ is:

$$\begin{aligned}
&A_0(t)=q_{01}(t)\oplus A_1(t)+q_{03}(t)\oplus A_3(t)+M_0(t)\\
&A_1(t)=q_{10}(t)\oplus A_0(t)+q_{12}(t)\oplus A_2(t)+\left[q_{11.4}(t)+q_{11.(45)^n}(t)\right]\oplus A_1(t)+M_1(t)\\
&A_2(t)=\left[q_{21}(t)+q_{26.(76)^n}(t)\right]\oplus A_1(t)+M_2(t)\\
&A_3(t)=\left[q_{31}(t)+q_{31.(54)^n}(t)\right]\oplus A_1(t)
\end{aligned} \tag{12.8}$$

where,

$$M_0(t)=e^{-\lambda t}, M_1(t)=\overline{G(t)}e^{-(\mu+\lambda)t}, M_2(t)=\overline{F(t)}e^{-\lambda t} \tag{12.9}$$

Using LT of the above relation (12.8), then we get

$$A_0(\infty)=\lim_{s\to 0} sA_0^*(s)=\frac{N_A}{D'}, \tag{12.10}$$

where, $N_A=[\mu_0 p_{10}+\mu_1+\mu_2 Pp_{12}]$ and $D'=\left[\mu_0 p_{10}+\mu_1'+\mu_2' p_{12}+\mu_3' p_{03} p_{10}\right]$

12.7 Busy Period of the Server Due to Repair of the Failed Unit

Suppose $B_i(t)$ represent the server business probability due to repair of the particular unit which is not in the position of working at a time (t) while the unit of the system $S_i \in R$ at a time (t = 0). Then the repetitive interface for $B_i(t)$ is:

$$\begin{aligned}
B_0(t) &= q_{01}(t) \oplus B_1(t) + q_{03}(t) \oplus B_3(t) \\
B_1(t) &= q_{10}(t) \oplus B_0(t) + q_{12}(t) \oplus B_2(t) + \left[q_{11.4}(t) + q_{11.(45)^n}(t)\right] \oplus B_1(t) + W_1(t) \\
B_2(t) &= \left[q_{21}(t) + q_{26.(76)}(t)\right] \oplus B_1(t) \\
B_3(t) &= \left[q_{31}(t) + q_{31.(54)^n}(t)\right] \oplus B_1(t) + W_3(t)
\end{aligned} \tag{12.11}$$

where, $W_1(t) = \overline{G_1(t)} e^{-(\mu+\lambda)t} + \overline{G_1(t)} \lambda e^{-(\lambda+\mu)t} \oplus \overline{G_1(t)} e^{-\mu t} + \ldots$
$W_3(t) = \overline{G_1(t)} e^{-\mu t} + \overline{G_1(t)} \mu e^{-\mu t} \oplus f_1(t) \oplus \overline{G_1(t)} e^{-\mu t} + \ldots$

Using LT of relations (12.11), then we get

$$B_0^R = \lim_{s \to 0} sB_0^*(s) = \frac{N_3}{D'}, \quad \text{where} \quad N_B = W_1^*(0) - W_3^*(0) p_{03} p_{10} \tag{12.12}$$

and D' is earlier declared.

12.8 Busy Period of the Server Due to Refreshment

Suppose $B_i^{rf}(t)$ represent the server business probability due to refreshment at a time (t) while the unit of the system performs at $S_i \in R$ a time (t = 0). Then the repetitive interface for $B_i^{rf}(t)$ is:

$$B_0^{rf}(t) = q_{01}(t) \oplus B_1^{rf}(t) + q_{03}(t) \oplus B_3^{rf}(t)$$
$$B_1^{rf}(t) = q_{10}(t) \oplus B_0^{rf}(t) + q_{12}(t) \oplus B_2^{rf}(t) + \left[q_{11.4}(t) + q_{11.(45)^n}(t)\right] \oplus B_1^{rf}(t)$$
$$B_2^{rf}(t) = \left[q_{21}(t) + q_{21.(76)^n}(t)\right] \oplus B_1^{rf}(t) + W_2$$
$$B_3^{rf}(t) = \left[q_{31}(t) + q_{31.(54)^n}(t)\right] \oplus B_1^{rf}(t) \quad (12.13)$$

where, $W_2(t) = \overline{G_1(t)} \oplus \lambda e^{-\lambda t} \oplus \overline{F_1(t)} \oplus f_1(t)$

Using LT of the above relations (12.13), then we get

$$B_0^{rf} = \lim_{s \to 0} sB_0^{rf*}(s) = \frac{N_B^{rf}}{D'}, \quad \text{where} \quad N_B^{rf} = W_2^*(0) p_{12} \quad (12.14)$$

and D' is earlier declared.

12.9 Estimated Visits Made by the Server

Assume that $V_i(t)$ is the estimated visits to repair the failed unit by the server at (0, t], while the unit of the system deals with the state $S_i \in R$ at a time (t = 0). Then the repetitive interface for $V_i(t)$ is:

$$V_0(t) = Q_{01}(t) \otimes [1 + V_1(t)] + Q_{03}(t) \otimes [1 + V_3(t)]$$
$$V_1(t) = Q_{10}(t) \otimes V_0(t) + Q_{12}(t) \otimes V_2(t) + \left[Q_{11.4}(t) + Q_{11.(45)^n}(t)\right] \otimes V_1(t)$$
$$V_2(t) = \left[Q_{21}(t) + Q_{21.(76)^n}(t)\right] \otimes V_1(t)$$
$$V_3(t) = \left[Q_{31}(t) + Q_{31.(54)^n}(t)\right] \otimes V_1(t) \quad (12.15)$$

Using LST of the above relations (12.15), then we get

$$V_0(\infty) = \lim_{s \to 0} sV_0^{**}(s)$$
$$V_0 = \frac{V_n}{D'}, \quad \text{where} \quad V_n = p_{10} \quad (12.16)$$

and D' is earlier declared.

12.10 Particular Cases

Suppose that $f_1(t) = \theta e^{-\theta t}, g_1(t) = \phi e^{-\phi t}, p_{01.3} = \dfrac{b\phi}{(\phi+\mu)}$,

$p_{01.3(54)^n} = \dfrac{b\mu}{(\phi+\mu)}, p_{11.4} = \dfrac{\lambda\phi}{(\phi+\mu+\lambda)(\phi+\mu)}$,

$p_{11.(45)^n} = \dfrac{\lambda\mu}{(\phi+\mu+\lambda)(\phi+\mu)}, p_{21.(76)} = \dfrac{\lambda}{(\theta+\lambda)}, p_{31.(54)^n} = \dfrac{\mu}{(\phi+\mu)}$

Also, $M_0 = \dfrac{1}{\lambda} = \mu_0, M_1 = \dfrac{1}{(\phi+\mu+\lambda)} = \mu_1, M_2 \dfrac{1}{(\theta+\lambda)} = \mu_2$

Also,

$$W_1(t) = \mu_1', W_2(t) = \mu_2', W_3(t) = \mu_3',$$

$$MTSF = \frac{\theta\phi + \lambda\theta + \phi\lambda + \mu\lambda + \lambda^2 + a\lambda\theta + a\lambda^2 + a\mu\lambda}{\theta\phi\lambda + \lambda^2\theta + \phi\lambda^2 + \mu\lambda^2 + \lambda^3 - a\theta\phi\lambda - a\lambda^2\phi} \tag{12.17}$$

$$(A_0) = \frac{\theta\phi^2(\lambda+\theta) + \lambda\theta\phi(\theta+\lambda) + \mu\lambda\theta\phi(\lambda+\phi+\mu)}{\theta^2\phi^2 + \theta\lambda(\phi^2+\lambda\mu) + \lambda(\theta+\lambda)[(\theta+\mu)(\phi+b\phi+\lambda)]} \tag{12.18}$$

$$(B_0^R) = \frac{[\theta\phi + (\lambda+b\phi)(\theta+\mu)b]\lambda(\theta+\lambda)}{\theta^2\phi^2 + \theta\lambda(\phi^2+\lambda\mu) + \lambda(\theta+\lambda)[(\theta+\mu)(\phi+b\phi+\lambda)]} \tag{12.19}$$

$$(B_0^{rf}) = \frac{\lambda\mu[\theta\phi + \lambda(\phi+\theta)]}{\theta^2\phi^2 + \theta\lambda(\phi^2+\lambda\mu) + \lambda(\theta+\lambda)[(\theta+\mu)(\phi+b\phi+\lambda)]} \tag{12.20}$$

$$V_0 = \frac{\lambda\theta\phi^2(\theta+\lambda)}{\theta^2\phi^2 + \theta\lambda(\phi^2+\lambda\mu) + \lambda(\theta+\lambda)[(\theta+\mu)(\phi+b\phi+\lambda)]} \tag{12.21}$$

12.11 Profit Analysis

The profit function may be used to do a profit analysis of the system.

$$P = T_0 A_0 - T_1 B_0^R - T_2 B_0^{rf} - T_3 V_0 \quad (12.22)$$

where $T_0 = 15000$ (Price tag per unit up-time).

$T_1 = 1500$ (When a server is busy due to repair then charge per unit time).

$T_2 = 800$ (When a server is busy due to refreshment then charge per unit time).

$T_3 = 500$ (One visit server charge).

12.12 Discussion

It can be seen in Figure 12.2, Figure 12.3, and Figure 12.4 that the trend of MTSF, availability and profit function increase smoothly with respect to refreshment rate (θ); however, other parameters $\lambda = 0.55$, $\mu = 0.4$, $\phi = 0.5$, a = 0.8, b = 0.2 have fixed values. Curves L_2, L_3, and L_4 show that when the value λ changes from 0.55 to 0.65, and μ changes from 0.4 to 0.6 then MTSF, availability, and profit function decrease but when ϕ changes from 0.5 to 0.7, then MTSF, availability, and profit function increase.

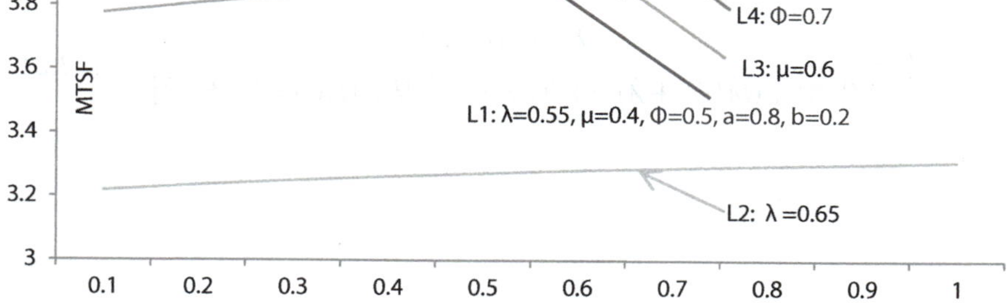

Figure 12.2 MTSF vs. Refreshment rate (θ) →.

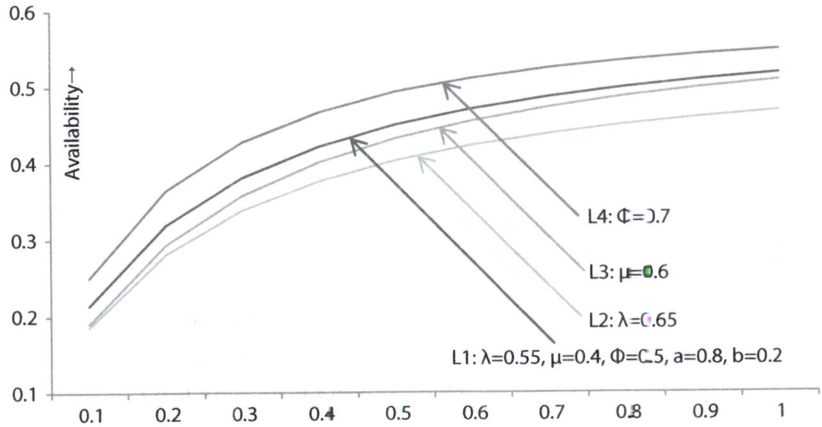

Figure 12.3 Availability vs. Refreshment rate (θ) →.

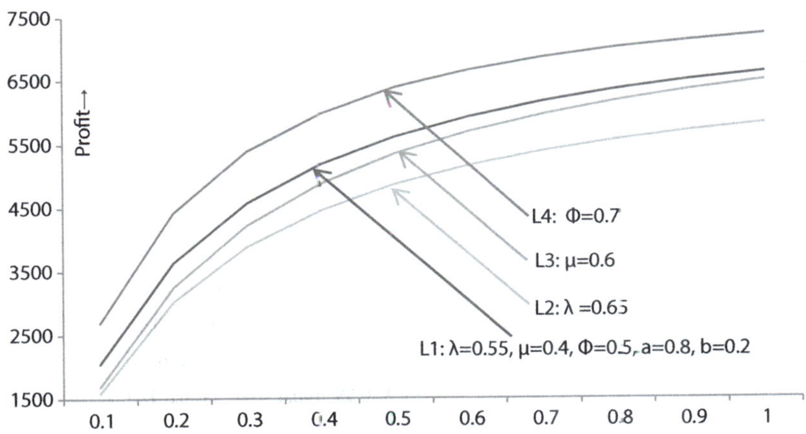

Figure 12.4 Profit vs. Refreshment rate (θ) →.

The behavior of availability, and profit function corresponding to repair rate (ϕ) increases when other parameters $\lambda = 0.55$, $\mu = 0.4$, $\theta = 0.5$, $a = 0.8$, $b = 0.2$ are assumed to be constant. It can be observed from Figure 12.5 and Figure 12.6 that when λ changes from 0.55 to 0.65 and μ changing from 0.4 to 0.6 then the availability and profit function decreases but when θ changes from 0.5 to 0.7 then the availability, and profit function increase.

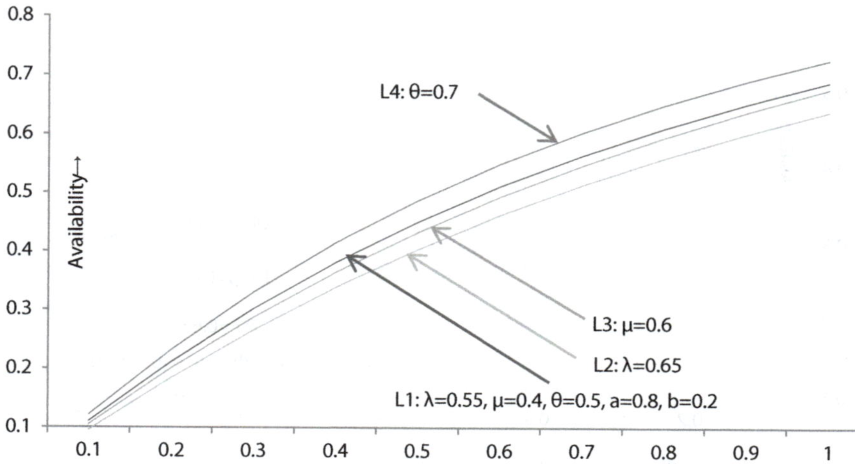

Figure 12.5 Availability vs. Repair rate (θ) →.

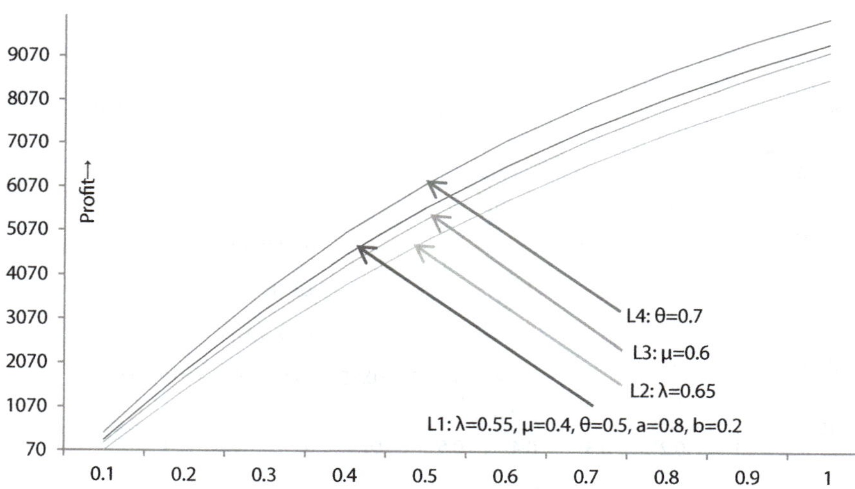

Figure 12.6 Profit vs. Repair rate (θ) →.

12.13 Conclusion

According to the findings of the study, offering refreshments to the server while he is on the job improves his aptitude, which is important in any repairable system and for the smooth operation of high-quality products. Corporate sectors, industries, cybercafés, and education, such as the university system, all use the concept of refreshment.

12.14 Contribution of Refreshment

From the above discussion, the sharply increasing trends of the availability and profit of the system with respect to refreshment as indicated in Figure 12.3 and Figure 12.4, respectively, clearly indicated that the server had to override the emotional and tried-and-tested state of the server. The idea to provide refreshment to the server is beneficial to all servers, the system, and the owner of the system. Thus, the idea of giving refreshments to the server plays an important role during his performance.

12.15 Future Scope

Refreshment to the server plays an important role in the water boosting station system where one unit is operative and another is kept on cold standby.

References

1. Osaki, S. and Nakagawa, T., On a two-unit standby redundant system with standby failure. *Oper. Res.*, 19, 2, 510–523, 1971.
2. Nakagawa, T. and Osaki, S., Stochastic behavior of a two-unit priority standby redundant system with the repair. *Microelectron. Reliab.*, 14, 3, 309–313, 1975.
3. Subramanian, R., Venkatakrishnan, K.S., Kistner, K.P., Reliability of a repairable system with standby failure. *Oper. Res.*, 24, 1, 169–176, 1976.
4. Gopalan, M.N. and Nagarwalla, H.E., Cost-benefit analysis of a one-server two-unit cold standby system with repair and age replacement. *Microelectron. Reliab.*, 25, 5, 977–990, 1985.
5. Dhillon, B.S., Reliability and availability analysis of a system with warm standby and common cause failures. *Microelectron. Reliab.*, 33, 9, 1343–1349, 1993.
6. Dhankhar, A.K. and Malik, S.C., Cost-benefit analysis of system reliability models with server failure during inspection and repair. *Int. J. Stat. Anal.*, 1, 3, 265–278, 2011.
7. Kumar, P., Bharti, A., Gupta, A., Reliability analysis of a two non-identical unit system with repair and replacements having correlated failures and repairs. *J. Inf. Math. Sci.*, 4, 3, 339–350, 2012.
8. Malik, S.C. and Barak, S.K., Reliability measures of a cold standby system with preventive maintenance and repair. *Int. J. Reliab. Qual. Saf. Eng.*, 20, 06, 1350022(1–9), 2013.

9. Dhall, A., Malik, S.C., Munday, V.J., A stochastic system with possible maintenance of standby unit and replacement of the failed unit subject to inspection. *Int. J. Comput. Appl.*, 97, 8, 26–30, 2014.
10. Bhardwaj, R.K. and Singh, R., Semi markov approach for asymptotic performance analysis of a standby system with server failure. *Int. J. Comput. Appl.*, 98, 3, 9–14, 2014.
11. Bhardwaj, R.K., Kaur, K., Malik, S.C., Performance analysis of a stochastic system with standby failure and maintenance. *Int. J. Comput. Appl.*, 125, 1, 06–11, 2015.
12. Bhardwaj, R.K. and Singh, R., A cold-standby system with server failure and delayed treatment. *Int. J. Comput. Appl.*, 124, 17, 31–36, 2015.
13. Kumar, A. and Barak, M.S., Sensitivity analysis of reliability measures of harvesting systems under different environmental conditions. *World Appl. Sci. J.*, 33, 6, 1010–1016, 2015.
14. Kumar, A., Baweja, S., Barak, M., Stochastic behavior of a cold standby system with maximum repair time. *Decis. Sci. Lett.*, 4, 4, 569–578, 2015.
15. Malik, S.C., Stochastic analysis of a parallel system with failure of service facility. *J. Reliab. Stat. Stud.*, 8, 2, 97–106, 2015.
16. Barak, A.K. and Barak, M.S., Impact of abnormal weather conditions on various reliability measures of a repairable system with the inspection. *Thailand Stat.*, 14, 1, 35–45, 2016.
17. Barak, S.K., Barak, M.S., Malik, S.C., Cost-benefit analysis of a cold standby system with preventive maintenance and repair subject to inspection. *J. Math. Stat. Sci.*, 01, 274–285, 2016.
18. Garg, V. and Kadyan, M.S., Profit analysis of a two-unit cold standby system subject to preventive maintenance. *Int. J. Stat. Reliability Eng.*, 3, 1, 29–39, 2016.
19. Kumar, J. and Goel, M., Availability and profit analysis of a two-unit cold standby system for general distribution. *Cog. Math.*, 3, 1, 1262937, 2016.
20. Yadav, D. and Barak, M.S., Stochastic analysis of a cold standby system with server failure. *Int. J. Math. Stat. Invent.*, 4, 6, 18–22, 2016.
21. Barak, M.S., Yadav, D., Barak, S.K., Cost-benefit analysis of a redundant system with the server having refreshment facility subject to inspection. *Int. J. Adv. Eng. Res. Sci.*, 4, 6, 24–31, 2017.
22. Barak, M.S. and Barak, S.K., Profit analysis of a two-unit cold standby system model operating under different weather conditions. *Life Cycle Reliab. Saf. Eng.*, 7, 3, 173–183, 2018.
23. Barak, M.S., Yadav, D., Barak, S.K., Stochastic analysis of two-unit redundant system with priority to inspection over repair. *Life Cycle Reliab. Saf. Eng.*, 7, 2, 71–79, 2018.
24. Kumar, A. and Saini, M., Profit analysis of a computer system with preventive maintenance and priority subject to maximum operation and repair times. *Iran J. Comput. Sci.*, 1, 3, 147–153, 2018.

25. Yadav, R.K. and Malik, S.C., Stochastic analysis of a computer system with unit wise cold standby redundancy and failure of service facility. *Int. J. Math Eng. Manage. Sci.*, 5, 3, 529–543, 2020.
26. Malik, S.C. and Yadav, R.K., Stochastic analysis of a computer system with unit wise cold standby redundancy and priority to hardware repair subject to failure of service facility. *Int. J. Reliab. Qual. Saf. Eng.*, 28, 02, 2150013, 2021.
27. Malik, S.C., Yadav, R.K., Barak, M.S., Stochastic analysis of a unit wise redundant computer system with priority to hardware repair and failure of service facility during software up-gradation. *Int. J. Stat. Reliability Eng.*, 7, 3, 342–349, 2021.
28. Saini, M., Devi, K., Kumar, A., Stochastic modeling of a non-identical redundant system with priority in repair activities. *Thailand Stat.*, 19, 1, 155–162, 2021.
29. Saini, M., Kumar, A., Devi, K., Stochastic investigation of two-unit redundant systems with the provision of different repair schemes, in: *Innovations in Computational Intelligence and Computer Vision*, pp. 3–9, Springer, Singapore, 2021.
30. Saini, M., Kumar, A., Sinwar, D., Parameter estimation, reliability, and maintainability analysis of sugar manufacturing plant. *Int. J. Syst. Assur. Eng. Manage.*, 13, 01, 1–19, 2021.
31. Saini, M. *et al.*, Availability and performance optimization of urea decomposition system using genetic algorithm and particle swarm optimization. *Life Cycle Reliab. Saf. Eng.*, 10, 3, 285–293, 2021.

13

Stochastic Modeling and Availability Optimization of Heat Recovery Steam Generator Using Genetic Algorithm

Monika Saini, Nivedita Gupta and Ashish Kumar*

Department of Mathematics & Statistics, Manipal University Jaipur, Jaipur, Rajasthan, India

Abstract
In steam turbine power plants generator, turbine, cooling towers, and heat recovery steam generators (HRSG) are the main equipment. The failure of HRSG resulted in the failure of the steam cycle that reduces the power production in the plant. The HRSG technology is quite recent, so its performance analysis through reliability measures become necessary to evaluate and enhance capacity of power plants. The present study proposed a novel mathematical model for the HRSG system to assess its availability. For this purpose, a state transition diagram is developed using the Markov birth-death process by considering all time-dependent failure and repair rates as exponentially distributed. The Chapman-Kolmogorov differential-difference equations are derived for the proposed model. The availability of the proposed model is optimized using genetic algorithm to attain the global solution. The numerical result of the present study is appended to highlight the importance of the study.

Keywords: Heat recovery steam generator, genetic algorithm, Markov birth death method, exponential distribution

**Corresponding author*: ashishbarak2020@gmail.com

S. C. Malik, Deepak Sinwar, Ashish Kumar, S. R. Gadde, Prasenjit Chatterjee and Bui Thanh Hung (eds.) Computational Intelligence in Sustainable Reliability Engineering, (269–286) © 2023 Scrivener Publishing LLC

13.1 Introduction

The performance of steam turbine power plant depends on the availability of its components like turbine, cooling tower, turbine, and heat recovery steam generator (HRSG). The availability of these components ensures the operation of the plant. Every component is designed and installed to perform a preassigned specific task. In steam turbine power plant, HRSG is used for conversion of exhaust gas energy into steam. The whole process from preheating to separation of water and steam are performed in economizer, drum, and evaporator. In most of the plants, two or three levels of steam generation are used extensively. The major consequence of HRSG failure is loss of power generation under following conditions: (a) the gas turbine attached with HRSG is shutdown, (b) it may be operated in open cycle. Keeping in mind the importance of HRSG in plant functionality, the analysis of its availability becomes very crucial. The availability of HRSG, like complex systems, is highly influenced by the reliability of components or maintenance strategies adopted. Most of the plant's availability measures as the ratio of uptime and uptime + down time. The availability is used to propose effective maintenance strategies. Several studies have been conducted by researchers to investigate the availability of process industries, like manufacturing, and power generation. Gupta and Tewari [1] developed a mathematical model for the thermal power plant and conducted the simulation study. Gupta and Tewari [2] developed a Markov model for the thermal power plant to evaluate the reliability of plant. Vora *et al.* [3] proposed a simulation model for stochastic analysis of steam generator used in thermal power plants. Jagtap *et al.* [4] optimized the availability of a turbo generator by particle swarm optimization to enhance the performance of the plant. Gupta *et al.* [5] suggested a mathematical model for investigating the operational availability of generator unit of thermal plant. Gupta *et al.* [6] analyzed a simulation model to depict the availability and performance of coal handling unit that is utilized in traditional thermal power plants. First time, Carazas *et al.* [7] tried to analyze the availability of a heat recovery steam generator by using reliability block diagram. Chaibakhsh [8] developed a mathematical model and performed simulation of heat recovery steam generator. Sindareh-Esfahani *et al.* [9] performed the thermodynamical modeling-based optimization of heat recovery steam generator and concept of standby unit is also used. Najjar *et al.* [10] studied the effect of degradation in heat recovery steam generator under full and partial load. Mert *et al.* [11] did the sensitivity and exergy analysis of triple pressure HRSG through developing a model. Sabouhi *et al.* [12] developed a reliability model for combined cycle power plant to analyze the availability of its units. Sabouhi *et al.* [13] used reliability centric maintenance methodology in identification of most critical component in power plants. Gupta *et al.* [14] performed the reliability, availability, and maintainability investigation of generator unit of steam turbine power plant. Radin [15] studied a model for enhancing the reliability and flexibility of steam power plants.

The whole manuscript is divided into six sections including present introductory section. System description, notations and assumptions included in section 13.2 while mathematical model is proposed and analyzed in section 13.3. Section 13.4 presented the availability optimization methodology and results and discussion shown in section 13.5. Concluding remarks and conclusion appended in section 13.6.

13.2 System Description, Notations, and Assumptions

13.2.1 System Description

The heat recovery steam generator is an arrangement of following components as shown in Figure 13.1:

- Subsystem A (low-pressure system): It consist of one unit of low-pressure system. Failure of this cause complete system failure as it is connected in series with subsequent unit.
- Subsystem B (intermediate-pressure system): It consist of one unit of intermediate pressure system. Failure of this

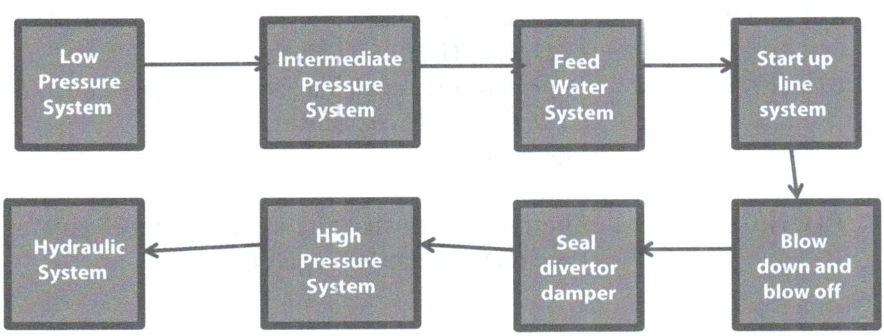

Figure 13.1 Configuration diagram of HRSG.

cause complete system failure as it is connected in series with subsequent unit.

- Subsystem C (feed water system): It consist of one unit of feed water system. Failure of this cause complete system failure as it is connected in series with subsequent unit.
- Subsystem D (Startup line system): It consist of one unit of startup line system. Failure of this cause complete system failure as it is connected in series with subsequent unit.
- Subsystem E (Blow down and blow off): It consist of one unit of blow down and blow off system. Failure of this cause complete system failure as it is connected in series with subsequent unit.
- Subsystem F (seal divertor damper): It consists of one unit of seal divertor damper. Failure of this cause complete system failure as it is connected in series with subsequent unit.
- Subsystem G (high-pressure system): It consists of one unit of high-pressure system. Failure of this cause complete system failure as it is connected in series with subsequent unit.
- Subsystem H (hydraulic system): It consists of one unit of hydraulic system. Failure of this cause complete system failure as it is connected in series with subsequent unit.

13.2.2 Notations

○:	Indicates system is in full capacity
□:	Indicates system is in failed states
A, B, C, D, E, F, G, H:	Represent units working with full capacity
a, b, c, d, e, f, g, h :	Represent failed units
$S_i(0 \leq i \leq 8)$:	Represent the different states
$\alpha_i(1 \leq i \leq 8)$:	The constant failure rates of the subsystems *A, B, C, D, E, F, G, H* respectively
$\beta_i(x)(1 \leq i \leq 8)$:	Represent the repair rates of the subsystem *A, B, C, D, E, F, G, H* respectively
$P_0(t)$:	The probability that system is in full capacity at initial time t
$P_i(x, t)(0 \leq i \leq 10)$:	The probability that the system is in i^{th} state at time t has an elapsed repair time

13.2.3 Assumptions

- The failure rates follow exponential distribution and repair rates follow arbitrary distribution and are independent to each other.
- There are no simultaneous failures among the subsystems.
- There are sufficient repair and replacement facilities. Repairmen always present in plant and performance wise repaired system is as good as new.

13.3 Mathematical Modeling of the System

In this section, a mathematical model is developed as shown in Figure 13.2 using Markov birth death process and simplified using supplementary variable technique.

$$P_0(t+\Delta t)=(1-\alpha_1\Delta t-\alpha_2\Delta t-\alpha_3\Delta t-\alpha_4\Delta t-\alpha_5\Delta t-\alpha_6\Delta t-\alpha_7\Delta t-\alpha_8\Delta t)P_0(t)$$
$$+\int_0^\infty \beta_1(x)P_1(x,t)\Delta t dx+\int_0^\infty \beta_2(x)P_2(x,t)\Delta t dx+\int_0^\infty \beta_3(x)P_3(x,t)\Delta t dx$$
$$+\int_0^\infty \beta_4(x)P_4(x,t)\Delta t dx+\int_0^\infty \beta_5(x)P_5(x,t)\Delta t dx+\int_0^\infty \beta_6(x)P_6(x,t)\Delta t dx$$
$$+\int_0^\infty \beta_7(x)P_7(x,t)\Delta t dx+\int_0^\infty \beta_8(x)P_8(x,t)\Delta t dx$$

Dividing both the sides of above equation by Δt, we get

$$\frac{P_0(t+\Delta t)-P_0(t)}{\Delta t}=(-\alpha_1-\alpha_2-\alpha_3-\alpha_4-\alpha_5-\alpha_6-\alpha_7-\alpha_8)P_0(t)$$
$$+\int_0^\infty \beta_1(x)P_1(x,t)dx+\int_0^\infty \beta_2(x)P_2(x,t)dx+\int_0^\infty \beta_3(x)P_3(x,t)dx$$
$$+\int_0^\infty \beta_4(x)P_4(x,t)dx+\int_0^\infty \beta_5(x)P_5(x,t)dx+\int_0^\infty \beta_6(x)P_6(x,t)dx$$
$$+\int_0^\infty \beta_7(x)P_7(x,t)dx+\int_0^\infty \beta_8(x)P_8(x,t)dx$$

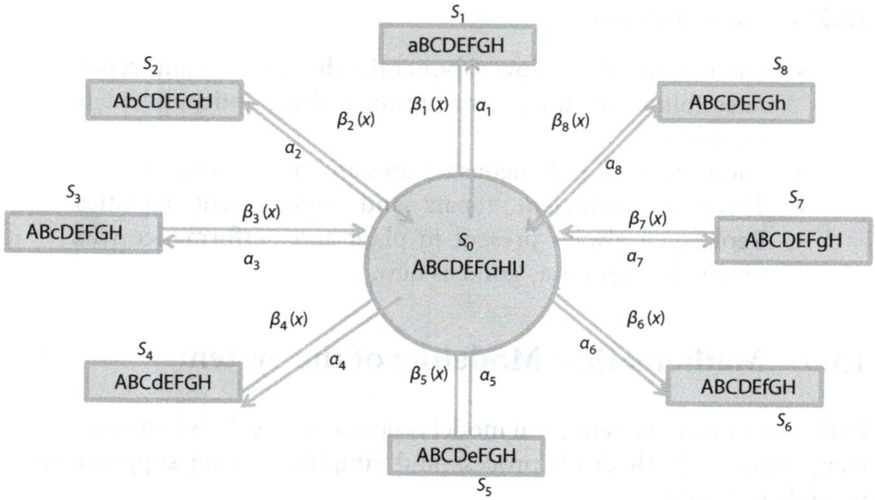

Figure 13.2 State changeover diagram of HRSG in steam turbine power plant.

As $\Delta t \to 0$

$$\lim_{\Delta t \to 0} \frac{P_o(t+\Delta t) - P_0(t)}{\Delta t} = (-\alpha_1 - \alpha_2 - \alpha_3 - \alpha_4 - \alpha_5 - \alpha_6 - \alpha_7 - \alpha_8)$$

$$P_0(t)$$

$$+ \int_0^\infty \beta_1(x)P_1(x,t)dx + \int_0^\infty \beta_2(x)P_2(x,t)dx + \int_0^\infty \beta_3(x)P_3(x,t)dx$$

$$+ \int_0^\infty \beta_4(x)P_4(x,t)dx + \int_0^\infty \beta_5(x)P_5(x,t)dx + \int_0^\infty \beta_6(x)P_6(x,t)dx$$

$$+ \int_0^\infty \beta_7(x)P_7(x,t)dx + \int_0^\infty \beta_8(x)P_8(x,t)dx$$

$$\Rightarrow \frac{dP_o}{dt} + (\alpha_1 + \alpha_2 + \alpha_3 + \alpha_4 + \alpha_5 + \alpha_6 + \alpha_7 + \alpha_8) P_0(t)$$

$$= \int_0^\infty \beta_1(x) P_1(x,t) dx + \int_0^\infty \beta_2(x) P_2(x,t) dx$$

$$+ \int_0^\infty \beta_3(x) P_3(x,t) dx + \int_0^\infty \beta_4(x) P_4(x,t) dx$$

$$+ \int_0^\infty \beta_5(x) P_5(x,t) dx + \int_0^\infty \beta_6(x) P_6(x,t) dx$$

$$+ \int_0^\infty \beta_7(x) P_7(x,t) dx + \int_0^\infty \beta_8(x) P_8(x,t) dx$$

$$\Rightarrow \frac{dP_o}{dt} + \sum_{i=1}^{8} \alpha_i P_o(t) = \sum_{i=1}^{8} \int_0^\infty \beta_i(x) P_i(x,t) dx$$

$$\Rightarrow \left(\frac{d}{dt} + u_0 \right) P_o(t) = f_0 \tag{13.1}$$

where

$$u_0 = \sum_{i=1}^{8} \alpha_i \qquad f_0 = \sum_{i=1}^{8} \int_0^\infty \beta_i(x) P_i(x,t) dx$$

Now

$$P_1(x + \Delta x, t + \Delta t) = \alpha_1 \, \Delta t \, P_0(t) + (1 - \beta_1(x) \, \Delta x) P_1(x, t) \tag{13.2}$$

Differentiating equation (13.2) with respect to x and t partially, we get

$$\Rightarrow \left(\frac{\partial}{\partial x} + \frac{\partial}{\partial t} + \beta_1(x) \right) P_1(x,t) = \alpha_1 P_0(t) \tag{13.3}$$

Similarly, we can obtain

$$\left(\frac{\partial}{\partial x}+\frac{\partial}{\partial t}+\beta_i(x)\right)P_i(x,t)=\alpha_i\,P_0(t) \qquad ;i=2,3,4,5,6,7,8 \tag{13.4}$$

The boundary conditions are given as

$$\begin{aligned} &P_1(0,t)=\alpha_1 P_0(t) && P_2(0,t)=\alpha_2 P_0(t)\\ &P_3(0,t)=\alpha_3 P_0(t) && P_4(0,t)=\alpha_4 P_0(t)\\ &P_5(0,t)=\alpha_5 P_0(t) && P_6(0,t)=\alpha_6 P_0(t)\\ &P_7(0,t)=\alpha_7 P_0(t) && P_8(0,t)=\alpha_8 P_0(t) \end{aligned} \tag{13.5}$$

and initial conditions are given as

$$\begin{aligned} &P_0(0)=1\\ &P_i(0)=0 \qquad ;i=1\ to\ 8 \end{aligned} \tag{13.6}$$

Equations (13.1), (13.3), and (13.4) together with boundary and initial conditions constitute Chapman-Kolmogorov differential difference equation. In a particular case, to show the importance of results based on availability of the system and profit analysis, we assume repair rates to follow exponential distribution. Then, the above sets of equations, boundary, and initial conditions reduce as follows:

$$\begin{aligned} \left(\frac{d}{dt}+u_0\right)P_o(t)=\beta_1 P_1(t)+\beta_2 P_2(t)+\beta_3 P_3(t)+\beta_4 P_4(t)+\beta_5 P_5(t)\\ +\beta_6 P_6(t)+\beta_7 P_7(t)+\beta_8 P_8(t) \end{aligned} \tag{13.7}$$

$$\left(\frac{d}{dt}+\beta_i\right)P_i(t)=\alpha_i\,P_0(t) \qquad ;\ i=1,2,3,4,5,6,7,8 \tag{13.8}$$

and initial conditions are given as

$$P_0(0) = 1$$
$$P_i(0) = 0 \quad ; i = 1 \text{ } to \text{ } 8 \tag{13.9}$$

The steady-state probabilities and long-time availability of the system are required, and here, this is calculated by taking $\frac{d}{dt} = 0$ as $t \to \infty$, $P_i(t) = P_i$ in equations (13.7) and (13.8), we get

$$\beta_1 P_1 = \alpha_1 P_0 \qquad \beta_2 P_2 = \alpha_2 P_0 \qquad \beta_3 P_3 = \alpha_3 P_0$$
$$\beta_4 P_4 = \alpha_4 P_0 \qquad \beta_5 P_5 = \alpha_5 P_0 \qquad \beta_6 P_6 = \alpha_6 P_0$$
$$\beta_7 P_7 = \alpha_7 P_0 \qquad \beta_8 P_8 = \alpha_8 P_0$$

$$that \text{ } is \text{ } P_j = \frac{\alpha_j}{\beta_j} P_0 \quad ; j = 1,2,3,4,5,6,7,8 \tag{13.10}$$

Solving set of equations (13.7) and (13.1) together, we get the probability of being in states S_1 *to* S_8 in terms of P_0. Using normalizing condition $\sum_{i=0}^{8} P_i = 1$ we have

$$\sum_{i=0}^{8} P_i = 1$$
$$\Rightarrow P_0 + P_1 + P_2 + P_3 + P_4 + P_5 + P_6 + P_7 + P_8 = 1$$
$$\Rightarrow P_0 + \frac{\alpha_1}{\beta_1} P_0 + \frac{\alpha_2}{\beta_2} P_0 + \frac{\alpha_3}{\beta_3} P_0 + \frac{\alpha_4}{\beta_4} P_0 + \frac{\alpha_5}{\beta_5} P_0 + \frac{\alpha_6}{\beta_6} P_0 + \frac{\alpha_7}{\beta_7} P_0 + \frac{\alpha_8}{\beta_8} P_0 = 1$$
$$\Rightarrow P_0 = \frac{1}{\left(1 + \frac{\alpha_1}{\beta_1} + \frac{\alpha_2}{\beta_2} + \frac{\alpha_3}{\beta_3} + \frac{\alpha_4}{\beta_4} + \frac{\alpha_5}{\beta_5} + \frac{\alpha_6}{\beta_6} + \frac{\alpha_7}{\beta_7} + \frac{\alpha_8}{\beta_8}\right)} \tag{13.11}$$

The steady state Availability of the system is given as

$$A_v = P_0$$

$$\Rightarrow A_v = \frac{1}{\left(1+\frac{\alpha_1}{\beta_1}+\frac{\alpha_2}{\beta_2}+\frac{\alpha_3}{\beta_3}+\frac{\alpha_4}{\beta_4}+\frac{\alpha_5}{\beta_5}+\frac{\alpha_6}{\beta_6}+\frac{\alpha_7}{\beta_7}+\frac{\alpha_8}{\beta_8}\right)} \tag{13.12}$$

13.4 Availability Optimization of Proposed Model

Instead of a local solution, it is always recommended to find the global solution for the real-world problems so that most accurate predictions can be made. For this purpose, computational intelligence-based algorithms are extensively used. Genetic algorithm is one such method. Genetic algorithm is an algorithm that is most frequently used for complex engineering optimization problems. It is a population-based search algorithm. Here, the new population is reproduced by parental population. The genetic operators, namely crossover and mutation, are used to select the off springs. Over successive generations, the fitness function converges to the optimal solution in the search space. The performance of any industrial system is based on its decision parameters namely failure and repair rates. The estimation of optimum values of these decision parameters is very crucial, and hence, here genetic algorithm is used. The failure and repair rates parameter constraint of HRSG subsystem are as follows: (α_1, β_1), (α_2, β_2), (α_3, β_3), (α_4, β_4), (α_5, β_5), (α_6, β_6), (α_7, β_7), (α_8, β_8)

The range of parameter constraints are:

For HRSG unit: $\alpha_1 \in [0.0009, 0.55]$, $\alpha_2 \in [0.0001, 0.08]$, $\alpha_3 \in [0.0002, 0.06]$, $\alpha_4 \in [0.0003, 0.04]$, $\alpha_5 \in [0.00025, 0.035]$, $\alpha_6 \in [0.00075, 0.04]$, $\alpha_7 \in [0.00045, 0.02]$, $\alpha_8 \in [0.00035, 0.055]$, $\beta_1 \in [0.05, 1.5]$, $\beta_2 \in [0.02, 1.1]$, $\beta_3 \in [0.03, 1.15]$, $\beta_4 \in [0.05, 1.05]$, $\beta_5 \in [0.01, 2.25]$, , $\beta_6 \in [0.025, 2.05]$, $\beta_7 \in [0.048, 1.65]$, $\beta_8 \in [0.02, 1.35]$

In this study, genetic algorithm technique is used to maximize the long run availability given by Eq. (13.12). The steps of solving GA are value encoding is used to encode the chromosome values, random selection technique is used to select the parents from the population to operate cross-over and uniform cross-over is used to generate new off-springs. These operators are described below: (i) encoding, (ii) fitness function, (iii) selection, (iv) cross-over, (v) mutation.

(i) Encoding

Encoding is an operator in GA problem for mapping chromosome, which is a set of different values. It highly depends on the problem under study. A sample chromosome value is given below:

HRSG unit	$\alpha_2 = 0.00423$, $\alpha_3 = 0.0055$, $\alpha_4 = 0.00215$, $\alpha_5 = 0.007$, $\alpha_6 = 0.0062$, $\alpha_7 = 0.008$, $\alpha_8 = 0.0015$, $\beta_1 = 0.6$, $\beta_2 = 0.35$, $\beta_4 = 0.145$, $\beta_5 = 1.05$, $\beta_6 = 0.9$, $\beta_7 = 0.86$, $\beta_8 = 0.52$, $\alpha_1 = 0.0095$, $\beta_3 = 0.45$

(ii) Fitness function: Fitness function is described as the quality of obtained solution near to the prescribed set of values. A sample chromosome with fitness value is given as below:

HRSG unit	$\alpha_2 = 0.00423$, $\alpha_3 = 0.0055$, $\alpha_4 = 0.00215$, $\alpha_5 = 0.007$, $\alpha_6 = 0.0062$, $\alpha_7 = 0.008$, $\alpha_8 = 0.0015$, $\beta_1 = 0.6$, $\beta_2 = 0.35$, $\beta_4 = 0.145$, $\beta_5 = 1.05$, $\beta_6 = 0.9$, $\beta_7 = 0.86$, $\beta_8 = 0.52$, $\alpha_1 = 0.0095$, $\beta_3 = 0.45$

(iii) Selection: In this, the parent population are selected randomly to generate offspring population.

(iv) Cross-over: Cross-over operator is a technique in which the child population is generated from the paired parent population.

Sludge digestion processing unit	$\alpha_2 = 0.00423$, $\alpha_3 = 0.0055$, $\alpha_4 = 0.00215$, $\alpha_5 = 0.007$, $\alpha_6 = 0.0062$, $\alpha_7 = 0.008$, $\alpha_8 = 0.0015$, $\beta_1 = 0.6$, $\beta_2 = 0.35$, $\beta_4 = 0.145$, $\beta_5 = 1.05$, $\beta_6 = 0.9$, $\beta_7 = 0.86$, $\beta_8 = 0.52$, $\alpha_1 = 0.0095$, $\beta_3 = 0.45$

(v) Mutation: The mutation operator is useful to maintain the diversity from one generated population to the next.

HRSG unit	$\alpha_2 = 0.00423$, $\alpha_3 = 0.0055$, $\alpha_4 = 0.00215$, $\alpha_5 = 0.007$, $\alpha_6 = 0.0062$, $\alpha_7 = 0.008$, $\alpha_8 = 0.0015$, $\beta_1 = 0.5$, $\beta_2 = 0.35$, $\beta_4 = 0.145$, $\beta_5 = 1.05$, $\beta_6 = 0.9$, $\beta_7 = 0.86$, $\beta_8 = 0.52$, $\alpha_1 = 0.0095$, $\beta_3 = 0.45$

13.5 Results and Discussion

In this section, at first stage impact on availability of HRSG unit is investigated by changing various failure and repair rates. The system availability is obtained w.r.t to repair rate (β_3) of subsystem C, i.e., feed water system by varying its values between α_3 = [0.0055, 0.0145] and constant value of failure rate $\beta_3 = 0.95$. The failure and repair rates of other subsystems have been taken as:

$\alpha_2 = 0.00423$, $\alpha_3 = 0.0055$, $\alpha_4 = 0.00215$, $\alpha_5 = 0.007$, $\alpha_6 = 0.0062$, $\alpha_7 = 0.008$, $\alpha_8 = 0.0015$ and $\beta_1 = 0.6$, $\beta_2 = 0.35$, $\beta_4 = 0.145$, $\beta_5 = 1.05$, $\beta_6 = 0.9$, $\beta_7 = 0.86$, $\beta_8 = 0.52$

From Figure 13.3, it is investigated that any increment in the failure rate of feed water system resulted a deep decline in availability of system.

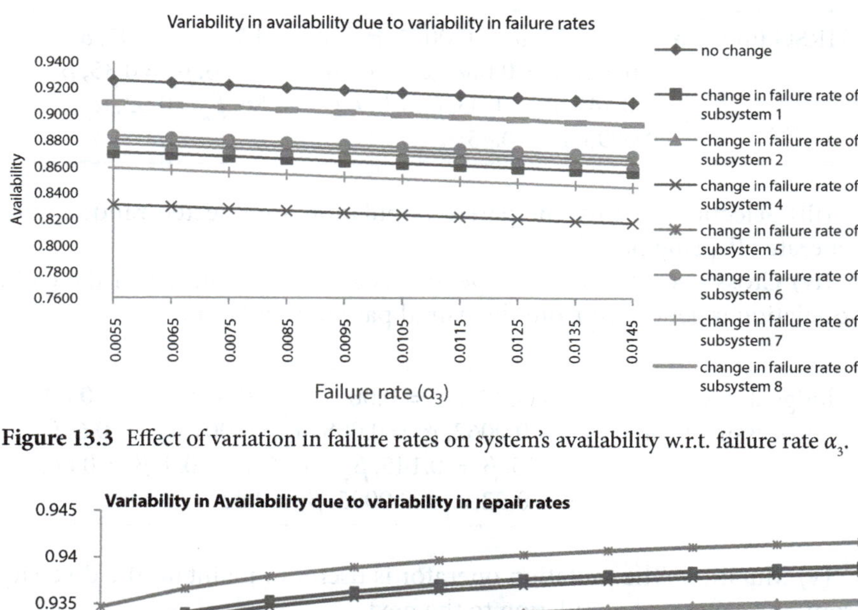

Figure 13.3 Effect of variation in failure rates on system's availability w.r.t. failure rate α_3.

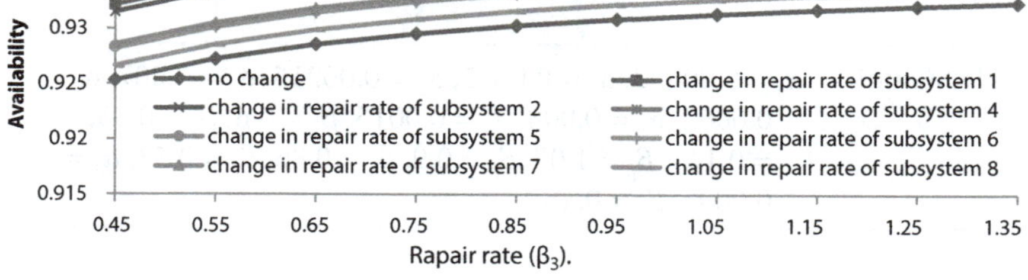

Figure 13.4 Effect of variation in repair rates on system's availability w.r.t. repair rate β_3.

Table 13.1 Impact of number of generations on HRSG availability having evolution size = 350, crossover probability = 0.85 and mutation probability = 0.65.

Number of generations	10	20	30	40	50	60	70	80	90
Availability	**0.9805**	**0.9741**	**0.9796**	**0.981**	**0.9793**	**0.9747**	**0.9758**	**0.9828**	**0.9811**
α_1	0.0065	0.0004	0.0051	0.0005	0.0051	0.0031	0.0034	**0.0066**	0.0114
α_2	0.0077	0.0255	0.0082	0.095	0.0172	0.0473	0.014	**0.0093**	0.0167
α_3	0.0015	0.0286	0.0041	0.011	0.0438	0.0016	0.0001	**0.0026**	0.0007
α_4	0.0004	0.0005	0.0011	0.0002	0.0008	0.0001	0.0016	**0.0014**	0.0001
α_5	0.0016	0.0028	0.0087	0.0179	0.0113	0.0058	0.0045	**0.0025**	0.0088
α_6	0.0004	0.0517	0.029	0.0066	0.0018	0.0403	0.0054	**0.0018**	0.0201
α_7	0.0232	0.053	0.0295	0.0079	0.0057	0.0046	0.0092	**0.02**	0.027
α_8	0.0017	0.0131	0.0011	0.0167	0.0017	0.0143	0.0035	**0.0032**	0.0049
β_1	0.6411	1.5254	1.2962	2.3231	1.2777	1.6894	1.5788	**2.3245**	1.9591
β_2	0.3343	1.2865	0.2552	3.451	0.3265	1.3891	1.0763	**0.7263**	2.1622
β_3	1.3814	0.1209	0.1666	0.4077	1.8569	2.5054	0.528	**1.8738**	0.0777
β_4	0.5651	0.3528	0.3933	0.3497	0.6393	0.3306	0.7652	**0.345**	0.3542
β_5	0.9801	0.4362	0.4951	3.1735	0.5092	0.9494	1.4948	**0.5657**	1.1524
β_6	0.5608	1.4766	1.6625	4.2456	0.6343	0.7424	1.8831	**0.9751**	6.8281
β_7	0.7415	2.0362	0.9425	0.6715	3.0612	2.3108	1.495	**1.9311**	1.0471
β_8	0.8503	0.8869	0.1804	1.8845	2.3702	1.8417	0.4656	**0.6807**	0.7199

Table 13.2 Impact of number of evolutions on HRSG availability having population size = 70, crossover probability = 0.85 and mutation probability = 0.65.

Evolution	70	140	210	280	350	420	490	560	630
Availability	**0.9665**	**0.9579**	**0.9731**	**0.9894**	**0.9758**	**0.9818**	**0.9881**	**0.9623**	**0.9773**
α_1	0.0034	0.0013	0.0078	**0.002**	0.0034	0.0064	0.0018	0.0028	0.0044
α_2	0.0055	0.0148	0.0802	**0.0004**	0.014	0.0515	0.0191	0.0013	0.0138
α_3	0.0016	0.0035	0.0009	**0.0066**	0.0001	0.0003	0.0056	0.0009	0.003
α_4	0.0013	0.0002	0.0005	**0.0003**	0.0016	0.0005	0.0005	0.0023	0.0009
α_5	0.0056	0.001	0.0222	**0.0005**	0.0045	0.0142	0.0041	0.0117	0.017
α_6	0.0072	0.0017	0.0067	**0.0077**	0.0054	0.0148	0.0055	0.04	0.0056
α_7	0.0108	0.0124	0.0022	**0.0001**	0.0092	0.0007	0.0324	0.0057	0.0009
α_8	0.0065	0.0091	0.0115	**0.009**	0.0035	0.0032	0.0073	0.0055	0.005
β_1	6.0348	1.3157	2.2004	**1.7775**	1.5788	0.9149	1.5857	4.7344	1.3615
β_2	0.3231	0.8441	3.7138	**1.754**	1.0763	1.815	0.6546	2.1146	0.5002
β_3	0.1963	0.1148	0.6105	**0.348**	0.528	0.1346	0.1622	0.6887	0.498
β_4	0.594	0.1613	0.2611	**0.0933**	0.7652	0.6546	0.298	0.332	0.2689
β_5	1.3357	3.045	0.1759	**1.7849**	1.4948	1.6921	0.2707	2.8401	0.8001
β_6	0.8184	0.3212	1.8704	**3.7902**	1.8831	1.1374	0.6279	1.5369	2.5743
β_7	0.6046	0.4693	0.275	**0.8929**	1.495	0.5839	2.2795	2.5866	5.3576
β_8	0.4194	3.034	2.1307	**1.5531**	0.4656	2.1604	0.4853	0.8324	1.1764

Table 13.3 Impact of crossover probability on HRSG availability having evolution size = 350, population size = 70 and mutation probability = 0.65.

Crossover	0.09	0.18	0.27	0.36	0.45	0.54	0.63	0.72	0.81
Availability	**0.9735**	**0.977**	**0.9685**	**0.9819**	**0.955**	**0.9862**	**0.9888**	**0.9906**	**0.9689**
α_1	0.0123	0.0234	0.0027	0.0035	0.0027	0.0293	0.0032	**0.0017**	0.0076
α_2	0.0229	0.0101	0.0434	0.009	0.0066	0.0271	0.0078	**0.0016**	0.0281
α_3	0.0115	0.0011	0.0026	0.0192	0.0083	0.035	0.0009	**0.0005**	0.0126
α_4	0.0008	0.0001	0.0003	0.0003	0.0005	0.0011	0.0018	**0.0013**	0.0006
α_5	0.0064	0.0012	0.0257	0.0019	0.0028	0.0141	0.0014	**0.0232**	0.0059
α_6	0.0168	0.0193	0.0109	0.0113	0.0282	0.0451	0.0091	**0.0019**	0.018
α_7	0.0104	0.0006	0.0106	0.0049	0.0202	0.0102	0.0052	**0.015**	0.0303
α_8	0.002	0.0028	0.0037	0.0026	0.0004	0.0056	0.0043	**0.0009**	0.0273
β_1	1.2617	1.5896	1.7067	5.4068	0.4184	8.3243	0.6426	**0.5456**	4.1375
β_2	2.8441	1.7485	2.8421	0.3966	0.7922	0.858	0.462	**0.6951**	1.4547
β_3	1.0059	1.3004	1.2371	0.4125	1.3933	1.1688	0.5724	**0.3566**	1.6954
β_4	0.9451	0.4248	0.0737	0.1744	0.1413	0.7687	0.9888	**0.2622**	0.3517
β_5	0.3237	0.3696	1.1978	1.4038	2.5169	3.7868	0.4613	**0.8137**	2.9554
β_6	2.1242	0.4942	0.595	2.3498	0.4688	4.4077	0.1578	**1.4793**	2.7157
β_7	0.7586	0.873	3.1722	3.3832	2.3087	0.426	1.3099	**0.4432**	0.8576
β_8	0.6408	0.9319	0.3544	2.5277	0.2473	1.0988	0.6752	**1.8642**	3.0537

Table 13.4 Impact of mutation probability on HRSG availability having evolution size = 350, population size = 70 and crossover probability = 0.85.

Mutation	0.1	0.2	0.3	0.4	0.5	0.6	0.7	0.8	0.9
Availability	**0.9676**	**0.9606**	**0.9821**	**0.984**	**0.97**	**0.9868**	**0.9834**	**0.969**	**0.9746**
α_1	0.0006	0.0003	0.0031	0.0106	0.0079	**0.0064**	0.0058	0.0007	0.0015
α_2	0.006	0.013	0.0085	0.0026	0.018	**0.01**	0.0547	0.0187	0.0103
α_3	0.0137	0.0089	0.0074	0.0042	0.0249	**0.01**	0.0063	0.0189	0.0226
α_4	0.0001	0.0009	0.0002	0.0004	0.0001	**0.0017**	0.0003	0.0006	0.0049
α_5	0.0098	0.0202	0.0032	0.0284	0.0052	**0.0074**	0.0041	0.0075	0.0074
α_6	0.0134	0.0039	0.0055	0.0106	0.0044	**0.052**	0.0027	0.0537	0.0145
α_7	0.0058	0.0067	0.0309	0.0024	0.0617	**0.0099**	0.0334	0.0053	0.0116
α_8	0.0057	0.0079	0.0015	0.0076	0.0021	**0.0017**	0.0143	0.0079	0.0002
β_1	1.3108	0.679	1.3696	2.5329	3.4373	**0.7691**	1.7356	1.0054	3.2506
β_2	0.9851	2.0875	0.1577	0.1867	0.7639	**0.5438**	1.0982	3.2223	2.1549
β_3	1.0238	0.8052	0.4604	0.2744	0.6131	**0.389**	6.8845	0.6829	2.5021
β_4	0.0201	0.1697	0.2809	0.2314	0.9017	**1.0343**	0.1523	1.0968	0.1904
β_5	0.7032	0.6008	3.9985	2.5565	2.1895	**2.4872**	1.2015	1.1947	0.4798
β_6	4.0023	1.9707	0.9182	0.3306	1.1633	**2.7519**	0.4395	3.8145	0.7499
β_7	0.5699	1.2305	0.9704	0.6802	3.047	**0.5437**	0.7192	0.8484	0.607
β_8	0.9488	1.4212	0.8759	0.3651	0.1846	**1.7298**	6.4088	1.6581	0.6809

Similarly, failure rate of all subsystems also showed the same trend. The failure rate of start up line subsystem showed the highest decline in availability of HRSG unit. The availability of system is derived with respect to repair rate of feed water system β_3 = [0.45, 1.35] and constant value of failure rate $\alpha_3 = 0.0095$. Rest of the parameters defined above. Figure 13.4 revealed that system availability increased with respect to repair rate of every subsystem. Start-up line subsystem showed the highest increment in availability.

In second stage, genetic algorithm is applied on the objective function (13.12) of proposed model to estimate the parmeters value to attain maximum availability. The numerical results of availability and estimated parameters is appended in Tables 13.1 to 13.4. It is observed that the optimum availability vlue is 0.9906 at crossover probability, 0.72; mutation probability, 0.65; population size, 70; and number of evolutions, 350.

13.6 Conclusion

In the present study, a novel stochastic model is developed, and performance of system is analyzed for a particular case. The availability of the HRSG system is also optimized using genetic algorithm. The availability increased with respect to repair rates and declined with failure rates. It is observed from GA results that the optimum availability vlue is 0.9906 at crossover probability of 0.72; mutation probability, 0.65; population size, 70; and number of evolutions, 350. The results can be used by system designers to develop the systems.

References

1. Gupta, S. and Tewari, P.C., Simulation modeling and analysis of a complex system of a thermal power plant. *J. Ind. Eng. Manag.*, 2, 2, 387–406, 2009.
2. Gupta, S. and Tewari, P.C., Markov approach for predictive modeling and performance evaluation of a thermal power plant. *Int. J. Reliab. Qual. Saf. Eng.*, 17, 01, 41–55, 2010.
3. Vora, Y., Patel, M.B., Tewari, P.C., Simulation model for stochastic analysis and performance evaluation of steam generator system of a thermal power plant. *Int. J. Eng. Sci. Technol.*, 3, 6, 5141–5149, 2011.
4. Jagtap, H.P., Bewoor, A.K., Kumar, R., Ahmadi, M.H., Chen, L., Performance analysis and availability optimization to improve maintenance schedule for the turbo-generator subsystem of a thermal power plant using particle swarm optimization. *Reliab. Eng. Syst. Saf.*, 204, 107130, 2020.

5. Gupta, N., Saini, M., Kumar, A., Operational availability analysis of generators in steam turbine power plants. *SN Appl. Sci.*, 2, 4, 1–11, 2020.
6. Gupta, S., Tewari, P.C., Sharma, A.K., An availability simulation model and performance analysis of a coal handling unit of a typical thermal plant. *S. Afr. J. Ind. Eng.*, 20, 1, 159–171, 2009.
7. Carazas, F.J.G., Salazar, C.H., De Souza, G.F.M., Availability analysis of heat recovery steam generators used in thermal power plants. *Energy*, 36, 6, 3855–3870, 2011.
8. Chaibakhsh, A., Modelling and long-term simulation of a heat recovery steam generator. *Math. Comput. Model. Dyn. Syst.*, 19, 2, 91–114, 2013.
9. Sindareh-Esfahani, P., Ghaffari, A., Ahmadi, P., Thermodynamic modeling based optimization for thermal systems in heat recovery steam generator during cold start-up operation. *Appl. Therm. Eng.*, 69, 1-2, 286–296, 2014.
10. Najjar, Y.S., Alalul, O.F., Abu-Shamleh, A., Degradation analysis of a combined cycle heat recovery steam generator under full and part load conditions. *Sustain. Energy Technol. Assess.*, 37, 100587, 2020.
11. Mert, S.O., Ozcelik, F.Z., Ceyda, K., Modelling, sensitivity and exergy analysis of triple-pressure heat recovery steam generator. *MANAS J. Eng.*, 8, 2, 106–114, 2020.
12. Sabouhi, H., Abbaspour, A., Fotuhi-Firuzabad, M., Dehghanian, P., Reliability modeling and availability analysis of combined cycle power plants. *Int. J. Electr. Power Energy Syst.*, 79, 108–119, 2016.
13. Sabouhi, H., Fotuhi-Firuzabad, M., Dehghanian, P., Identifying critical components of combined cycle power plants for implementation of reliability-centered maintenance. *CSEE J. Power Energy Syst.*, 2, 2, 87–97, 2016.
14. Gupta, N., Kumar, A., Saini, M., Reliability and maintainability investigation of generator in steam turbine power plant using RAMD analysis. *J. Phys. Conf. Ser. IOP Publishing*, 1714, 1, 012009, 2021.
15. Radin, Y.A., Improving the flexibility and reliability of steam power units at thermal power plants. *Therm. Eng.*, 68, 6, 481–489, 2021.

14

Investigation of Reliability and Maintainability of Piston Manufacturing Plant

Monika Saini[1], Deepak Sinwar[2] and Ashish Kumar[1*]

[1]*Department of Mathematics & Statistics, Manipal University Jaipur, Jaipur, India*
[2]*Department of Computer and Communication Engineering, Manipal University Jaipur, Jaipur, India*

Abstract
The main objective of the present study is to investigate the reliability and maintainability of a piston manufacturing plant. For this analysis, data on time to repair and the number of failures have been collected for 2 years. The descriptive analysis of the subsystems has been done along with trend and serial correlation testing. The best-fitted repair and failure time distributions among Weibull, normal exponential, and lognormal distributions have been investigated. The useful parameters corresponding to best-fitted distribution have been estimated using U-statistics methodology and NHPP-PLP. The reliability, availability, and hazard rates of the entire plant were calculated. The results are stored numerical and graphical concerning time to highlight the importance of the study. The model is useful to assess the anticipated time for planning a maintenance schedule of the plant. This study will be very helpful in terms of identifying the occurrence of failures in plants and solving them.

***Keywords*:** Reliability, availability, maintainability, goodness-of-fit, trend, serial correlation

**Corresponding author*: ashishbarak2020@gmail.com

S. C. Malik, Deepak Sinwar, Ashish Kumar, S. R. Gadde, Prasenjit Chatterjee and Bui Thanh Hung (eds.) Computational Intelligence in Sustainable Reliability Engineering, (287–310) © 2023 Scrivener Publishing LLC

14.1 Introduction

The piston is a tube-shaped formed segment firmly fitted inside another cylinder. Piston (cylinder) assumes a significant job in all vehicles, and this manner, the exhibition of the cylinder producing process, is similarly significant as that of cylinder quality and should be assessed. The current examination talks about the conduct investigation of a plant fabricating plant. Manufacturing plants are unpredictable repairable frameworks, and in this way, it is hard to assess their presentation metric, including unwavering quality, reliability, accessibility, and maintainability. As the piston manufacturing process is based on a complex structure of the arrangement of subsystems, availability and reliability assessments for such systems. So, it is very difficult for the management of such plants to arrange repair facilities and implement maintenance strategies. A lot of researchers tried to develop some methodologies for reliability and maintainability evaluation of complex industrial systems. Mahajan and Singh [1] Gupta [2], Gupta *et al.* [3], Gupta and Tiwari [4], Sharma and Kumar [5], Kumar *et al.* [6], Aggarwal *et al.* [7], Kumar and Malik [8], and Saini and Kumar [9] suggested some reliability evaluation techniques, including Lagrange's method, stochastic process, Laplace transformation, matrix method, and some advanced optimization techniques genetic algorithm, GABLT, fuzzy reliability, particle swarm optimization, and RAMD approach. Kumar and Lata [10] analyzed the reliability of a piston manufacturing plant using the fault tree method and adopted risk reduction worth as a measure of importance. Lal *et al.* [11] used numerical methods for solving two stochastic models of the piston manufacturing plant to analyze the behavior of the plant. Modgil *et al.* [12] performed a simulation study of the sole lasting unit in the shoemaking industry for availability analysis. Goyal *et al.* [13] carried out the performance analysis of multiple component manufacturing systems of a rubber tube production industry using the concept of priority in repair disciplines. Reliability optimization of a series-parallel k-out-of-n system whose failure rate is based on components has been studied by Sharifi *et al.* [14]. Lin *et al.* [15] evaluated the system reliability for a freeway system using stochastic concepts. Kumar *et al.* [16] proposed a stochastic model for a system of non-identical units with failure of the repairman. During the last few decades, RAM analysis of industrial systems attracted researchers and system developers. Blischke and Murthy [17] illustrated reliability analysis, failure analysis, availability, and other descriptive statistical measures based on 26 cases. Saraswat and Yadava [18] focused on RAMS attributes to improve the performance of engineering systems. A two-phase methodology incorporating particle swarm optimization metaheuristic algorithm has been proposed by Garg and Sharma [19]. The key objective behind this methodology is to identify the optimized value for RAM indices at the global level useful for industries. Tsarouhas *et al.* [20] analyzed statistically the time between failure and time to repair of the strudel production line at subsystem and component level. The best-fitted probability distribution for failure and repair at the line and subsystem level has been identified. The estimated parameters have been used to estimate the reliability and maintainability of the strudel line. Yingjie and Liling [21] presented a dependability examination technique for machining framework considering framework cost, where just the working long stretches of machines are viewed as when figuring mean-time-between-disappointment (MTBF).

Maintainability is a basic structure trademark that shows how well an item can be kept up. Zhou *et al.* [22] described maintenance time as a key parameter of merchandise maintainability layout. Tsarouhas [23] utilized viability examination to improve the exhibition of the yogurt generation line dependent on satisfactory support exercises. Tsarouhas and Makrygianni [24] concentrated on the use of practicality examination in a fly contender flying machine to call attention to shortcomings of support activities. In another investigation, Tsarouhas and Besseris [25] gave results to a total viability investigation using informational collections from a creation framework in a shaving cutting edges division of an enormous innovative razor producer. Ding and Kamaruddin [26] suggested a maintainability strategy that has been consistently used to examine accentuation for unwavering quality investigation of assembling frameworks. Duarte *et al.* [27] indicated that the dependability attributes of gear, which is assessed in the configuration stage, can somewhat change during the operational stage. The result of this is a colossal impact on the exhibition of generation offices. Henceforth, they referenced that all affecting operational conditions ought to be gathered in the maintenance database. Tsarouhas [28] tried to investigate the trend and serial correlation between successive repair and failures in the wine-producing industry.

The existing methodology that has been used for RAM analysis of gas and oil industry, production line, mining industry, and packaging line likely Barabady and Kumar [29], Louit *et al.* [30], Gustafson *et al.* [31], and Tsarouhas *et al.* [32], Barak *et al.* [33] to fulfil various objectives like identification of failure pattern and its theoretical probability distribution, TBF and TTR analysis, reliability and maintainability analysis.

The piston manufacturing plant is designed in a series structure by comprising 12 repairable subsystems. It is a universally known fact that reliability and maintainability are key characteristics of repairable systems to measure their performance. In a long time, these measures have been supported an upgraded part due to the competitive environment, quality administration, and generation (production) costs. After that, the maintenance agency of the piston manufacturing plants started recording failure-related data in their databases. The main objectives of the present are as follows:

- to investigate reliability and availability for the plant,
- to estimate the reliability and maintainability for different periods,
- to find out the most critical/sensitive component/subsystem,
- to analyze the trend and serial correlation between failures.

From the existing literature, it is revealed that RAM measures extensively used in the oil and gas industry, wine manufacturing, and agricultural products processing but applying it in piston manufacturing are unique. It is also helpful for other industries to propose maintenance strategies. The complete paper is organized into seven sections including the first introductory section. Section 14.2 thrusts on the system development and data collection while in the third section descriptive statistical analysis is performed. The power law process has been explained in section 14.4. Sections 14.5 and 14.6 are devoted to the trend and reliability analysis of the system. Also, the relationship between TBF and TTR is identified in section 14.6. Concluding remarks are made in the last section 14.7.

14.2 System Description and Data Collection

Piston manufacturing plant starts piston manufacturing with the help of semiautomatic die machines when a gravity die casting process is performed in foundry by controlling outside pressure. After that, consequential pistons are forwarded to risers and runners for cleaning and quality measures testing like hardness, strength, or any manufacturing defect identification. A lot of pistons are positioned in the piston space for their machining ventures afterward a fixture seat machine is utilized. By using a carbide tool tip two bumpy channels have been made on the piston for fixing a pin in them. A similar hole for oil passing is also made before finishing the channels and hole using the diamond tool and smoothness of the piston is ensured for smooth movement in the cylinder. In the next step, cavity machines have been used for putting a cavity and its finishing. On the corners of the piston, values are made and rounded off. In the absence of coolant and air pressure, circlip trenches are made on the piston. Then, final deburring, cleaning, surface evaluation, and operation test are performed before delivery. The complete process passes through twelve stages/subsystems explained as below: Subsystem (S_1) i.e. fixture seat machine is the first stage in a piston manufacturing plant. It is used for immobilizing the pistons. Rough grooving and turning machine are termed as subsystem S_2 and it is used for making grooves/channels on the piston. Subsystem S_3 i.e. rough pin hole boring machine is used for hole boring operations. This hole is utilized for connecting a rod to a crankshaft. Subsystem S_4, i.e., oil hole drilling machine is used for making holes for proper supply of oil. Finishing grooving machine i.e. subsystem (S_5) castoff to give wrap up to the bumpy channels. Subsystem S_6 is the finish profile turning operation. In this subsystem, the piston is modeled into an oval shape to conquer the issues raised during the working time of the piston in high temperatures. Subsystem S_7, S_8, and S_9 are respectively finished pinhole, finish crown, and cavity and valve milling machine respectively used for hole boring and valve slump. For round offing, the corners of piston subsystem S_{10} namely chamfering are used, and circlip channels are made through circlip grooving machine i.e. subsystem S_{11}. Finally, internal and external cleaning is made using subsystem S_{12}.

The piston manufacturing plant considered here operates into two shifts of eight hours each and all Sunday it remains closed. It comprises twelve subsystems arranged in a series structure. For the present study, failure data is collected from the maintenance sheet of the plant. A total number of 407 failures were recorded for 622 days. The number of failures on individual subsystems, time to repair, the time between failures was then collected manually and using MS-excel. The time between failures (TBF) is important for the identification of the system's key characteristic defined as reliability while time to repair (TTR) is important for maintainability analysis. For, planning of maintenance scheduling and appointment of repairman both are very important. The flow chart of the system model has been given in Figure 14.1. To gain knowledge about the theoretical distributions of TBF and TTR histogram of the plant's data has been drawn. The histogram of TBF and TTR are stored in Figures 14.2 to 14.3. From the Pareto chart appended in Figure 14.4, it is observed that subsystem S_7 is the most sensitive unit in the plant has 30% failures. After the subsystem S_{10} have 21.1% failure failures followed by S_3 having 73 failures. Only these three subsystems reported 69% of the total number of failures.

The total number of failures, total TBF, total TTR, and availability of subsystems and system are given in Table 14.1.

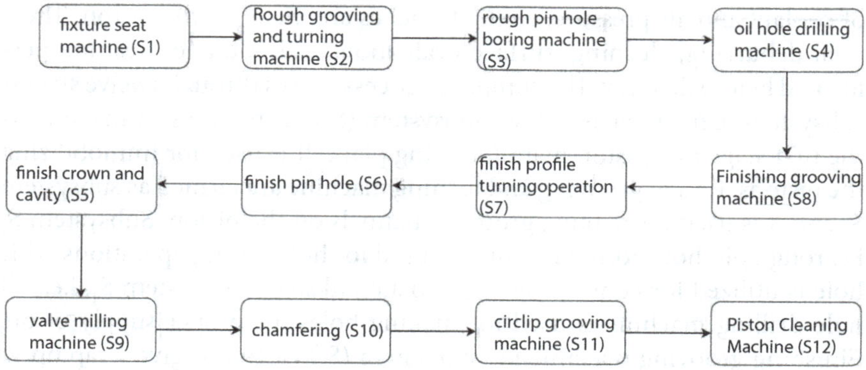

Figure 14.1 System description.

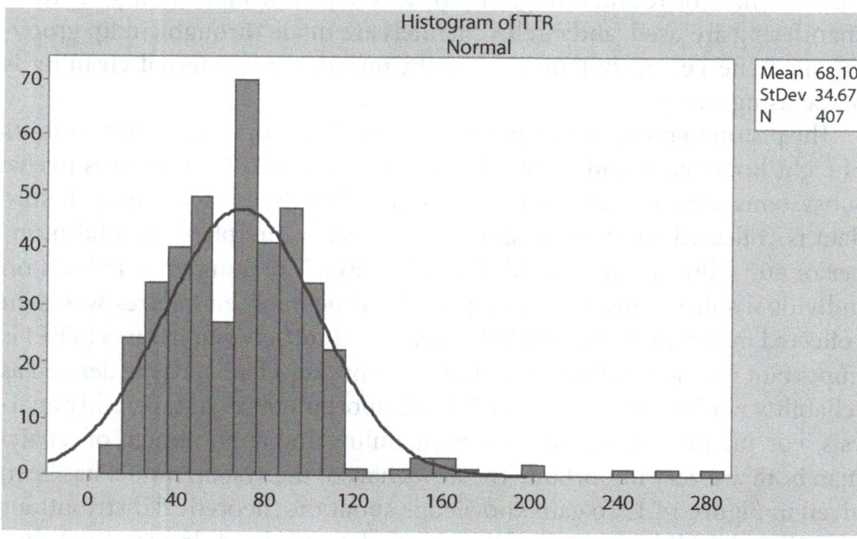

Figure 14.2 Histogram of TTR of plant.

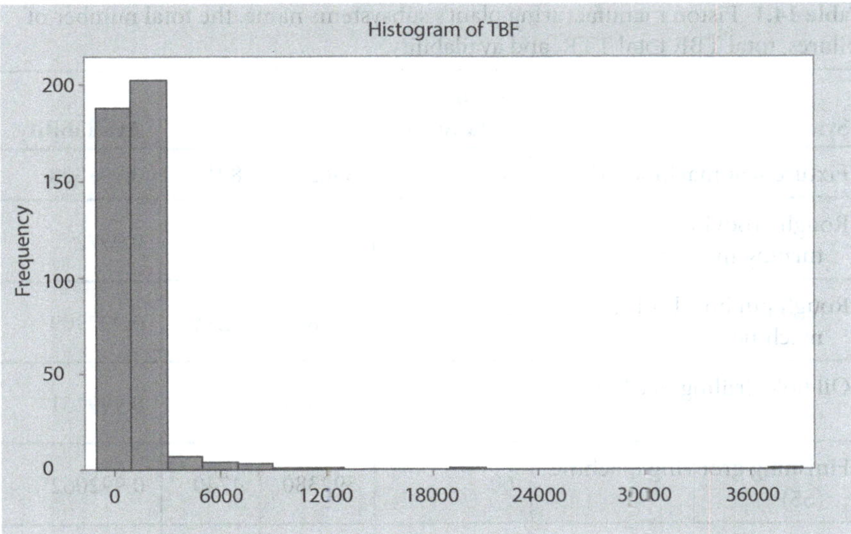

Figure 14.3 Histogram of TBF of plant.

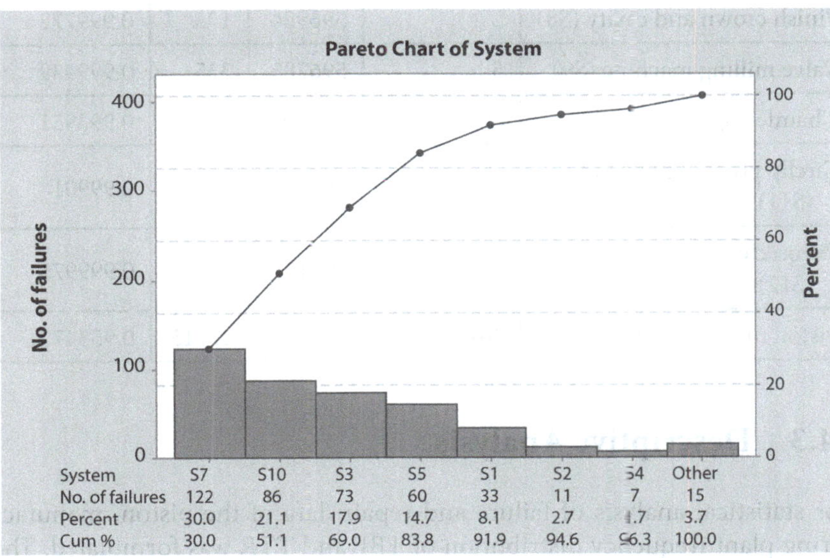

Figure 14.4 Pareto chart of plant.

Table 14.1 Piston manufacturing plant's subsystems name, the total number of failures, total TBF, total TTR, and availability.

System	No. of failures	TBF	TTR	Availability
Fixture seat machine (S1)	33	596229	891	0.998508
Rough grooving and turning machine (S2)	11	595448	1672	0.9972
Rough pin hole boring machine (S3)	73	592886	4234	0.992909
Oil hole drilling machine (S4)	7	596679	441	0.999261
Finishing grooving machine (S5)	60	592380	4740	0.992062
Finish profile turning operation (S6)	4	596808	312	0.999477
Finish pin hole (S7)	122	586384	10736	0.98202
Finish crown and cavity (S8)	2	596984	136	0.999772
Valve milling machine (S9)	5	596785	335	0.999439
Chamfering (S10)	86	593508	3612	0.993951
Circlip grooving machine (S11)	3	596529	591	0.99901
Piston cleaning machine (S12)	1	597105	15	0.999975
Piston manufacturing plant	407	569405	27715	0.953477

14.3 Descriptive Analysis

For statistical analysis of failure and repair data of the piston, manufacturing plant frequency distribution of TBF and TTR was formulated. The basic characteristics of the frequency distribution, including measures of location, measures of variation skewness, and kurtosis, have been obtained. In addition to the above, the minimum and maximum values of the TBF and TTR of plant and subsystem are also obtained. The measure of location gives us a single value, which represents the whole set of values, and the rest of the observations lies in the neighborhood of that value. Measures of central tendency do not provide any information regarding the variation among the observations from average, so the measure of dispersion is required. Dispersion provides information about the variation. But the information about the direction of variation is not given by dispersion. So, skewness provides the direction information. Skewness is also known as the measure f symmetry of the distribution. If data shows symmetry, then data follows the normal distribution. In the case of asymmetry, there will be either positive skewness or negative skewness. Kurtosis gives information about the shape of the data. If the coefficient of kurtosis is equal to three, then data follow a normal distribution (mesokurtic). If among observation variation is low, then data show a high peak then the curve is leptokurtic and high variation indicates the curve is platykurtic. The normal distribution has the characteristic that means, median, and mode all are equal while in the case of positive skewness median is less than mean. Negatively skewed data have mean that is less than median. The summary statistics of TBF and TTR of plant and all subsystems have been shown in Tables 14.2 and 14.3. It is revealed from the descriptive analysis that mean TBF of the plant is 1408 min that is 23 hours approximately. So, on average, in three consecutive shifts, plant faces a failure while the average TTR of the plant is 68.10 min. The most time-consuming repair is performed on subsystem S11. There are also other time-consuming repairs reported on subsystems S2 & S3. Standard deviation is a good measure of dispersion to measure the variation among TTR and TBF. The highest values of standard deviation for TTR are observed on S3 however on TBF it is observed on S11. The minimum standard deviations for TBF and TTR are S7 and S1, respectively. Failure modes of most of the subsystems and plant are positively skewed except S8, while TTRs of S1, S3 S11, S12, and the whole plant are positively skewed. The rest of the subsystems are negatively skewed. The highest value of TBF is reported on the whole plant and for TTR it is reported on S1. Figures 14.5 and 14.6 show the box plots of TBF and TTR of all subsystems and plant. Quartiles, median, and outliers are also presented in the box plot.

14.4 Power Law Process Model

The reliability and maintainability analysis of data based on the independent and identical distributed (IID) assumption of TBF and TTR in the time sphere. In a non-IID situation, fitting of theoretical distribution is not

Table 14.2 Descriptive analysis of TBF.

Variable	Count	Mean	StDev	CoefVar	Min	Median	Max	Skewness	Kurtosis
Plant	408	1408	2434	172.87	2	1121	38519	10.39	141.48
S1	34	17535	2368	13.50	13108	16830	25154	1.18	2.29
S2	12	49608	6617	13.34	39595	47779	66575	1.51	3.75
S3	74	8011	2135	26.65	4475	7585	19449	3.91	16.77
S4	8	74577	14577	19.55	60872	68162	100997	1.20	0.05
S5	61	9710	2953	30.41	7218	8871	22250	2.25	5.63
S6	5	119346	12968	10.87	108832	114525	141786	1.88	3.82
S7	123	4767	1289	27.03	2097	4391	11029	2.48	6.94
S8	3	198972	18441	9.27	177917	206744	212255	1.56	*
S9	6	99453	22145	22.27	67741	96904	135232	0.39	1.5
S10	87	6821	1391	20.39	5606	6383	13720	2.50	7.51
S11	4	149083	26943	18.07	124537	144087	183622	0.72	-1.53
S12	1	597105	*	*	597105	597105	597105	*	*

Table 14.3 Descriptive analysis of TTR.

Variable	Count	Mean	StDev	CoefVar	Minimum	Median	Maximum	Skewness	Kurtosis
S1	33	24.18	12.84	53.11	7.00	22.00	87.00	3.74	18.51
S2	11	129.5	66.0	50.99	15.0	155.0	201.0	-0.90	-0.69
S3	74	114.4	485.4	424.40	15.0	57.5	4232.0	8.59	73.87
S4	7	64.86	23.53	36.28	19.00	68.00	94.00	-1.22	2.54
S5	60	77.77	13.94	17.93	17.00	75.50	116.00	-0.83	5.64
S6	4	97.50	12.71	13.04	81.00	98.50	112.00	-0.47	1.59
S7	122	88.29	18.45	20.90	17.00	88.50	151.00	-0.28	3.12
S8	2	60.5	72.8	120.38	9.0	60.5	112.0	*	*
S9	5	80.4	22.9	28.53	52.0	86.0	102.0	-0.39	-2.63
S10	86	42.49	13.23	31.13	24.00	40.00	112.00	2.42	9.25
S11	3	262.7	19.6	7.44	244.0	261.0	283.0	0.38	*
S12	2	15.000	0.000000	0.00	15.000	15.000	15.000	*	*
Plant	407	68.10	34.67	50.91	7.00	68.00	283.00	1.70	7.29

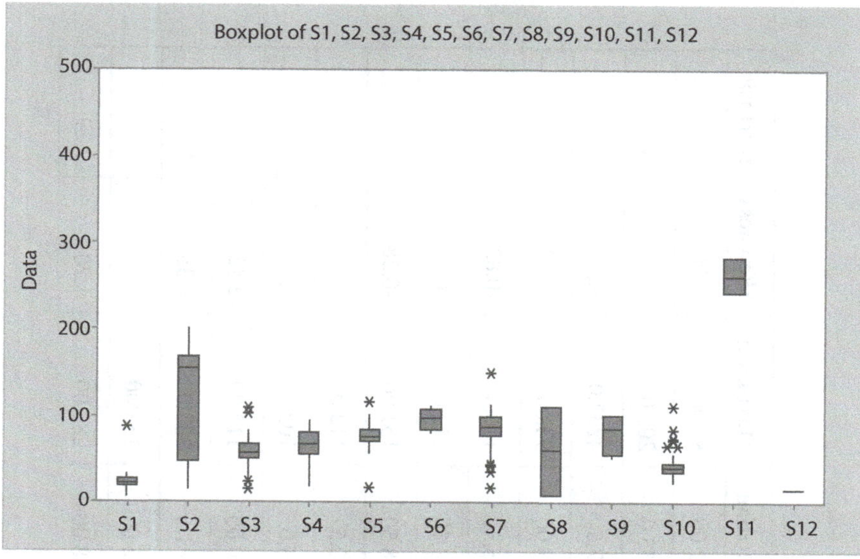

Figure 14.5 Box plot of TTR of plant.

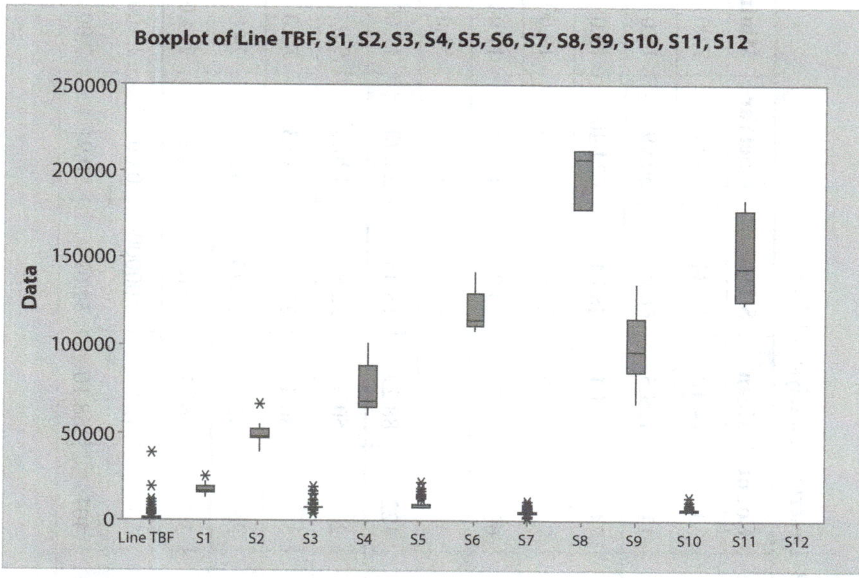

Figure 14.6 Box plot of TBF of plant.

justified and the nonhomogeneous Poisson process, i.e., power-law process (PLP) is the appropriate distribution for analysis. In degraded or improving situation PLP is also proved a useful tool. The probability density function of PLP model and failure rate are defined as follows

$$\lambda(t) = \frac{\beta}{\theta}\left\{\frac{T}{\theta}\right\}^{\beta-1} \tag{14.1}$$

$$f(t) = \frac{\beta}{\theta}\left\{\frac{T}{\theta}\right\}^{\beta-1} e^{[-(t/\theta)\beta]} \tag{14.2}$$

Having scale parameter = θ, shape parameter = β and running time = T

The parameter appended into equation (14.1) can be estimated by the methods proposed by Crow [34] noted in equation (14.3).

$$\hat{\beta} = \frac{n}{\sum_{i=1}^{n-1} L_n(T_n / T_i)}, \hat{\theta} = \frac{T_n}{n^{1/\hat{\beta}}} \tag{14.3}$$

After performing the goodness-of-fit test and trend analysis if the trend is observed then the PLP tool should be applied for reliability evaluation. In this case, reliability and maintainability functions are defined as

$$\mathrm{R(t)} = \exp[-(T/\theta)^{\beta}] \;\&\; m(t) = 1 - \exp[-(T/\theta)\beta] \tag{14.4}$$

MIL-HDBK-189 [35] suggested the goodness-of-fit as

$$U = 2\sum_{i=1}^{n-1} Ln(T_n/T_i) \tag{14.5}$$

when data is time truncated, U-statistic is equivalent to χ^2 distribution having a 2n degree of freedom. For trend analysis, i^{th} *and* $(i-1)^{th}$ TBF is plotted on a two-dimensional scale.

14.5 Trend and Serial Correlation Analysis

The present section incorporated the graphical and analytical/statistical methods of trend and serial correlation analysis. A quantitative analysis has been performed here to identify the trend or serial correlation significantly present in the data or not in the piston manufacturing plants. The validation of the independently and identically distributed (IID) nature of TBF/TTR data for every subsystem and entre plant has been performed. The most common methods for identifying the IID nature of data are the trend and serial correlation. If the IID assumption is validated, then the use of classical statistical techniques for hypothesis testing is appropriate and in case of failure of IID phenomena such statistical methods are not applicable for reliability and maintainability analysis. In such situations, non-stationary distribution is the best fit for the data. So, a non-homogenous

Table 14.4 Trend analysis of TBF and TTR of the piston manufacturing plant.

Variable	Df	U-statistics (MIL-HDBK-189, 1981)	Chi-square	Decision
Line TTR	810	930.1205	877.321	Not rejected (IID)
S1	64	53.878	83.67	Rejected (non-IID)
S2	20	15.245	31.41	Rejected (non-IID)
S3	144	144.81	173.004	Rejected (non-IID)
S4	12	9.880	21.006	Rejected (non-IID)
S5	118	115.843	144.35	Rejected (non-IID)
S7	242	263.202	279.28	Rejected (non-IID)
S10	170	158.001	201.423	Rejected (non-IID)
S1	66	59.398	85.96	Rejected (non-IID)
S2	22	18.534	33.92	Rejected (non-IID)
S3	146	141.855	175.19	Rejected (non-IID)
S4	14	13.058	23.68	Rejected (non-IID)
S5	120	106.808	146.56	Rejected (non-IID)
S7	244	237.456	281.437	Rejected (non-IID)
S10	172	156.985	203.601	Rejected (non-IID)
Line TBF	812	774.472	879.403	Rejected (non-IID)

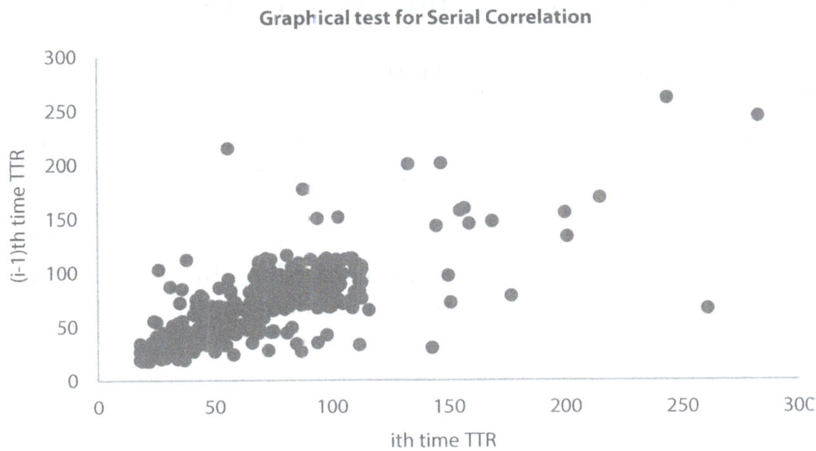

Figure 14.7 Graphical test for serial correlation.

Poisson process is recommended. To analyze the trend U-statistics given in equation (14.5) has been utilized using the null hypothesis that no trend presents in the data. At a 5% level of significance, the U-statistic is compared with χ^2 tabulated value having 2(n-1) degree of freedom. The degree of freedom represents the number of linearly independent observations in the set. It is determined by using the number of parameters and sample observations. If U-statistic $\geq \chi^2$ tabulated value then a null hypothesis is not rejected, and no trend present in the TBF/TTR data while U-statistic $\leq \chi^2$ tabulated value null hypothesis rejected. The validation of trends for TBF/TTR of the piston manufacturing plant is shown in Table 14.4. At a 5% level of significance, it is observed that only TTR data of plant shows IID` behaviour rest all the subsystems and plant TBF are non-IID. Now serial correlation of plant's TTR is observed by plotting the ith-order data against (i-1)th-order data set and correlation is observed. The coefficient of correlation is a relative measure that provides information about the linear relationship between two variables, and it varies from −1 to +1. Zero values represent the absence of a linear relationship. After observing Figure 14.7 and Table 14.4, it is concluded that except TTR for all subsystems NHPP-PLP is the best-fitted distribution for reliability and maintainability analysis.

14.6 Reliability and Maintainability Analysis

Ebeling [36] defined reliability as a probability that a system/unit/ product will perform its intended function for a given period when used under stated environmental conditions. In other words, it is the probability of non-failure over time. According to Ebeling [36], maintainability is defined to be the probability that a failed unit/component will be repaired or reinstalled/restored to a specific condition with a period when maintenance is performed under the prescribed procedure. The plant TTR shows no trend and serial correlation in analyzed data and an effort has been

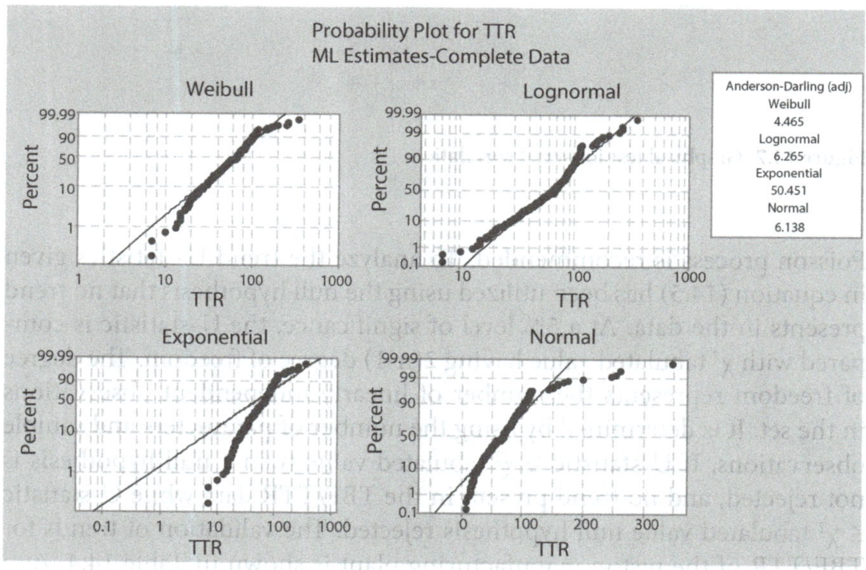

Figure 14.8 Probability curve of plant's TTR.

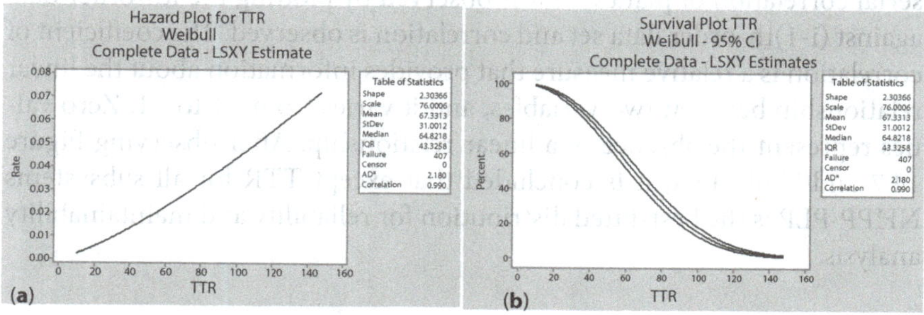

Figure 14.9 (a) Hazard rate plot of TTR. (b) Survival rate plot of TTR.

Table 14.5 Best fitted distribution, shape, and scale parameters for TBF & TTR of the piston manufacturing plant.

Variable	Distribution	Shape	Scale
Plant TTR	Weibull	2.30366	76.0006
S1	NHPP-PLP	1.224992	45.96128
S2	NHPP-PLP	1.442629	270.1709
S3	NHPP-PLP	1.008234	60.03981
S4	NHPP-PLP	1.416961	114.9847
S5	NHPP-PLP	1035881	89.61593
S7	NHPP-PLP	0.053191	0.000643
S10	NHPP-PLP	1.0886	61.05439
S1	NHPP-PLP	1.144818	27393.29
S2	NHPP-PLP	1.294892	87361.28
S3	NHPP-PLP	1.114561	12468.87
S4	NHPP-PLP	1.225274	109305.8
S5	NHPP-PLP	1.142235	15200.46
S7	NHPP-PLP	1.047451	5927.704
S10	NHPP-PLP	1.108382	10556.77
Plant TBF	NHPP-PLP	1.053621	1895.064

made to find the best-fitted distribution. The Anderson darling goodness-of-fit test was used on the theoretical distributions Weibull, exponential, lognormal, and normal. The minimum AD value as shown in Figure 14.8 is 4.465 associated with Weibull distribution. So, the best-fitted distribution is the Weibull distribution for the plant's TTR. For the rest of the subsystems, TBF/TTR have been found Non-IID and NHPPLP is the best-fitted distribution. Figures 14.9(a) & 14.9(b) plot the hazard rate and survival function of TTR of the plant using the least square estimation method. The values of shape and scale parameters are shown in Table 14.5. It is observed that the probability of repair to failure increases up to 150 minutes. Table 14.6 shows the reliability of various subsystems and plants with respect to time. It is observed that in the initial 500 mins the plant reliability is 0.7573, and in 20 hours reliability becomes 0.5131. The highest reliability reported

Table 14.6 Reliability of Subsystems and piston manufacturing plant.

Time	S1	S2	S3	S4	S5	S7	S10	Plant
0	1	1	1	1	1	1	1	1
500	0.979321	0.992616	0.956290305	0.994411	0.965361	0.915438	0.948858	0.757304
1200	0.951086	0.982371	0.898287604	0.986639	0.918873	0.808928	0.881623	0.513154
1800	0.927534	0.973673	0.851379346	0.980025	0.880812	0.727554	0.827798	0.367597
2400	0.904565	0.965052	0.80692062	0.973456	0.844327	0.654365	0.771837	0.263328
3000	0.882165	0.956507	0.764783513	0.96693	0.809354	0.588539	0.72092	0.188634
3600	0.86032	0.948039	0.724846791	0.960449	0.775829	0.529334	0.673362	0.135128
4200	0.839016	0.939645	0.68699555	0.954011	0.743693	0.476085	0.628941	0.096798
4800	0.818239	0.931325	0.651120887	0.947616	0.712889	0.428193	0.587451	0.069341
5400	0.797976	0.923079	0.617119587	0.941264	0.68336	0.385119	0.548698	0.049673
6000	0.778216	0.914906	0.584893822	0.934954	0.655054	0.346378	0.512501	0.035583
6600	0.758945	0.906806	0.554350875	0.928687	0.627921	0.311534	0.478692	0.02549
7200	0.740151	0.898777	0.52540287	0.922462	0.601911	0.280195	0.447113	0.018259

Table 14.7 Maintainability of subsystems and piston manufacturing plant.

Time	S1	S2	S3	S4	S5	S7	S10	Plant
0	0	0	0	0	0	0	0	0
500	0.999998	0.930737	0.999774	0.997891	0.99691	1.00000	0.999866	0.91364103
1200	1.00000	0.998351	1.00000	1.00000	0.999999	1.00000	1.00000	0.99720013
1800	1.00000	0.999933	1.00000	1.00000	1.00000	1.00000	1.00000	0.999851848
2400	1.00000	0.999997	1.00000	1.00000	1.00000	1.00000	1.00000	0.999992161
3000	1.00000	1.00000	1.00000	1.00000	1.00000	1.00000	1.00000	0.999999585
3600	1.00000	1.00000	1.00000	1.00000	1.00000	1.00000	1.00000	0.999999978
4200	1.00000	1.00000	1.00000	1.00000	1.00000	1.00000	1.00000	0.999999999
4800	1.00000	1.00000	1.00000	1.00000	1.00000	1.00000	1.00000	1.00000
5400	1.00000	1.00000	1.00000	1.00000	1.00000	1.00000	1.00000	1.00000
6000	1.00000	1.00000	1.00000	1.00000	1.00000	1.00000	1.00000	1.00000
6600	1.00000	1.00000	1.00000	1.00000	1.00000	1.00000	1.00000	1.00000
7200	1.00000	1.00000	1.00000	1.00000	1.00000	1.00000	1.00000	1.00000

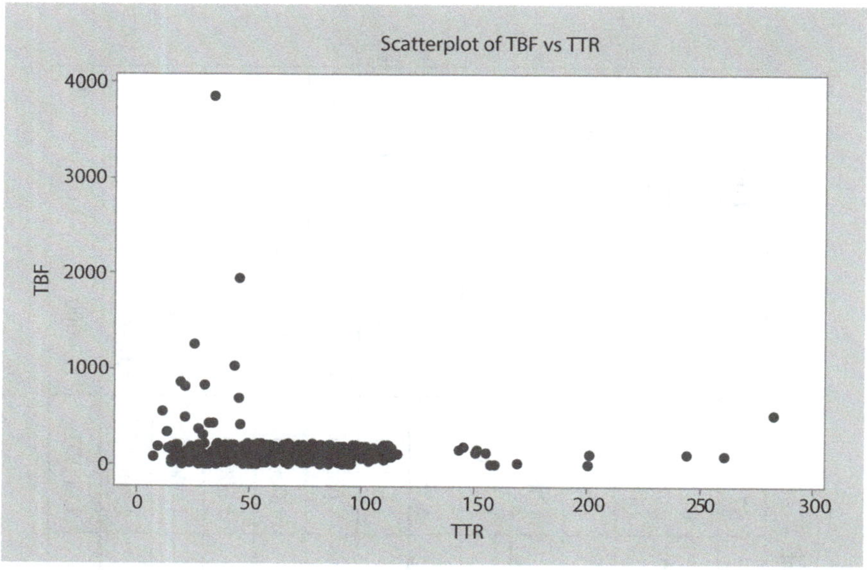

Figure 14.10 Scatter plot of TBF vs. TTR

on subsystem S4 followed b S2 and S1. The reliability diagram has been shown in Figure 14.10. After 3600 minutes, the management must plan any maintenance because the chance of successful operation is 13.51%. The maintainability procedure includes the availability of resources like a skilled repairman, spare parts. From Table 14.7, it is revealed that any failure that happened in the plant can be repaired in 20 hours probability is 99.72%. Subsystem S2 depicts the lowest maintainability. There are 100% chances that in 4800 min every failure of the plant must be repaired. The maintainability graphs of subsystems and plant are shown in Figure 14.9. Finally, Figure 14.10 shows the correlation between TBF and TTR data using the scattered diagram approach, and no correlation is visualized.

14.7 Conclusion

The main findings of the present work are noted as follows:

- System availability is 95.34% and don time is 4.66%.
- Subsystem Finish crown and cavity (S_8) is the highly available component and the least available component is the finish pinhole (S_7).

- No correlation is visualized between time between failure and time to repair.
- The time to repair of the plant follows the Weibull distribution while for all other subsystem's TBF/TTR the best-fitted distribution is NHPP-PLP.
- The finish pinhole (S_7) needs high attention to improve the reliability of the plant.
- The average TBF and TTR of the plant were reported as 1408 min and 68.10 min.

Based on the above analysis, it is recommended to improve the reliability, availability, and maintainability special attention must be given to less reliable and available components like finish pinhole (S_7). After some specific time, intervals maintenance policies must be planned to enhance the operational capacity of the plant.

References

1. Mahajan, P. and Singh, J., Reliability of utensils manufacturing plant. *Opsearch*, 36, 260–269, 1999.
2. Gupta, P., *Mathematical Analysis Of Reliability And Availability Of Some Process Industries*, Dissertation, Thapar University, India, 2003.
3. Gupta, P., Lal, A.K., Sharma, R.K., Singh, J., Analysis of reliability and availability of serial processes of plastic-pipe manufacturing plant: A case study. *Int. J. Qual. Reliab. Manag.*, 24, 404–419, 2007.
4. Gupta, S. and Tiwari, P.C., Simulation modeling and analysis of a complex system of a thermal power plant. *J. Ind. Eng. Manag.*, 2 387–406, 2009.
5. Sharma, S.P. and Kumar, D., Stochastic behavior and performance analysis of an industrial system using GABLT technique. *Int. J. Ind. Syst. Eng.*, 2, 1–23, 2010.
6. Kumar, S., Tewari, P.C., Kumar, S., Gupta, M., Availability optimization of CO-shift conversion system of a fertilizer plant using genetic algorithm technique. *Bangladesh J. Sci. Ind. Res.*, 45, 133–140, 2010.
7. Aggarwal, A., Kumar, S., Singh, V., Performance modeling of the skim milk powder production system of a dairy plant using RAMD analysis. *Int J. Qual. Reliab. Manag.*, 32, 2, 167–181, 2014.
8. Kumar, N. and Malik, S.C., Stochastic modeling of a system of non-identical units with conditional failure of repairman. *Int. J. Stat Reliability Eng.*, 5, 2, 112–120, 2019.
9. Saini, M. and Kumar, A., Performance analysis of evaporation system in sugar industry using RAMD analysis. *J. Braz. Soc. Mech. Sci. Eng.*, 41, 4, 1–10, 2019.

10. Kumar, A. and Lata, S., Reliability analysis of piston manufacturing system. *J. Reliab. Stat. Stud.*, 4, 2, 43–55, 2011.
11. Lal, A.K., Kaur, M., Lata, S., Behavioral study of piston manufacturing plant through stochastic models. *J. Ind. Eng. Int.*, 9, 24, 1–10, 2013.
12. Modgil, V., Simulation, modeling and availability analysis of sole lasting unit in shoe making industry. *Int. J. Ind. Eng. Theory*, 19, 2, 2012. https://doi.org/10.23055/ijietap.2012.19.2.664
13. Goyal, A., Sharma, S.K., Gupta, P., Performance analysis of a multiple component manufacturing subsystem of a rubber tube production plant under pre-emptive resume priority repair. *Int. J. Ind. Eng. Theory*, 16, 4, 2009. https://doi.org/10.23055/ijietap.2009.16.4.113
14. Sharifi, M. *et al.*, Reliability optimization of a series-parallel k-out-of-n system with failure rate depends on working components of system. *Int. J. Ind. Eng. Theory*, 22, 4, 438–453, 2015.
15. Lin, Y.-K. *et al.*, Evaluation of system reliability for a freeway system with stochastic speed. *Int. J. Ind. Eng. Theory*, 26, 1, 12–33, 2019.
16. Kumar, A., Goyal, D., Sinwar, D., Saini, M., Stochastic modeling and performance optimization of sludge digestion processing system using genetic algorithm. *Int. J. Inf. Technol.*, 1–10, 2022. https://doi.org/10.1007/s41870-022-00908-x
17. Blischke, W.R. and Murthy, D.N.P., *Case Studies in Reliability and Maintenance*, pp. 351–445, John Wiley and Sons, Inc., Hoboken, NJ, 2003.
18. Saraswat, S. and Yadava, G.S., An overview on reliability, availability, maintainability and supportability (RAMS) engineering. *Int. J. Qual. Reliab. Manage.*, 25, 3, 330–344, 2008.
19. Garg, H. and Sharma, S.P., A two-phase approach for reliability and maintainability analysis of an industrial system. *Int. J. Reliab. Qual. Saf. Eng.*, 19, 3, 19, 2012.
20. Tsarouhas, P., Reliability, availability and maintainability (RAM) analysis for wine packaging production line. *Int. J. Qual. Reliab. Manage.*, 35, 3, 821–842, 2018.
21. Yingjie, Z. and Liling, G., Reliability analysis of machining systems by considering system cost. *Int. J. Comput. Integr. Manuf.*, 28, 8, 836–843, 2015.
22. Zhou, D., Kang, L., Chuan, L., A virtual reality-based maintenance time measurement methodology for complex products. *Assembly Autom.*, 33, 3, 221–230, 2013.
23. Tsarouhas, P., Application of statistical approaches for analyzing the reliability and maintainability of food production lines: A case study of mozzarella cheese, in: *Mathematical and Statistical Methods in Food Science and Technology*, D. Granato, and G. Ares, (Eds.), pp. 491–510, John Wiley & Sons, Chicago, IL, 2014.
24. Tsarouhas, P. and Makrygianni, M., A framework for maintenance and combat readiness management of a jet fighter aircraft. *Int. J. Syst. Assur. Eng. Manag.*, 8, 2, 1895–1909, 2017.

25. Tsarouhas, P. and Besseris, G., Maintainability analysis in shaving blades industry: A case study. *Int. J. Qual. Reliab. Manag.*, 34, 4, 581–594, 2017.
26. Ding, S.H. and Kamaruddin, S., Maintenance policy optimization–literature review and directions. *Int. J. Adv. Manuf. Technol.*, 76, 5-8, 1263–1283, 2015.
27. Duarte, J.C., Cunha, P.F., Craveiro, J.T., Maintenance database. *Proc. CIRP*, 7, 551–556, 2013.
28. Tsarouhas, P., Performance evaluation of the croissant production line with reparable machines: A case study. *J. Ind. Eng. Int.*, 11, 1, 101–110, 2015.
29. Barabady, J. and Kumar, U., Reliability analysis of mining equipment: A case study of crushing plan at jajarm bauxite mine in Iran. *Reliab. Eng. Syst. Safe.*, 93, 4, 647–653, 2008.
30. Louit, D.M., Pascual, R., Jardine, A.K., A practical procedure for the selection of time-to-failure models based on the assessment of trends in maintenance data. *Reliab. Eng. Syst. Safe.*, 94, 10, 1618–1628, 2009.
31. Gustafson, A., Schunnesson, H., Kumar, U., Reliability analysis and comparison between automatic and manual load haul dump machines. *Qual. Reliab. Eng. Int.*, 31, 3, 523–531, 2015.
32. Tsarouhas, P., Arvanitoyannis, I., Varzakas, T., Reliability and maintainability analysis of cheese (feta) production line in a Greek medium-size company: A case study. *J. Food Eng.*, 94, 3-4, 233–240, 2009.
33. Barak, M.S., Garg, R., Kumar, A., Reliability measures analysis of a milk plant using RPGT. *Life Cycle Reliab. Saf. Eng.*, 10, 3, 1–8, 2021.
34. Crow, L.H., Reliability analysis for complex repairable systems, in: *Reliability and Biometry*, F. Proschan and R.J. Serfling (Eds.), pp. 379–410, SIAM, Philadelphia, PA, 1974.
35. *MIL-HDBK-189 Reliability Growth Management*. Headquarters, US Army Communication Research and Development Command, ATTN, DRDCO-PT, Fort Monmouth, NJ, 1981.
36. Ebeling, C., *An Introduction to Reliability and Maintainability Engineering*, 10th ed, Tata McGraw-Hill, New Delhi, 2013.

Index

Also of Interest

Check out these published and forthcoming titles in "Sustainable Computing and Optimization" series from Scrivener Publishing

Optimization Techniques in Engineering Advances and Applications
Edited by Anita Khosla, Prasenjit Chatterjee, Ikbal Ali and Dheeraj Joshi
Forthcoming 2022. ISBN 978-1-119-90627-8

Computational Intelligence in Sustainable Reliability Engineering
Edited by S. C. Malik, Deepak Sinwar, Ashish Kumar, S. R. Gadde, Prasenjit Chatterjee and Bui Thanh Hung
Forthcoming 2022. ISBN 978-1-119-86501-8

Decision Support Systems for Smart City Applications
Edited by: Loveleen Gaur, Vernika Agarwal and Prasenjit Chatterjee
Forthcoming 2022. ISBN 978-1-119-89643-2

Fuzzy Computing in Data Science Applications and Challenges
Editors: Sachi Nandan Mohanty, Prasenjit Chatterjee and Bui Thanh Hung
Published 2022. ISBN: 978-1-119-86492-9

Handbook of Intelligent Computing and Optimization for Sustainable Development
Edited by: Mukhdeep Singh Manshahia, Valeriy Kharchenko, Elias Munapo, J. Joshua Thomas and Pandian Vasant
Published 2022. ISBN 978-1-119-79182-9

Machine Learning Algorithms and Applications
Edited by Mettu Srinivas, G. Sucharitha and Anjanna Matta
Published 2021. ISBN 978-1-119-76885-2

Smart Sustainable Intelligent Systems
Edited by Namita Gupta, Prasenjit Chatterjee and Tanupriya Choudhury
Published 2021. ISBN 978-1-119-75058-1

www.scrivenerpublishing.com

Printed in the USA/Agawam, MA
September 16, 2026

910661.088